Agassiz's Legacy

Louis Agassiz, founder (in 1873) and "Master" of the Anderson School of Natural History on Penikese Island, Massachusetts. Photo courtesy Woods Hole Historical Society.

Agassiz's Legacy

Scientists' Reflections on the Value of Field Experience

Elizabeth Higgins Gladfelter

2002

OXFORD
UNIVERSITY PRESS

Oxford New York
Auckland Bangkok Buenos Aires Cape Town Chennai
Dar es Salaam Delhi Hong Kong Istanbul Karachi Kolkata
Kuala Lumpur Madrid Melbourne Mexico City Mumbai Nairobi
São Paulo Shanghai Singapore Taipei Tokyo Toronto

Published by Oxford University Press, Inc.
198 Madison Avenue, New York, New York 10016

www.oup.com

Library of Congress Cataloging-in-Publication Data
Gladfelter, Elizabeth H.
Agassiz's legacy : scientists' reflections on the value of field experience / Elizabeth Higgins Gladfelter.
p. cm.
ISBN 0-19-515441-X—ISBN 0-19-515442-8 (pbk.)
1. Scientists—Interviews. 2. Science—History—20th century.
3. Research—History—20th century. I. Title.
Q141 .G728 2002
509.04—dc21 2002070129

1 3 5 7 9 8 6 4 2

Printed in the United States of America
on acid-free paper

To all my "students" and "teachers,"

who I have discovered are one and the same,

and especially to Bill

Preface

Il est tres simple: on ne voit bien qu'avec le coeur.

L'essentiel est invisible pour les yeux.

Antoine de Saint-Exupery, "Le secret du renard," Le Petit Prince

Through the years I have developed the overriding belief that *people* are what is important. People, not institutions or governments, make decisions; people get things done; people influence other people's lives. My husband once articulated my professional interests for me (as I was attempting to write an essay for a fellowship). They are a fundamental need to be in nature; an interest in discovering as much as I can about how nature works; and an interest in facilitating the opportunities for others to learn about nature. Fortunately, I have had ample opportunities to exercise all of these interests, not the least of which was preparing this book. I have written it because:

1. I know what being in nature means to a student and to a natural science scholar: nature is the source of inspiration and the site of "truth." The skills necessary to make new discoveries in nature take time to develop, just as do most educational skills. Observational skills, needed by all field scientists, are useful in many other life applications as well. Although extensive educational research has been conducted on when a child is best prepared mentally to develop language ability, learn music, and learn how to read, for examples, the developmental process of learning and refining field skills is an unexplored area.

2. I also know what it takes to support and keep scientists and their students in the field. Field institutions, access, and logistics take energy, commitment, planning, and financial resources (although, ironically, not really very much in the big scheme of things). Convincing university administrations and government bureaucracies about the value of

field experience is becoming increasingly difficult, time consuming, and energy draining. Fewer and fewer administrators and bureaucrats have sufficient field experience themselves to appreciate the necessity of it in our future progress in understanding how the oceans, Earth, atmosphere, and living world work.

I want to demonstrate the vital importance of field experience in the lives and careers of natural scientists by presenting their personal accounts. I have interviewed a number of natural scientists who pursue fundamental questions about how the natural world works. Through their life stories, which I have arranged by the generation in which they were undergraduates, we can examine how the process of doing science and how the process of developing future scientists has changed. My study is qualitative, not quantitative—an approach that would shock funding agencies but is well understood by anyone who has done field science. I did not choose my interviewees by random selection. My baseline criteria for an interview were that I had to find the person interesting and I had to respect the person's contributions to field research. I also wanted to span spatial and temporal scales in scientists' approaches to investigating the world, to present a variety of geographical and ecosystem settings, and to cover all of the major groups of organisms: animals, plants, fungi, protists, and bacteria.

Several people have asked why I haven't interviewed more women or ethnic minorities. My answer is that I have attempted to present a reasonable representation of the natural scientists I have known over the past 35 years. There are few ethnic minorities in natural science; women have become professionals in increasing numbers only since the 1960s. Science is an international endeavor, and I have included some foreign scientists. About half of the scientists I've interviewed I knew before through my teaching, research, and administrative activities—or as friends met when I was a child. The other half are those whose seminars inspired me or who were recommended to me by trusted colleagues.

The people behind this book are numerous, and I thank them for their many tangible and intangible contributions to it. My husband, Bill Gladfelter, has always encouraged me to follow my visions and turn them into concrete realities; he also continues to share with me explorations of the natural world, which we love. Through our travels, observations, and musings, we have pondered how nature works. Through our teaching and research, we have tried to share some of those insights with others. Through his editing of some of this book, he has tried to help you understand my thoughts, by clarifying how I expressed them. My discussions with Arthur Gaines about people, the natural sciences, and institutions have helped me better understand my vague uneasiness about the future of field experience. When Gaines suggested a position as a guest investigator at the Woods Hole Oceanographic Institution (WHOI), I seized the opportunity to begin these interviews and produce this book. Steve Wainwright ("I support people, not ideas or projects") and Mitch Baumeister have provided financial support and encouragement.

During the course of writing this book, I have interviewed a number of other scientists whose stories are equally fascinating, but due to a ruthless but realistic editor, I do not have room to include them, but I acknowledge them below. I have also enlisted the help of relatives, colleagues, and friends to read and comment on all or part of the manuscript, and their input has been invaluable. I have tried to include most of these "teachers" and "students" in the lists below; if I have forgotten someone, I sincerely apologize. My sisters, Anne Porter and Mary Etta Higgins, my mother, Elizabeth Higgins, and Jennifer Gaines, Dan Curran, Anne Cohen, Harry Greene, Andy Beet, Doris Kretschmer, Keith Benson, and E. O. Wilson have read and given constructive comments on portions of the text. Peter Klopfer, George Middendorf, and a third anonymous reviewer made some valuable suggestions to improve the text, which I have incorporated. Discussions with Dennis Willows, Charlie McClennen, Jack Pearce, Rich Aronson, Bruce Selleck, Russ Schmidt, Harry Roberts, John Grotzinger, Chuck Greene, Craig Taylor, Carl Wirsen, Jeff Miller, Ken Sebens, Jeannette Baiardi, Howard Towner, Susannah Porter, Jamie Kellogg, John Porter, Hillel Gordin, David Stoddart, John Bythell, John Ebersole, Aaron Adams, Rick Grigg, Sarah Gaines, and John Ogden provided critical ideas for my research, although they may not have realized it at the time we talked. To all the scientists I interviewed and those other contributors I have noted, my sincere thanks and appreciation.

I thank the staff of the WHOI, especially the Marine Policy Center (MPC) and its director, Andy Solow. The MPC provided not only the logistics but a collegial atmosphere within which to do this work. I also want to thank my Oxford editors, particularly Kirk Jensen, who has graciously supported this work since its infancy, and Stacey Hamilton, who enthusiastically carried it along to final publication.

Finally, I want to issue instructions about how this book is to be read: in any order you wish. If you are curious about your generation, start there; if you are interested in the stories of the women, start there; if you want to start at page one and proceed in an orderly fashion, that's perfectly OK also. The point is to be inspired by these stories; to learn something new about nature or about yourself; and to do something that only you can do to ensure that new field scientists will be adequately educated for the twenty-first century.

Contents

Agassiz's Legacy

1

Natural Sciences at the End of the Twentieth Century

The truth is, the Science of Nature has been already too long made only a work of the brain and the fancy. It is now high time that it should return to the plainness and soundness of observations on material and obvious things.

Robert Hooke, 1665

In 1982, near the end of his career and at the beginning of hers, Dr. D. P. "Don" Abbott stood knee-deep in the intertidal off China Point on the Monterey Peninsula and spoke with Gaby Nevitt about her future. She had taken his courses invertebrate zoology and special problems in marine biology while a student at Stanford University's Hopkins Marine Station. Don told Gaby that he felt this was the most exciting time to be a scientist. But, if she chose to go on in graduate school, she should prepare herself by developing her skills in the use of some of the specialized tools becoming available to sense and measure processes in nature. He also told her that she would derive a deep satisfaction from a teaching and research career, but that she would never get rich.

The way Don practiced science was not very different from that of the natural historians of the nineteenth century. He donned hip boots and carried a bucket to collect invertebrates from the intertidal; he used compound and dissecting microscopes to observe the specimens he dissected using surgical tools; he made hand drawings of his observations. He specialized in the taxonomy of ascidians (sea squirts), yet Donald Abbott "knew" invertebrates. His comprehensive knowledge of these "animals without backbones" had established him as one of the world's experts. The questions he inspired in his students, the first dating back to the early 1950s, are still being investigated today, a half century later. As is true of many pure or academic science investigations, while the subjects may appear trivial or irrelevant to people outside the field, the answers many times eventually lead to practical applications, for example, in natural resource management or in environmental or medical fields.

In contrast, Gaby is a scientist at the dawn of the twenty-first century. She took Don's advice and became an expert in the use of some sophisticated tools. Her research today is determining how Antarctic seabirds sense their prey; it is conducted on shipboard (often in 50-foot seas), on remote southern ocean islands, and in the laboratory back in California. The answers to her questions are addressed with techniques ranging from observations made with binoculars to remote sensing of areas of high productivity in the southern oceans; from microelectrode recordings of nerve signals to detection of minute traces of gases in the atmosphere; from transmission electron micrographs to biochemical studies of changing hormone production; from recording thermometers placed in seabird nesting burrows to satellite tagging of foraging birds. Despite the level of sophistication of some of her investigative techniques, Nevitt could not do this work without going into the field. Her story, how she got into this interdisciplinary research, and where it is leading in the future is in chapter 6.

Don and Gaby met as professor and undergraduate student at Hopkins Marine Station (HMS). Because she was fascinated with marine animals, she had chosen to go to Stanford University so that she could take courses at Hopkins. HMS was built in 1892 as a place to conduct summer courses and research on the biology of marine organisms and biology as learned from marine organisms. Over the years, the small seaside campus has brought together undergraduates, graduate students, postdoctoral fellows, and senior scientists in a relaxed yet focused atmosphere conducive to learning and practicing the tools of marine biology. The relationships between teacher and student, while often formal as Don and Gaby's was, were also as mentor and friend. Looking at the picture of the encounter in the intertidal when they talked about her future, Gaby states, "DPA is just how I remember him to be: focused, a little grumpy, and at home in the intertidal. But it could be any student. Probably not any teacher though!" In such small settings as marine labs or field expeditions, the enthusiasm and frankness of students could meld with the experience and dedication of senior scientists to synergistically lead to the production of good science and good scientists.

Don retired shortly after that conversation with Gaby and died a couple of years later. His departure left a significant gap in the educational resources not only of Hopkins Marine Station and Stanford University, but also of the country. His legendary invertebrate zoology course was an intense ten weeks of field trips, laboratory work examining the great diversity in the rich marine fauna of the central California coast, and inspiring lectures oriented to the functional and evolutionary adaptations of the animals observed daily. It provided not only a comprehensive knowledge of the organisms themselves but also a way of approaching their biology that could be applied to any organism and the environment in which it lives. This course, given for more than 30 summers at HMS, was always fully subscribed with students from around the country. Don worked throughout the year on modernizing his course, reading the literature, and continuing to do research on his own special group, the tunicates. Abbott's former students, now at colleges and universities throughout the country, even today use his lecture notes and class organization as the bases for their own teaching. Unfortunately, invertebrate zoology, although perhaps the best subject to introduce a student to the diversity of life forms, not only is not taught on many university campuses any more, it is not even taught at most marine laboratories. The special problems course, an undergraduate independent research quarter focusing on the biology of a selected marine species, was created in the early 1960s by Abbott and his colleagues at HMS. It was initially supported by the National Science Foundation (NSF), which recognized what was, for its time, an innovative approach to educating future scientists.

Ironically, Don Abbott would probably not be hired as a university professor today. His teaching potential might have helped him gain a position at a good, small, liberal arts school, but his record of relatively few published research papers and funded grants would not have survived the present tenure process at a major university.

Gaby did go on to become a modern natural scientist, now associate professor at the University of California at Davis. Unlike many of her contemporaries, however, she is broadly trained: a behavioral scientist with expertise in neurobiology and olfactory physiology, morphology and ecology, invertebrates and vertebrates. Her research on seabird olfaction and salmon homing is carried out at sites around the globe, but somehow (through her 80-hour work weeks) she also manages to pursue research money through grants; communicate results through publications and oral presentations; be a mentor to her own graduate students; and teach undergraduate courses.

When Don gave Gaby advice on her future if she chose to be a scientist, he left out two important ingredients, perhaps because he intuitively had faith that Gaby already possessed them. A natural scientist should know the elements of the system in which he works; he should be a generalist in the sense that he can place his research in the context of the broader system; if he is a biologist, he should understand organismal biology and respect systematics. Also, a natural scientist must learn to sense the environment that he studies, not by visual observation alone (e.g., of behavior of animals; types of rock; numbers and sizes of organisms, crystals, or rock strata; instrument-recorded data) but also by all of the other senses a human possesses. Only then does one truly begin to gain an understanding of nature and allow oneself to be inspired by its magnificence. Only then can one study it effectively in a rigorous scientific manner.

Hooke's Caution: The Pendulum Swings

The pendulum swings between observations and theory as humans struggle to learn more about the natural world. Both are crucial to the advancement of science in the production of new research and of a new generation of scientists. Unfortunately, when the theoretical side of a science dominates the social atmosphere in which science is conducted, the observational side is often belittled. Today, in the halls of academia, natural history is considered passé, a hobby for amateurs. However, if a scientist has never had firsthand experience of a direct interaction with nature, his work will remain only "of the brain and the fancy."

Science is a human endeavor, a product both of individual effort and social interactions; it is "organized curiosity." Science is a mixture of observation, intuition, deduction, conversation, reading, arguments, guessing, and last, but certainly not least, chance. People make observations; people detect patterns; people formulate hypotheses, models, concepts, and theories. A scientist may find himself more adept at one of these skills than at the others, but he should be aware of how his work contributes to that of others (both those who came before him and his contemporaries) and toward the advancement of science. As a scientific discipline evolves, a body of knowledge grows, which is initially primarily descriptive, a natural history. As time passes, more and more theory is developed. Yet, as UCLA professor emeritus George Bartholomew stated in an essay on scientific innovation and creativity, "A logically derivable conclusion is just a logically derivable conclusion, nothing more. Logic and reasoning alone, and all the rules developed for them, cannot stand as science without empirical testing or the establishment of detailed consonance with a body of systematically organized observations" ("Scientific Innovation and Creativity: A Zoologist's Point of View," *American Zoologist* 22 [1982]: 229).

The natural sciences deal with matter and energy, their relationships and transformations. Some natural sciences, physics and chemistry, for example, lend themselves to a deductive investigative approach. Their simplicity and regularity, often stated in mathematical terms, portray an orderliness about nature that belies the confusion, complexity, and frustration encountered by scientists whose questions deal with the spatial and temporal scales to which we are accustomed in our everyday lives. Millimeters to kilometers, seconds to human lifetimes are the scales of human experience. Direct observation in the field is essential not only to determine the basic relationships among things as they exist in nature and the degree of variability that exists but also to provide a source of inspiration, a motivation for pursuing the study. Direct observation underlies the science conducted on and in the Earth, the oceans, the atmosphere, and the living world. It forms the base of the pyramid of knowledge of the sciences of geology, oceanography, biology, ecology, meteorology, and their ever-growing subdisciplines.

As we begin the twenty-first century, environmental concerns dominate national and international news coverage—for instance, earthquakes, hurricanes, wildfires, sea-level changes, disease, conservation, pollution, land-use questions, and loss of biodiversity—but, ironically, emphasis is now placed, in both teaching and research, on theory and modeling as opposed to developing observational skills and experience. The latter are critical factors to ensure that one is capable of not only collecting good data from the field but also optimally analyzing that data by integrating it with personal past experience and the results of other scientists. The educational and research bureaucracies are thwarting rather than enhancing the ability of field scientists and their students to work in the field and to pass the art of field science to the next generation. Knowing where and how to collect valid data, learning how to judge the quality of the data collected by others, and learning how to formulate good scientific questions are special skills, which will be lost if not taught to and practiced by the current generation of students, from which will come our future scientists. It is vitally important that we continue to produce the scientists who can maintain the flow of new information and interpret it correctly in terms of prior knowledge, so that we meet not only the known but also the presently unknown challenges of the future.

Study Nature, Not Books

The philosophical basis of the modern approach to natural science education and research in the United States can be traced back to a Swiss-born natural historian, Louis Agassiz. His contributions to the field of natural history include a broad range of topics spanning fields as diverse as glacial geology and fish biology. He lectured at Harvard College in the mid-1800s, entertaining his audiences with dazzling performances, but was disturbed with the method used to teach American students about nature: rote memorization of textbook or lecture material. Eventually, he developed a new field-based approach, which he implemented by attracting financial support to build a seaside laboratory in 1873 on Penikese Island located off Cape Cod, Massachusetts. Agassiz envisioned this laboratory to be a sister institution of Harvard's Museum of Comparative Zoology, a place where his students, 28 men and 16 women, all of whom were preparing for teaching careers, could "study nature, not books."

In contrast to a lecture or a textbook, in which a natural phenomenon is presented in a logical progression and as clear, often dry, fact, in the field, one is faced with the confusing complexity of nature in its entirety. The lecture or textbook version of a natural phenomenon has undergone a scientist's interpretation and simplification. It is a social product of interactions among individuals through both space and time. It is the scientist's approximation of some truth in nature. For instance, the geologic map that so clearly portrays the spatial relations among various rock formations of different eras is the result of meticulous measurements made by a skilled geologist; the untrained observer will find these same formations little more than a confused, albeit perhaps aesthetically pleasing, landscape. Science does not usually proceed in a linear fashion as is portrayed in the scientific method. While this is often taught to students to illustrate how science is done, it is actually the means by which scientific results are communicated to others. The steps have been put in a logical order, but they are not in the sequence in which they may actually have occurred.

Agassiz's students were urged to find, make observations on, and formulate questions about sea creatures living in their own environment. The seaside laboratory offered easy logistical access to the marine organisms. Each person had ample time and opportunity to focus his attention on nature. In addition to providing firsthand field experience for each student, the marine lab setting offered a supportive social climate. The location of the lab on a small, remote island effectively isolated all of the students and instructors from other, distracting interests for a period. There were the active interchange of ideas and chances to teach others about what one had observed and how one interpreted it, a key element in the development of self-confidence. On Penikese Island, a student had the opportunity to interact on a daily basis with Agassiz, other distinguished natural historians of the day, and his own peers in a positive setting where new observations and discoveries, made in the field, were greeted with enthusiasm and critically discussed. This, in turn, helped prepare the developing scientific mind for future study and the generation of new ideas.

Science is an interactive field. Although an idea, the "right question," comes from an individual creative mind, there are many things in the lifetime of an individual that lead up to such an idea: his introduction to nature, training in the basics of science, inspiration to pursue an area of inquiry, discussions of observations and interpretations, and learning to accept and to make critical evaluations. These are things that come about through direct interaction with other people and through reading the literature.

Agassiz's contributions to the natural sciences were not only as a noted researcher and a distinguished teacher of both university students and the public, but also as an innovative administrator. His vision of a seaside laboratory was brought to fruition through his energy, his creativity, and his ability to convince philanthropist John Anderson of the importance of this educational experiment, which resulted in the donation of land and a generous financial contribution to build the laboratory and its infrastructure. Agassiz's legacy is embodied in the famous quote attributed to him: "Study nature, not books." To advance our knowledge of our natural world, we, the present generation of Agassiz's heirs, must continue to study nature, and we must ensure that future generations have the skills and the opportunities to study nature as well.

Agassiz died after that first summer session of the Anderson School of Natural History. His son Alexander continued the popular and successful field course on Penikese Island the next summer, but he too succumbed to illness shortly thereafter. Anderson withdrew his financial support when the school no longer had the dominating personality of an Agassiz, and the seaside laboratory was closed permanently. The lessons learned during those summers were not forgotten, however. Many of the students from those initial classes went on to teaching careers, and some became distinguished researchers. In addition, a number of them were also instrumental in the founding and fostering of other marine laboratories, including the oldest extant laboratories on the East and West coasts of the United States: the Marine Biological Laboratory (MBL), founded in Woods Hole, Massachusetts, in 1888, and Hopkins Marine Station (HMS) of Stanford University, started in Pacific Grove, California, in 1892. Another student of Agassiz, Nathaniel Shaler, whose ideas initially inspired Agassiz to establish a marine field school, founded the first geology field camp, long a mainstay of a geology degree.

The Agassiz tradition, the direct study of nature, has continued for more than one and a quarter centuries at marine laboratories, terrestrial biological stations, geology field camps, marine and terrestrial field expeditions, and science field trips all over the world. The study of nature firsthand is enriching for all students whether they continue on to become professional scientists or not. Field institutions, programs, and camps have served the intertwined role of education and research. They facilitate access to nature, and they also foster a social climate that encourages rigorous investigation of the natural world through personal experience across a continuum of generations. As we will see, some individuals and institutions have played a particularly significant role in maintaining these traditions. A scientist's contribution to the advancement of knowledge is not only through his own research. It is also through the students and colleagues whom he has taught and inspired and through the institutions (departments, field stations, and camps) that thrive with the help of his support or that, perhaps, were conceived and developed by him.

It is important for a student, especially if he may be thinking of becoming a field scientist, to learn why and how a professional scientist works in the field. As you will see, field experiences are intellectually, and sometimes physically and/or emotionally, demanding. The contact that students have with experienced field scientists, graduate students, and their own peers can be crucial to their development as scientists. They learn that their task is a serious one: what they learn about the world around them is intrinsically important, and it requires dedication, commitment, and experience. In contrast to the clarity of a concept in a textbook, students learn that nature itself is far more complex and confusing. One also learns about the frustrations of fieldwork: things aren't as one thought they would be; experiments and equipment don't work; observations are hard to make and, once made, difficult to interpret; the weather gets bad; one gets seasick; one gets lost; personalities may clash. Conversely, and just as important, one also discovers the joys of fieldwork. An observation leads to a flash of inspiration; one realizes the professor is just human and shares one's sense of humor; people can argue passionately about a scientific interpretation but remain the best of friends; and there are those memories of special moments, shared or experienced alone, like watching a sunset or the green flash at dawn.

Agassiz's Heirs

This book contains oral histories of the lives and careers of some of the people who have continued the Agassiz tradition, from the 1930s to the present day. Their ages when I interviewed them ranged from 83 to 31. Their fields span the natural sciences at the scale of human experience: biology, geology, ecology, oceanography, and anthropology. Each has personal attributes that I feel are crucial to doing good science: creativity; the capacity to translate that into observations, discrete questions, and completed thoughts; and the use of direct interactions with the natural environment to catalyze the process. What each scientist has done in the field obviously will differ from case to case, but all present strong arguments that their presence in the field was critical to how they have done science. They have all learned to make observations in a particular systematic way and, by integrating their previous knowledge and experiences with current observations, to ask meaningful scientific questions. Good science is generated by asking the right questions and by maintaining a tight link between the questioner and the person in the field (who is making observations, gathering both qualitative and quantitative data). As you will discover, observations and insights made while in the field have often led these scientists to their best research.

In addition to doing research, most of these scientists have been teachers and administrators as well. The advancement of science includes producing the next generation of scientists, and both teaching and research require good administrative support. As they relate how they became professional scientists, our subjects recall people and experiences that have influenced their careers. Not all were born scientists; most have followed a convoluted pathway on the road to a career. As children, those of the earlier generations grew up without television and superhighways, but in a world they investigated firsthand and through books and their own imaginations. These stories relate not only successes but also disappointments, self-doubts, emotions, humor, and especially their passion and enthusiasm both for work and for the people with whom these scientists have worked.

The book is divided into chapters chronologically so that we can examine how the pendulum has swung in natural science research and education during the past half century from observation to theory, from breadth in training to early specialization. Scientists do not develop in a social vacuum. Expectations of society, wars, national and global economies, educational and job opportunities, periods of scientific tunnel vision and scientific imagination, and interactions among government, academia, and the private sector have all interwoven to influence their careers. The chapters are delineated by the approximate decade in which the scientists were undergraduates. For most of them, that was the point in their formal education when their interests became focused and when they took college courses in the basics of natural science. Combining discipline and hard work, they tentatively began to acquire the tools of the trade. Each made a conscious decision to pursue graduate studies to become a professional field scientist.

Obviously, the scientists I interviewed in the different generations are at different stages in their careers. The oldest are retired, while the youngest are still searching for a permanent position. The oldest can look back at 50 years of changes, which they personally experienced; the youngest have only a most recent knowledge and experience of an academic career as their standard. At its ideal, science is a scholarly pursuit and a search for truth. However, we will see that the selective pressures for success in an academic career in the natural sciences, in particular the balance of teaching and research (and how one is prepared and rewarded for each), have altered in the latter half of the twentieth century. Institutions have become more bureaucratic, characterized by a specialization of functions, an adherence to strict rules, and a hierarchy of authority. This affects not only the teaching and research organizations but the funding agencies as well.

The natural science community is international. The majority of the scientists whom I have interviewed are American, but some are not. Their experiences and comments about "the American way of doing science," often contrasting it to the style of their home country, help define what is good and what is not in our present way of doing science in the United States. The ethnic and gender composition of the people whom I have interviewed approximately reflects that of the scientists with whom I have interacted professionally in the past 35-plus years. The scientists who tell their stories in this book are all well known within some subset of the scientific community, and although they are not average household names, many have been awarded the highest honors in their fields. As one example, every four years, one coral reef scientist is selected to receive the Darwin Medal, awarded to recognize lifetime achievements in that field. To date, there have been four medals; three of them were awarded to scientists interviewed in these pages: Ian Macintyre, Peter Glynn, and Yossi Loya. I asked about half the scientists whom I interviewed to give me a list of the three most influential people ever in their field, alive or dead, and also a list of the three most influential alive today. Several of the people I interviewed were cited on both lists.

I interviewed most of the scientists in their offices, their homes, or my office. The conversations were taped and then transcribed. It is important to bear in mind that the oral interviews presented within this book are greatly shortened and edited portions of those conversations; these are not autobiographical sketches written and reworked by the scientist. My own comments made during the interviews are indicated in italics. The first time people I interviewed are mentioned in other scientists' interviews, their names appear in small capital letters.

The principal goal of this book is to focus attention on the gradual diminution of the natural history approach in teaching and in research during the second half of the twentieth century. The crucial question we ask is: Are we presently educating our students so that we can continue to develop scientists who will be capable of producing the high quality of research and needed growth of field science to meet our future challenges and environmental needs?

Learning How Nature Works: The Tools, the Data, the Theory, and the Scientists

Nature exists. Humanity's attempt to understand how nature works is through scientific investigation. Technological advances of the late twentieth century have provided a bewildering array of new tools, both equipment and techniques, with which to measure the natural world at unprecedented spatial and temporal scales. Yet tools, no matter how sophisticated they are, do not make new discoveries. The tools are only as foolproof as the people employing them. Where, when, how, and what data to collect are key considerations. Even when the data are available and the measurements are "correct," they are meaningful only in the context of the question being asked. Someone has to interpret the data. And someone needs to articulate the next question.

Much of what we currently understand about how nature works is the result of a complex interaction of qualitative and quantitative observations, conceptual models to account for those observations in a broader context, and mathematical models (i.e., equations) to theoretically account for natural phenomena. More observations may lead to a change in the conceptual model and may prove the theoretical mathematical model totally false. For example, one of the subjects discussed in some of the oral histories (see in particular, Nagle and Molnar) is the rather rapid acceptance during the mid-1960s of the theory of plate tectonics.

In the 1960s, an abundance of observational data was gathered on shipboard by geophysicists (Bunce, Bowin) sampling the world's ocean depths. These included data on the thickness of ocean sediments, temperatures of the ocean crust, magnetic reversal, topography, and earthquakes; each set had a pattern of distribution. When overlapped and interpreted, the patterns were strongly suggestive of sea-floor spreading. Almost overnight, a model emerged: the earth's crust was made up of a series of movable plates, which accounted for the observed patterns.

This new model appeared after decades of almost arrogant denial of the concept of continental drift by some of the same geophysicists. Continental drift— the idea that the continents were once connected but had drifted apart through time—was first proposed in 1912 by Alfred Wegener and others. It was supported by good observational data. Geologists had matched rock strata between North America and Europe; paleontologists had discovered patterns of fossils suggesting lands now separated by oceans were once connected; and biogeographers had found striking patterns in the distribution of animals and plants among the continents, which supported the idea that they were once connected but had moved apart through time.

However, geophysicists of that era could demonstrate with their mathematical models that there was no possible way that continents could move, given the mechanism first proposed, that the continents drifted through the ocean crust. The continental rocks were soft relative to the oceanic crust. As one scientist put it, it would be like "pushing spaghetti through steel." Thus, the geophysicists strongly rejected the concept, and their influence was so strong among their colleagues that many geologists of that period also dismissed the idea. However, with the surge in interest in the oceans during and after World War II, principally due to military needs, new techniques were developed and new data were collected, which led to a plausible explanation. Consequently, in the mid-1960s, there was a dramatic acceptance of plate tectonics as a unifying theory in the earth sciences. Moreover, both the interdisciplinary nature of the data that supported this theory and the relevance of the theory to a wide assortment of geological, oceanographic, and biological disciplines ushered in a period of intense exchanges of scientific ideas across disciplinary boundaries, in contrast to the sharp division of the natural sciences that had preceded it, during which biology and geology were considered as almost mutually exclusive disciplines. Unfortunately, in today's world of specialization and reductionist approach to problems, the sciences are reverting to the idea-stifling compartmentalization mentality again.

The advent of the computer age has glorified the role of modeling in scientific research. While models are an integral part of how a scientist does science, the number crunching and elegant three-dimensional graphics of modern computers are but one stage in learning something new about nature, a tool to use but not an end in themselves. While this is obvious to scientists who first had field experience and later incorporated computer modeling into their repertoire (Bowin, Wiebe), some modern students are brought up with computers as their main source of information about nature, rather than the natural world. One factor characteristic of nature (and part of its appeal) is its variability. At some institutions, biology degrees are awarded to students who have never seen a whole organism in a laboratory; geology degrees are awarded to students who have never been on a field trip. The former may have done dissections on a computer; the latter may have learned to map using a computer. But neither has experienced the reality of nature.

Nature is often unpredictable and, at the scale of the earth and biological sciences, the chance event may have a big effect. For example, in the 1950s the ecological dogma of the day declared that tropical ecosystems, such as coral reefs and tropical rainforests, were complex, diverse, yet stable systems. The problem was that no one had looked at reef systems (or tropical forest systems) over time. When some people (Glynn, Loya, Bak) began to examine coral reefs throughout the world (and other systems as well, e.g., polar; Stirling), they made repeated (and, most important, repeatable) measurements over time. On reefs, we now recognize that chance events, be they natural (hurricanes), or manmade (oil spills), or perhaps a little of each (e.g., disease), occur. When systematically gathered data collected before and after such chance events were analyzed, a different and more complex picture began to emerge. As the modern story of the historical evolution of the Earth (Burchfiel, Hoffman) and of reefs (Macintyre, D. Hubbard) developed by geologists continues to be combined with the ecological changes documented over the last 30 years of the twentieth century, it is clear that coral reefs are indeed changing today, and their presence and distribution represent an unusual period for the Earth when viewed over geological time. Recent reefs date from the sea-level rise after the last glaciation. Nature has recorded climate change on island shores (Nunn), in glacial and wind deposits (Porter), in river deltas (Stanley), and in polar ice caps (Alley). It is the work of scientists to develop the means to decipher those records.

Often, it is an unforeseen piece of evidence, uncertain as to when it will occur in time and space, that leads to a major new discovery. For example, oceanographers followed up an observation of anomalously high water temperatures in the ocean depths by revisiting the site using the deep-diving submersible *Alvin*. No one expected that this exploration would reveal a rich, complex ecosystem, the Galapagos Vent Community, which has no dependence on the sun for its energy. Instead, at the base of its food web it has chemosynthetic bacteria (Jannasch), rather than photosynthetic organisms, like green plants or phytoplankton, which we know form the base of the overwhelming majority of terrestrial and aquatic ecosystems. It and other subsequently found vent communities are filled with a rich assortment of large-bodied, but little known or previously undescribed invertebrates. Nature has surprised us once again. It is not simple; it is complex. To discover a key piece of evidence, to unravel another of nature's mysteries, one must be in the right spot, one must be able to make and to recognize the significance of the right observation, and one must then know the correct way to proceed. Chance plays a role, but one must be prepared.

The Modern Scientist

The actual mechanics of the educational process by which we became biologists, geologists, or oceanographers has not really changed much in more than 50 years. After an undergraduate degree, usually but not always in a science, the student eventually enters a graduate program, sometimes obtaining a master's degree before proceeding in a Ph.D. program. The process of achieving one's doctorate in a science is a pivotal period in one's life when one demonstrates the confidence and the ability to do original work and the ability to communicate the results to others. Unlike students in professional schools of medicine and law, a future field scientist has no prescribed series of courses to take. Nor does he have a standard examination to pass to qualify for certification. Rather, the student meets certain broad standards by passing written and oral qualifying exams (designed by the graduate faculty in each institution) and conducts a piece of original research. *Original* is the operative word. The research, communicated in proper written form as a dissertation and often defended in an oral presentation, is deemed acceptable for a doctoral candidate, as assessed by his graduate committee. These rather ambiguous hurdles are what have been required for more than a half century to obtain one's union card of academia: the Ph.D. Prior to that time, bright, innovative, observant students who learned in the field alongside experienced people were able to become good scientists and to gain academic employment without a doctorate.

With doctorate in hand, the new scientist continues his career. Sometimes, the next stop is a postdoctoral position, where he contributes his skills to an ongoing project or, if lucky, develops a new set of skills to broaden and enhance his repertoire. If the new Ph.D. wishes to emphasize research in his career, an independent research facility, such as Woods Hole Oceanographic Institution (WHOI), or a major university offer the best opportunities, but a small liberal arts college is more conducive to combining research with teaching undergraduates. At top institutions of either sort, competition for faculty positions is intense, and in some cases, hundreds of qualified applicants seek a single position.

An academic job at a major university, finally obtained, may involve some teaching, but increasingly through the years, and much more important to the institution, is published, grant-supported research. For the most part, most academic institutions still have the tenure system as part of the promotional process. After several years as an assistant professor, one's record is critically reviewed both within the department and the institution and by outside academics in the same field. Nominally, teaching and service contributions (such as committee work) are counted in the tenure decision (or any promotional decision, for that matter), but at most institutions that emphasize research, publications resulting from grant-supported research and the amount of grant support are the most influential factors.

To get funding for a project, a scientist applies through his institution to a monetary source, usually by means of a grant or contract proposal. In the United States, most of the funding for basic research in the natural sciences—geology, oceanography, ecology, and integrative and comparative biology—has come from the National Science Foundation (NSF), the National Oceanic and Atmospheric Administration (NOAA), the Office of Naval Research (ONR), the National Institutes of Health (NIH), or the Department of Energy (DOE). Private funding from oil companies has been an important source of both research and educational monies, particularly for geology. A variety of other public and private sources have been periodically available through the years, including various foundations, the U.S. National Park Service, the U.S. Geological Survey (USGS), and state geological surveys, which have employed students for mapping projects, for example, that contributed to degrees.

Research results are communicated to the scientific community through oral presentations at scientific meetings and through publications in the scientific literature. A startling new discovery may be published as a short article in the weekly journals *Science* or *Nature*. Lengthier communications are usually published in journals specific to a field, for example, *Ecology*, *Journal of Sedimentary Research*, *Quaternary Research*, or *Coral Reefs*. There are hundreds of journals in the subspecialties.

Publication and funding both require peer review. This is a process by which a study written up for communication in a scientific journal or a research proposal is sent out to several anonymous reviewers (who work in similar research areas and presumably can judge the quality of the work or of the proposed work). Articles are written by the scientists who did the research and are sent to the editor of the journal that the scientist deems is most appropriate for the topic, most prestigious, and most widely read. The editor sends out the manuscript for review. Based on the opinion of the editor and the reviewers, the article is either accepted as written (not common), accepted with suggested revisions (which the author then proceeds to incorporate or argue against their inclusion), or rejected. In many cases, funding proposals are sent to a program manager, who sends out copies for anonymous review, later convenes a panel of scientists to rank the submitted proposals, and recommends which should receive funding. Reviewing papers and grant proposals, as well as writing recommendations for students and colleagues for career-advancing opportunities, such as graduate school, jobs, or promotion, are time-consuming processes, but they are taken seriously by those scientists who realize the function these activities play in the advancement of science.

Research does require money. A scientist has a social obligation both to the people who fund his work (often ultimately the public) and to his scientific colleagues to communicate his results in a timely manner. However, increasingly today, research dollars also support a complex bureaucracy that is far removed from the science being done. The budget proposed in any research grant or contract includes significant overhead to support the institution's administrative activities. Many scientists are dependent on grant and contract support for their salaries, which is referred to as "soft money," as opposed to "hard money," salaries paid for by the institution's own funds.

Science at major universities has become a big business, and academic success often goes to the scientist who can best contribute to the institution's financial bottom line. Thus the traditional components of an academic position—teaching, service, and research—have become highly skewed in favor of research. As competition for grant funds is fierce, faculty members have become more specialized; they focus on small subsets of a major discipline, in which they are most likely to be successful in continuing to secure grant funds. Many departments in the natural sciences are now dominated by a reductionist or a theoretical—as opposed to a natural history—approach because these are the types of research that attract the largest proportion of research dollars; often they can turn out quicker results than field studies, which require the investigator to operate on nature's time schedule. As examples, in the biological sciences, cellular and molecular biology have successfully captured the biggest grants and now control the undergraduate as well as the graduate curricula in many departments. In geology, some graduate students can now obtain doctorates by determining the chemical makeup of a single rock (collected by someone else); one can graduate with a degree in oceanography without ever setting sail on a ship. Over time, a faculty becomes more specialized. Specialists are hired in place of generalists (who would be unsuccessful in securing large grants); generalists are growing old and retiring. Breadth of knowledge within a department (or a field facility) decreases, which leads to more specialized courses. Consequently the breadth of the background given an undergraduate science major decreases as well, which leads to a vicious spiral. At some point in the education of a scientist, he must be exposed to theory and facts, to breadth in a field, and he must develop his unique set of skills in order to become expert in some area of inquiry.

The trend has been for departments to hire more and more specialists, who can compete successfully for grant monies. Specialization is important and an inevitable consequence of the growth of a science, but it should not be at the expense of general knowledge. Generalists are often not only the best teachers (they simultaneously present the basics of a science and inspire students), they also provide broadly based expertise, guidance, and inspiration for graduate students and senior scientists within a department. At the scale of human experience, the ability to synthesize and integrate material from a variety of disciplines and approaches and the ability to put that knowledge in a wider context is important. Unfortunately, the strong trend toward specialization, at the exclusion of generalists, has occurred not only on university campuses but at field institutions as well. Not only have research and course offerings become more specialized, but this trend has also resulted in the demise of some field labs. Universities grew after World War II and then needed to consolidate in the 1980s; the remote field station or summer field camp was often an easy target to eliminate.

A large university requires its science faculty members to support not only themselves, but also their graduate students, their departments, and the institution itself with monies from research grants. Obtaining this funding is an endless process. Writing and submitting, rewriting and resubmitting grant proposals; publishing papers; keeping up with the relevant literature; generally making oneself well known by giving papers at meetings and seminars; and serving on peer review committees and editorial boards—all leave precious little time for actual research work, let alone for teaching. Scientists add quick publications to their resumes by presenting partially completed stories again and again. At many institutions today, scientists are writing more funding proposals than research papers. In both their papers and their proposals, their time is spent rehashing the interpretations of "how nature works" written by other people, rather than examining how nature works firsthand, through data collected by themselves for this purpose.

Collectively, the scientists whom I have interviewed have been doing research, teaching students, and administering programs and institutions continuously for a half century. Some of the retired scientists continue to contribute to the advancement of science, conducting research and advising students, as professors emeriti. The research results of the people whom I have interviewed, and that of people whose careers they have influenced, play an important role in our current understanding of how the natural world works. Government bureaucrats may talk about biodiversity and climate change, but it is the scientists who have the knowledge and the skills needed to gather and interpret field data, to integrate their results with those of other scientists, and to address the questions that will arise in the future.

How I Inherited Agassiz's Legacy

I am fortunate to have been a student during the golden age of the field sciences, from the early 1960s until the early 1980s, when funding was at its most generous level, especially for those research activities that brought together scientists of different disciplines and different generations. Like Gaby Nevitt, I was a Stanford undergraduate. My first field course was also at HMS; in the summer of my sophomore year in 1966 I took Dr. Isabel Abbott's marine algae course. As a college senior, I also took special problems at HMS with the help of a tuition grant from the NSF. I found independent scientific research to be simultaneously daunting and fascinating.

My family was academically oriented. Our dinner table was often the site of spirited discussions; the unabridged dictionary and the world atlas were consulted frequently. We spent summers together, hiking and backpacking in the Tetons and the Canadian Rockies. As the youngest child of four, I followed in my siblings' footsteps and took courses in biology, geology, anthropology, world history, and literature; these subjects still form the bases of our mutual interests.

While I learned how to do formal scientific research at Stanford, my strength (according to my graduate advisor, Len Muscatine at UCLA) was my field experience and knowledge of organisms. My husband, Bill, whom I had married at the beginning of my sophomore year in college, is especially intrigued with the diversity of life. Taking advantage of the natural marvels of California, we took field trips to the rocky intertidal, to the deserts, to the mountains, to mudflats, to the coastal redwoods, and to the marshes, finding and identifying the animals, plants, and fungi in each. Bill did most of the work in identifying the organisms, using guides and taxonomic keys; I absorbed the knowledge from him.

While Bill was in graduate school, we spent four summers at Friday Harbor Laboratories (FHL), the University of Washington marine lab founded in 1904 in the San Juan Islands. FHL held graduate classes each summer and hosted scientists from around the world. I would visit the various labs and ask people questions about their research. I also took classes, including invertebrate zoology and fish biochemistry. I loved the total immersion at the lab: learning about the animals, going out to collect specimens, keeping them in the lab tables supplied with running seawater for observation and experimentation, interacting with scientists doing interesting work, looking up background information in the library. It was fun, and it was exciting. This was a great way to learn!

Another important aspect of the time at Friday Harbor, one that was to play a role later in my life, was that I began to observe how the marine laboratory operated. Each summer more than 200 people—students, teachers, researchers, and their families—would gather on the 484-acre FHL campus. I was always interested to watch how the director, Dr. Robert Fernald, handled the inevitable problems that arise when resources (of one sort or another, including housing, boats, tank space, research space) become limiting. His firm yet patient and consistent manner in dealing with these minor crises stuck with me.

I taught high school biology and grammar school science at a prep school while Bill finished his Ph.D. In 1974, Bill took a job at the West Indies Laboratory (WIL), a teaching and research lab in the Virgin Islands. Only three years old, WIL had already distinguished itself by its combination of teaching and coral reef research. The students, upper-class science majors from colleges and universities throughout the United States, spent a semester at the lab. They took three consecutive, full-time courses: tropical marine biology, marine ecology, and marine geology, each combining lectures with extensive field and laboratory projects. This was followed by a five-week independent study, during which the students planned their own field schedules and laboratory work and completed a faculty-edited research paper.

Coral reef science was a rapidly emerging field by the early 1970s; most of the researchers who worked at WIL were coral reef scientists. Since coral reefs are produced by organisms and have a high diversity of organisms but are so large that they form geomorphic structures and the current ones have been developing over thousands of years, they are of interest to both biologists and geologists. Ancient reefs are a major reservoir for the world's oil supplies so oil companies have an interest in them. Howard Odum and Eugene Odum had published a landmark paper in 1955 on the trophic structure of the coral reef ("Trophic Structure and Productivity of a Windward Coral Reef Community on Eniwetok Atoll," *Ecological Monographs* 25 [1955]: 291–320); this work inspired many aspiring scientists of following generations. As in many scientific fields, in coral reef science, a technological advance, in this case the easy availability of scuba gear, permitted the rapid development of a new discipline. In the late 1950s through the 1960s, Tom Goreau, in a prodigious effort, made innumerable dives, described the geology of the reefs in Jamaica, and did some classic work on the physiology of reef corals. As important as his research, however, was his mentoring of a whole generation of coral reef scientists (Loya, Bak,

Fautin) from universities throughout the world at the Discovery Bay Laboratory on the north coast of Jamaica, where he did his work and which became the first famous interdisciplinary coral reef laboratory.

In the mid-1980s one comment, made by an anonymous peer reviewer of an NSF facilities grant proposal for WIL, stated that the list of researchers who had worked at the West Indies Laboratory read like a "who's who" of coral reef science. Visiting the lab to do research were carbonate geologists and fish ecologists, natural product chemists and physical oceanographers, coral physiologists and evolutionary biologists, structural geologists and taphonomists, behavioral ecologists and functional morphologists, phycologists, benthic ecological physiologists, and, on occasion, even desert plant ecological physiologists and lizard ecologists. Many WIL visiting scientists combined youthful enthusiasm and extensive field experience with a dynamic approach to asking questions in the environment. Having the opportunity to work in the field with them and to discuss observations and ideas in an open-ended forum changed lives, including my own. Scientists and students learned from one another. The interaction of coral reef scientists trained in many different disciplines, but all focusing on the integrative theme of the coral reef ecosystem, is a good model to keep in mind as we read the interviews in this volume, as we consider such topics as climate change, biogeology, and symbiosis, and as we ask how science might best be conducted in the future.

The WIL resident scientific staff provided not only site-specific knowledge to help scientists orient to a new area, but also a breadth of scientific disciplinary expertise and experience. My husband, Bill, a D. P. Abbott student, did research on the biology and ecology of a variety of invertebrates: cnidaria (sea anemones, hydroids, reef corals), crustaceans, echinoderms, sponges, mollusks, and coral reef fishes. He knew the taxonomy and biology of the local tropical marine organisms, plus he knew where to find them and how to keep them alive. John Ogden, the director, believed in a comparative field approach to questions, and he encouraged visiting specialists to collaborate and expand on his studies on behavioral ecology and plant-animal interactions. Dennis Hubbard brought a background in sediment dynamics to a modern carbonate environment. He combined that research interest with a study of how coral reefs grow, plus he introduced hundreds of biology students to the principles of marine geology. John Bythell completed the field research for his University of Newcastle (U.K.) Ph.D. at WIL and then returned to add his expertise in reef coral organismal physiology and ecology. This combination of academic interests and approaches to problems provided the WIL staff, as well as visiting students and scientists, with an interdisciplinary approach to the coral reef, which was fostered in formal and informal lab activities.

For the first few years, I did a variety of jobs around the lab, including teaching assistant (TA) and research technician. One evening at a lab party, an undergraduate said, "Betsy, when are you going to go to grad school?" When I replied, "Sometime," he asked me how old I was. When I answered, "30," he said, "You ain't getting any younger, you know." I applied to two schools and was encouraged by the response from Len Muscatine at UCLA.

He immediately answered my letter, stating that he thought the work I was currently doing (on the reef coral symbiosis with intracellular algae) sounded of interest and related to research being done by him and his graduate students. I had asked, Did the location of the coral (shallow or deep, exposed or protected from wave action) affect how much carbon is fixed by algal photosynthesis and how much carbon is deposited as skeleton?

In some ways, my graduate years at UCLA were "the best of times, the worst of times." Here I was again, back in the spring quarter independent research mode, finding research to be simultaneously daunting and fascinating. To my delight, at UCLA I found a number of active labs led by dynamic advisors and full of hard-working, creative graduate students, where I could find stimulating discussions about invertebrates, plants, and vertebrates, ecological physiology and ecology, functional morphology and physiochemical biology. Our lab studied symbioses between animal hosts and plantlike unicellular organisms. Len ran a very strict lab. We all went through the same steps; we were trained in a variety of techniques, such as using the radioactive substance C^{14} to determine organic productivity (see Ryther). My labmates took me in hand: discussions with one led to the idea for my Ph.D. research; another generously allowed me to duplicate his comprehensive reprint collection on coral biology, which dated back to the early 1800s.

This Muscatine lab didn't happen by accident. Len carefully crafted a formula to help students become independent investigators. While many of his students worked on the *Hydra* symbiosis and pursued basic biological questions (for example, cell recognition), Len insisted that each student accompany him at least one summer in the field, be it in Hawaii at the Coconut Island Lab, at the Discovery Bay Lab in Jamaica, or at Friday Harbor Labs or at Enewetak in the tropical Pacific. He felt field experience was a requisite for a doctoral degree in biology. Although I never did research with Len in the field, we did spend a memorable few months at Catalina Island teaching in the UCLA marine biology quarter.

After passing my orals at the beginning of my second year, my research went fairly smoothly, although I really had no idea of what I had until I needed to organize it into a three-page summary for an American Association of University Women (AAUW) dissertation fellowship for which I was applying. My question was: How does the presence of zooxanthellae (dinoflagellates living within the animal's cells) lead to light-enhanced calcification in reef corals? Like many Ph.D. students, I never answered my original question. Basically, what I did determine, however, was that the coral builds its skeleton in several phases. It deposits a framework and then, as other workers stated so succinctly, "fills in the bricks and mortar" by depositing the bulk of calcium carbonate during the daylight hours. I also worked out the mechanism by which metabolites and other materials were transported throughout the coral colony, through a technique developed by one of the undergraduates at WIL. I did my field studies in St. Croix, and I brought material collected at specific times in the daily cycle back to UCLA to examine differences in both skeleton and tissue using scanning electron microscopy (SEM) and other techniques. The beauty of the SEM images were a source of inspiration, which kept me going in the occasional times of graduate student despair.

Graduate school was more than just a place to learn how to do research, however. It was also the place to learn how to teach. This was another of those rather ambiguous phases of graduate school. No one ever actually explained how to teach; you picked up the skill by practicing as a teaching assistant. As you'll see in many of the interviews, it was only when confronted with having to teach a subject that people really began to understand things they had learned. It is like going into the field to do research. No matter what you've done to prepare yourself beforehand, the complexity and dynamics of a classroom are only learned through experience. When a student asks for an explanation, one's comprehension of a concept is truly tested.

Two days after I turned in my dissertation at UCLA in January 1983, I began to teach as an assistant professor at WIL. Although research activity was still high there, the number of students taking the semester courses was declining. This was partly a function of the period (the Reagan era when NSF funding for science education was at a record low), but also due to the fact that after the first few years of the lab, little thought had been put into attracting the serious preprofessional student we felt benefited most from our program. We began an active recruiting effort, which involved visits to many of the good, small, liberal arts colleges and private and public universities of the Northeast and Midwest. This activity fell largely to me. I also got involved with a number of other, more administrative jobs at the time. For instance, the National Science Foundation had a grant competition to secure facility funding. After the first WIL proposal, to which all of the resident staff contributed, I took the lead on the next five grants, all of which we were awarded. Through these administrative activities and through giving research seminars at major national and international meetings, I met a lot of other scientists in a number of fields. When John Ogden took another job, he suggested to the dean, Lance Blackshaw, at Fairleigh Dickinson (the university in New Jersey that owned WIL) that I be appointed the director of the West Indies Lab. I was.

All went well the first year. Our recruiting effort had resulted in full classes with topnotch students; I and several colleagues had just received funding for two big field research projects; the lab was bursting with visiting researchers. Then came Hurricane Hugo in September 1989. To make a long story short, the university that owned the lab dallied with its options for almost a year, while we continued research and teaching in buildings we had repaired. The summer following the hurricane, the university board of trustees finally decided to close the lab and cash in on the insurance money.

In 1994, I did a small research paper for a group at WHOI that was under contract to the Environmental Protection Agency (EPA) to assess the potential for establishing a regional protocol to deal with "land-based marine pollution in the Caribbean." There was a meeting in Barbados. I met Arthur Gaines, a scientist with the Marine Policy Center at WHOI; we talked about people, institutions, and research in the field sciences. He couldn't believe that I had never been to Woods Hole—the "mecca of marine science," I remember him calling it—and he invited me to do a project there sometime. In addition to the Marine Biological Laboratory (1888), the little village of Woods Hole is also the home of the Woods Hole Oceanographic Institution (1930; descended from MBL and thus a direct descendant of Agassiz's Anderson School of Natural History), a National Marine Fisheries Laboratory (1872); a USGS lab (1962); the Sea Education Association (1971; undergraduate oceanographic semesters); the Children's School of Science (1914; summer field courses for children); and several other scientific research institutions. A few years later, I accepted the invitation to be a guest investigator at WHOI and began this book. The concentration of scientists in Woods Hole and its proximity to the scholarly metropolis of Boston have provided an intellectually stimulating environment to contemplate the issues I discuss. It has afforded, as well, frequent contact with many of the individuals interviewed, who either work at one of the local institutions or have visited to give a seminar, attend a meeting, or confer with colleagues.

This book has evolved from a combination of my curiosity about how my friends first got interested in science, my love of Studs Terkel's books, and my concern, particularly by the late 1980s, with the background preparation of students who came to the West Indies Lab. They came for a semester at WIL, upper-class science majors from dozens of good colleges and universities throughout the country. While they were certainly intelligent, many had had little laboratory experience, and most had almost no field experience before coming to the lab. Hardly any of the biologists had had a geology course, and many of the geologists had not had a college biology course. Within those disciplines, it seemed as if they were taking courses in which they were learning more and more about less and less. We could give them the conceptual and practical tools they needed during their semester with us, however, because all of our courses had the integrative theme of the coral reef. The biology of the organisms and the ecological and geological principles that operate in a coral reef setting could be studied through field and laboratory investigations. This approach could then be applied to any new ecosystem a researcher might study later in life.

Are We Currently Producing the Next Generation of Agassiz's Heirs?

It is true that the technological advances of the late twentieth century have provided an array of new tools useful for investigating the natural world. One result is specialization, the creation of many more subdivisions of disciplines. In the mid–twentieth century, biology and geology were separate disciplines. Within biology, one was often either a botanist or a zoologist, but most students were taught about the classification of organisms and the morphology and physiology of some subgroup, such as plants, invertebrates, vertebrates, or a class of vertebrates (ornithology, mammalogy, herpetology, or ichthyology). In any case, one could say most were trained with the idea of the organism as a central, unifying theme. In geology, the course in stratigraphy, in which students were introduced to the unit "the rock stratum" provided a context in which to place other subjects, including mineralogy, paleontology, and petrology.

Ironically, at a time when our ability to observe and measure our surroundings is far greater than at any previous period, essential elements are threatened in our current way

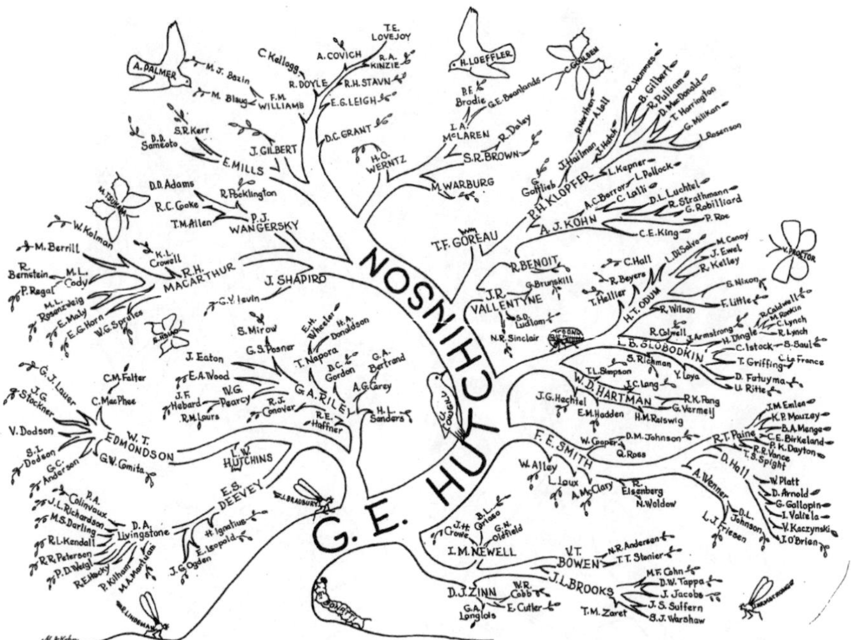

The Hutchinson Tree, showing the academic descendants (Ph.D.s) of G. Evelyn Hutchinson of Yale University as of 1971. Reprinted with permission from *Limnology and Oceanography* 16(2) (1971): 163.

of educating scientists in how to investigate the natural world. The "thorough knowledge of the elements of the system" is either neglected completely in many current undergraduate curricula or, in the best cases, severely diminished. Practical experience, whether in the field or in the laboratory, is often deemed too costly (or to pose too much liability) to maintain as a part of undergraduate curricula. Observational skills are not considered a basic requirement for a biology or a geology degree. It is important to go to the field to experience nature in nature, to see the big picture so that one has a better context in which to place one's own small piece of the puzzle. Also, the field is where the intangible element exists, where "excitement happens." Scientists of today are often so consumed in raising money to support their research that their office and laboratory doors remain firmly shut; few students get the opportunity to work alongside practicing scientists, to learn how to do science by watching how an expert does it. John Grotzinger, a biogeologist from MIT, recently told me that in academia today, "We are all so busy running around and chasing ants, we wouldn't know if there was an elephant among us."

I used to consider coral reef science as the proverbial "small world," but I have found through doing research for this book that the natural sciences, including the earth sciences, oceanography, and integrative and comparative biology, is a small world too. In 1988 Daphne Fautin and others published their "academic family tree" to honor Professor R. I. Smith from UC Berkeley upon his retirement. Ralph Smith was a Harvard Ph.D., and

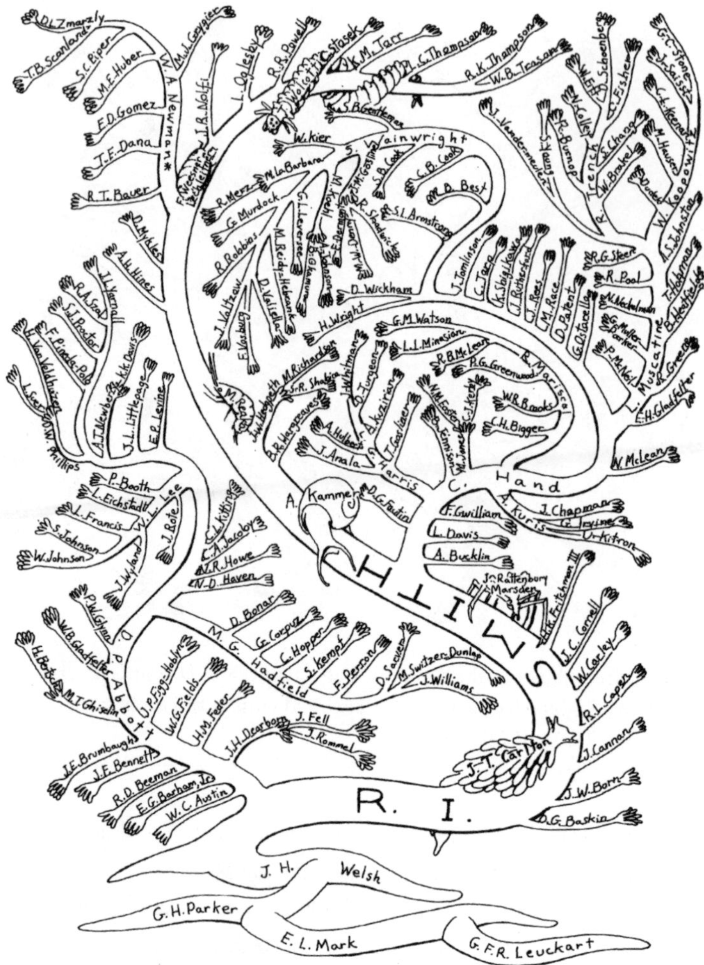

The Smith Colony, showing the academic descendants (Ph.D.s) of Ralph I. Smith of the University of California at Berkeley as of 1988. Reprinted with permission from *The Veliger* 31 (1988): 137.

his lineage dates back to G. H. Parker, one of the professors who taught in the first class at Agassiz's lab on Penikese Island. I, too, am such a descendant; I am a "great-granddaughter" of Smith, as it were. D. P. Abbott, who began this chapter, was in the second generation of the Smith tree.

In 1971, a decade before the Smith tree appeared, a Hutchinson tree had been similarly published in a tribute to G. Evelyn Hutchinson by his academic descendants. Professor Hutchinson, a Yale faculty member who did not obtain a Ph.D. himself, produced an academic lineage that dominates the fields of terrestrial and marine ecology today. In the year 2002, many of the little leaves on those trees are sturdy branches themselves. For instance, on the Hutchinson tree, Yossi Loya has produced 13 Ph.D.s and Alan Kohn 21

Ph.D.s to date; and, on the Smith tree, Steve Wainwright has produced 27 Ph.D.s. There are second (Kohn), third (Loya), fourth (Harvell, Shulman, Dacey), and fifth (Nowlis) generations of the Hutchinson tree among the individuals I have interviewed, and third- (Wainwright, Glynn) and fourth- (Fautin, Koehl) generation members of the Smith tree.

Another instructor in the first Penikese class of 1873 was David Starr Jordan, who had been a student of Louis Agassiz. Jordan became president of Stanford University and was instrumental in the establishment of Hopkins Marine Station. He also produced influential academic descendants, including Carl Hubbs (who taught at Scripps Institution of Oceanography) and Bill Gosline (who taught at the University of Hawaii). Hubbs's students included Boyd Walker (who taught at UCLA), and among Walker's students are Bill McFarland and Dick Rosenblatt, who was Bob Warner's major professor and Maggie McFall-Ngai's ichthyology professor. At Hawaii, Peter Sale was a graduate student of Bill Gosline, while Phil Lobel had him as an undergraduate advisor. The individuals on these trees not only have shared a tradition of how to do science, including the relevance and importance of fieldwork, but also have a ready-made family of contacts. As you will discover, many of the scientists interviewed have been influenced by these and other distinguished academic lineages. These lineages form webs, which weave among one another through collaborative efforts at the undergraduate, graduate, postdoctoral, and senior scientist levels. Furthermore, the scientists featured in this volume have been educated and have worked at institutions that have a long tradition of excellence in the natural sciences. Will the next generation have the benefit of the same traditions?

2

The World War II Generation

Science is an adventure, not a career.

Holger Jannasch

The Great Depression followed by the world conflict revolutionized society. Educational and career opportunities expanded to include people who would never have dreamed of them previously. Each of the World War II generation scientists (all more than 70 years of age) exudes warmth, humor, and a zest for life. They share with us their love for nature, for science, for teaching, and for people, including mentors, colleagues, and students whom they've worked alongside during their long careers. Most recall, in great detail, one or more influential mentors, who taught them not only about field science but about life as well. I believe it is these formal yet intense personal relationships that instilled in these scientists the need to provide similar opportunities to future students. It is clear that they think creatively not only about scientific research but also about the educational process. Experiences, particularly during the war years, gave many of these scientists a glimpse of a broader world. Not only did they meet people who grew up in other places and thus had different perspectives on issues, but many also traveled themselves as part of their wartime occupation. Perhaps this exposure to new people, new places, and new ideas helped initiate and then sustain the broad view they brought to their scientific disciplines and careers. Nevertheless, none strayed far from home during his undergraduate years (often starting in junior colleges) or even for graduate school. Almost every person interviewed recalled the struggle to find enough money to support himself in school and early in his career.

I still feel the best ideas I've pursued have come from observations of organisms in the field and asking, What are those animals doing?

William McFarland (b. 1925)

Professor Emeritus of Ecology and Systematics
Cornell University
Interviewed at his home 13 July 1998

In the mid-1970s, Bill McFarland visited the West Indies Lab to do research. His work on fish vision, including the adaptation of coral reef fish visual pigments to their light environment is considered classic (see MCFALL-NGAI). He frequently inveigled students (including me) to assist him in the field by arousing their curiosity about fish and the behavior of light in the sea; truthfully one was eager to go along because Mc's boundless enthusiasm is infectious. I interviewed him at his home on San Juan Island, where, although retired, he continues his research and helps guide graduate students at FHL.

So what you are asking me is how did I get into the fish business? I suppose it started when I was a little kid. I was raised in southern California, prior to freeways, prior to smog. Every weekend, at least during the spring and summer, my parents would picnic at the beach with friends. When I would surf with the older kids in 10- or 12-foot waves, it used to terrify my dad, but not my mother, who was much more laid back about it. The sea just became part of me. Just before World War II, when I was 12 or 13, I encountered ocean life in a firsthand manner. Bard, an older boy on the block and built like a beer barrel, used to collect lobster off the Palos Verdes peninsula. He invited Bob Carty, my childhood buddy, and me to go diving. We used funny-looking goggles and did not have swim fins. Ugh—the first thing I had to do was to swim out through the rocks over a bed of *Phyllospadix*. Algae was all over my body and scared the heck out of me until I got out a little farther and could see down. All of a sudden just a whole new world opened up to me. I had never seen this before. There was all the beauty of the invertebrates and the algae, and there were a few fish swimming around. I continued to snorkel throughout high school. Of course, by then we had regular masks and Owen Churchill swim fins.

During the war, I was trained as a marine engineer at California Maritime Academy, and I went overseas on a tanker delivering high-octane gasoline in the Pacific and Mediterranean theaters. I had also, while in high school, become a hotrod builder. I can't explain why, but I loved doing things with my hands. I built my own dragster and raced it on the dry lake that is now Edwards Air Force Base. The result of that and the academy training is that I gained experience in building things.

Nothing had really gelled as to where I was really going with my life. After the

Still studying fish vision in 2000, Bill McFarland now works at Friday Harbor Labs. Photo by William Calvin.

war, I went back to a junior college and then on to UCLA. As a chemistry major, I was told that I had to minor in two other sciences to get my degree. So I chose geology at first. This was long before plate tectonics. People knew about [Alfred] Wegener, but nobody paid any attention to the ideas. I took geology and I liked it a lot, but it wasn't modern; the excitement of continental drift just wasn't there. It was all: learn the minerals, learn the strata. I decided to take zoology as my second science. In those days, the first semester basically was invertebrate zoology and the second semester, vertebrate zoology. When I took it, my lecturer was Ted Bullock, who is now a very famous man, a leader in comparative neurobiology. I sat in the class with 400 other students, most of whom were pre-med. Bullock was unusual because he did not follow the textbook. He would come into class holding the latest journal. He had been reading it the night before, and he would tell us about the excitement of the biological research he had just finished reading about. One of the topics was the nature of the action potential; it was the period when [Sir Alan Lloyd] Hodgkin and [Andrew Fielding] Huxley were demonstrating the ionic basis of nerve conduction. The pre-meds went crazy with this guy; they were dismayed. I was fascinated; not so much about what he was telling me, but because here was a human being who you could tell loved what he was doing. Right then and there I changed my major from chemistry to zoology. And, yes, it took me another year to graduate with my B.A.

I ended up going to graduate school at UCLA because I didn't know that one should really get an undergraduate degree and then go on somewhere else. There was no NSF support. There was nothing in those days. When I graduated I ended up walking into the fish lab, because I was interested in fishes. I asked the fish guy, Dr. Boyd Walker, if he would consider sponsoring me for a master's degree. He said, "Sure." To make a long story short, it went from master's to a Ph.D. I fell in love with fishes, and I still am in love with fishes, to this day. The beauty of the training I got at UCLA as a graduate student was, in part, due to the fact that Boyd ingrained in his students the importance of knowing your animals. First their systematics and then their ecology, behaviors, and physiology. Boyd was never much of a publisher, even though he was a gifted researcher in systematics of fishes, but he was great with students. Another important factor was that I was privileged to attend grad school with a bunch of people who have become, quite frankly, famous fish biologists in our country. The interactions among us had as great an effect on our development as many of the courses we attended. The other professor who influenced my development, of course, was Dr. Fred Crescitelli. I was interested in function. I don't know why. What made animals work? Why did they do what they did? So I didn't work in systematics, even though Boyd insisted I be trained in it. I worked in physiology and morphology, adaptations of fishes. That just fascinated me.

But, while I was a graduate student, I also had to earn a living. At that time there was no NSF support, *Sputnik* was yet to fly, and TA-ships were very hard to come by. I was very fortunate to be hired as chemist-biologist at Marineland of the Pacific in Palos Verdes, the first West Coast oceanarium, in 1953 before it opened to the public. The other grad student who worked there was Ken Norris. The two of us made a team that was just incredible. Marineland had hired a commercial Italian fisherman, Frank Brocato, as head collector. He often took Ken and myself out as additional helpers. We collected all the initial fishes and all the original cetaceans. Frank has recently passed away. He turned out to be my *gumba,* which I believe in Italian means "big brother." The relationship among Frank, Ken, and me was really incredible, and it was truly important in my development as a fish biologist. Frank never had much formal education, but he had an immense practical experience about the sea and the fishes and cetaceans in it. When I put Frank's practical know-how together with my formal training at UCLA, it just worked wonders with my thinking about fishes.

He knew where to find things?

You betcha, Betsy. He had a love for what was going on out there. It is hard to explain, but I know that Frank Brocato and Ken Norris were as important to my training as my professors were.

I am basically a field biologist; I'm not a natural historian. What I do, however, when observing an organism, say, underwater, is ask, What is it doing? What is it doing? Ken Norris and I used to talk a lot about this approach. When we were at Marineland and looking at organisms, we would try to put ourselves inside their skin and think, in my case, like a fish, in his case, think like a cetacean.

In my case, think like a coral.

OK. Although my expertise is in fish, if I have expertise, it also lent itself to other organisms. The interaction of the organism with the environment has always fascinated me the most. The physiological adaptation, the morphological adaptations are often molded by environmental factors, be they abiotic, temperature or whatever, or biotic, predator-prey, for example. So my researches have tended to center on those factors.

Even though some of my ideas come from discussions with other scientists and reading the literature and the "wisdom" gathered over many years of experience, I still feel the best ideas I've pursued have come from observations of organisms in the field and asking, What are those animals doing? I'll give you a specific example. I'm kind of proud of this, because I was just a lowly little graduate student at the time. I had gone with Norris and Brocato, and we had been collecting fishes around Catalina Island for Marineland. In the evenings we'd come in and anchor in Catalina Harbor, which is a very protected area. While Frank Brocato was cooking dinner (my job was to cook breakfast) I would climb into the crow's nest to enjoy the setting sun. One evening, I noticed groups of fishes schooling at the surface. They were progressing around the curve of the harbor and heading for the open sea, but they did not proceed in a straight line. The schools were 20 to 30 feet across. They would swim in one direction for a minute or two and then suddenly the schools would change direction, so that the fish that had been at the back of the school became the leaders. I sat up there wondering, if they wanted to go out to sea, from point A to point B, why don't they just do it? Well, maybe they are feeding, but . . . then it suddenly dawned on me! When tightly packed, as they were, they would reduce the oxygen levels in the water. If they get swimming in the same direction, those in the back would be in lower oxygen levels than those in the front, therefore they would be at a metabolic disadvantage. That sudden insight stuck, and later we demonstrated in the field that in dense schools of anchovies and mullet, oxygen levels could be reduced to significantly low levels. I emphasize that those studies began from field observations and trying to put myself in the fish's position.

One should never, however, underrate the interactions with other students and scientists, for such interactions hone one's own ideas and, I think, fortify the excitement of doing science. In this modern day and age of publish or perish, sometimes ideas aren't freely exchanged because researchers are worried about being scooped. I've been scooped a couple of times. But that doesn't bother me much. Some people say all science has been done. What more can we do that is new? Or, a beginning student looks around and looks up at the gods, the big professors or Nobel laureates, and says, "Gee, there is nothing else to do." There is always something to do! Look what has happened in the last four decades. The scientific world has changed with a multitude of ideas and new ways to approach problems. Science, in my view, can never be any better than the questions asked. And good questions are hard to come by because they reside in the individual. Good questions, and therefore good science, don't come from money, nor from equipment, even though they are important to getting answers.

What then is a good question? That is hard to define. I think a good question must address an issue that has either yet to be explored or extends and examines issues that are unresolved by work that has gone before. I know that I've asked and pursued some dumb questions, and by that I mean insignificant questions, during my career. But I think I've asked a couple of good ones too, and I've pursued them and have found the answers personally satisfying, and so far at least [they] have stood the test of scientific time.

You've always intrigued me because of your approach to fish biology, which goes all the way from functional morphology and physiology at a very detailed level to the broader ecological perspective of these phenomena. When did you first get into actual studies of fish vision?

I was in Fred Crescitelli's lab. Even though I didn't work on vision, I learned all the lore. I couldn't help but be interested in vision, particularly with Fred's enthusiasm as a human being. I can still remember the day I walked into his lab, and he grabbed me to see the very latest visual pigment from a gecko. I mean, you knew Fred. As you said, he was a little leprechaun himself, an Italian leprechaun.

You were excited about the subject because of him, because he was excited about the subject?

Yes, it was that, but the correlation between visual pigments and the photic environment also fascinated me.

When you first came to West Indies Lab, it was with John and Gillian Lythgoe in 1977, the year I went off to UCLA. That's when I first remember you talking about vision, about Snell's window.

Oh, yes! John and I fooled the undergraduates with those wonderful stories of Snell's window and Rosenberg's hole. We actually conned them into assisting us at twilight, so they could observe them firsthand underwater. But that's another story. You want to know when and how I actually started research on fish vision? Fred Munz and I were fellow graduate students together in Fred Crescitelli's laboratory. As an aside, we became interested in hagfish osmoregulation. We rechecked some of the published findings that had been done on their osmoregulation. We were so excited about our results that, after graduating, when I eventually got to Cornell and he was at the University of Oregon, we wrote to NIH and got a grant to continue our study.

That first summer, Fred came to Cornell on the grant. Both being interested in fish, we visited Dr. [Dwight] Webster, the trout biologist at the Department of Conservation. We did some work on the visual pigments of trout; char and rainbow trout have different rod visual pigment separated by 10–12 nanometers. Webster suggested that we look at the hybrids among chars, rainbows, and lake trout. That got me involved with Fred Fry at the University of Toronto, Mr. Fish Physiology, a wonderful man. He really is the father of modern fish physiology in the organismal sense. So I called him on the phone because he had all the hybrids, stocks of the F1, F2, and backcrosses among the three species at his field station, and he graciously said I could sample as many as we needed. So I went up to his place on two consecutive weekends. I spent all Friday night and all Saturday night dark-adapting fish. Fred Munz and I did this study together and we published in *Science*. Fred Crescitelli would say, "Mc, I can't understand it; it is a totally unappreciated paper," because despite the fact that the paper was the first to ever show that visual pigments are genetically inherited in a codominant Mendelian manner, it was rarely cited.

Really?

Yeah, that's life! But it was at least appreciated by NIH, for it continued our funding.

How did you get from vision in trout to vision in coral reef fishes?

Hmmph! Serendipity. It really was an unplanned set of occurrences. We did the trout work, and NIH thought it was sufficiently important that it gave us money for 17 consecutive years, initially for osmoregulation and then for all the vision work. That led to the coral reef stuff. That happened in a funny way. When I was totally engrossed in doing vision work on trout, I had the instruments by this time in my lab to do it, I accepted a young lady who wanted to do a master's degree. In those days, not much was known about the visual pigments of cetaceans and I had an "in" to get eyeballs out of various cetaceans.

From Ken Norris?

From Ken Norris. So I said, "Why don't you do this, Karen? It is a great project." I initially had her help us with the trout work and had her learn the extraction techniques necessary to characterize the absorption spectrum of a visual pigment. Six months later, when Karen had mastered the techniques, she came in one day and said, "Mc, I'm getting married and can't continue."

That's why professors at UCLA wouldn't take women graduate students for a while.

But that's changed a lot. Anyway, I decided that if Karen cannot do cetaceans, I will. So I went to Hawaii. Ken Norris and Gregory Bateson, who was a social-type psychologist . . .

Sure, he is at Oregon and was Margaret Mead's . . .

Husband, right! We were all on this trawler off Keana Point, Oahu, following schools of spinner porpoise. Eventually they were going to collect some, and I was to study their eyes. But Ken said, "For seven days, Gregory doesn't want to spook the school by collecting any of them, so Mc, what do you want to do?" I said, "Why don't I go ashore and visit Wayne Baldwin at HIMB [Hawaiian Institute of Marine Biology]?" When I arrived there, Wayne and I walked around, and we passed fish pens filled with coral reef fishes. My eyes lit up! At that time, the visual pigments of very few coral reef fishes had been characterized. I asked, "Is there any chance I could sample these?" I had six days and six nights. What was I going to do? And I had all the stuff because I was going to sample cetaceans. Wayne took me to Lester Sucaron, the collector, who told me I could have any I wanted except those, those, and those. Basically, help yourself. Wayne got me set up with a little dark room. I slept during the day, and every night I worked and went through as many coral reef fishes as I could.

I went back to Cornell, where I spent a couple of months measuring the absorption spectra of each species of reef fish. The results were fascinating, but baffling. Dogma had it that when you extracted the visual pigments from a vertebrate, you only extracted the rod pigments and not the cone pigments. The only exception was iodopsin from the chicken. I used the method of partial bleaching. This is a technique that allows one to analyze the homogeneity of a visual pigment extract, only in every diurnal reef fish I had sampled the extracts were not homogeneous. There were multiple visual pigments, not just a single one. I wondered, could they be cone pigments?

OK. Fred Munz, my compadre, was going to England on sabbatical leave, taking a boat across the Atlantic. We went down to New York for a big champagne bon voyage, but I took all these baffling curves with me. Fred is leaving, mind you. It was the night before in a hotel. I said, "Fred, look at these things." He looked and he looked, and he just went bananas. He said, "My God, Mc, you've got to be extracting cone pigments. There is no other explanation. This isn't Vitamin A2 and Vitamin A1."

Whew! Our investigations literally exploded, but we worked for an additional six years before we published one word. I think there was a series of four papers that came out. You look back on your life and say, "What have I published that really was a study?" Those four papers with Fred Munz were. They are classics. I think they were our major contribution. Although I am excited today about our current investigations, those papers were a real synthesis of several different approaches.

At that time, you could wait that long to publish?

Well, yes. Oh, I was publishing other things on trout and on cetaceans to assist in promotion.

But what you really thought was important, you wanted to wait?

We sat on it. And, actually, it took us that long to bring the information together to make a coherent story. As I recall, we extracted and classified visual pigments from 169 species of coral reef fishes. From Hawaii initially, but we also went to New Guinea and Enewetak atoll. Other than those initial fish obtained in Hawaii, Fred and I did all of our own collecting. In fact, we did almost everything which was required. For example, we needed to evaluate the underwater spectrum of where the reef fishes lived. It also happened to jive at a time when instruments were starting to become more available. A commercial spectroradiometer became available that provided a direct spectral measure of photons. I welded together a steel case; it weighed over 100 pounds, which allowed us to measure light underwater in the areas where we were collecting the fish. That was fascinating. Through the generosity of NIH, we were able to obtain a second identical instrument. One went underwater and one measured in air at the same time. The resulting spectra were critical pieces of information, especially during twilight. So, we knew visual pigments from a wide variety of reef fishes, and we had the light fields characterized in which the fishes lived, but we did not have a handle on their critical behaviors. Oh, we kind of broadly knew, yeah, a chaetodon, a butterfly fish, eats on coral in the daytime and a squirrel fish eats crustaceans at night.

That's where Ted Hobson came into play. Ted was instrumental in this work. He added the behavioral information, and it became apparent that the major behavioral changes in reef fishes occurred during twilight. Things just suddenly came together: we had the pigments, the photic environment, and the critical behaviors. That is when we finally published.

Anyhow, that's how I got into science. You know how old I am. I'm 73. I'm retired from, not so much an illustrious career, but a long career. My days are still often involved in research. And I'll tell you, I still get excited when I'm pursuing something new. Like the work Ellis Loew and I are doing on UV vision in fishes. Perhaps only a few people in the world will care, but we care and that is important!

Of course, once a student graduates with a Ph.D., one enters the real world of science: the difficulties of finding funding, worrying about and supporting students, serving on committees, and all that. A lot of people aren't careful about it, and they get pulled away from the nitty-gritty of doing science hands on. They run big laboratories, and I'm not knocking it. It requires that they spend a lot of time writing grants to get money and politicking to get money to support postdocs, technicians, and graduate students, all of which are important. But I don't want to do that, never have done that. I've had my share of grants, and I've even several times tried my hand at administering scientific units, but it has always been the active participation in the field and lab that has driven me. I didn't spend my time worrying over grants to support a lot of people. If I can't do the research myself or be involved, I don't want my name on the paper, ever. That self-centered individuality has been something I've tried to instill in the students whom I have supervised. After all, what does Ph.D. mean? If you actually look it up, it means the ability to do original and creative research.

I do think, for a biologist, there is something important that I don't see happening much any more. I think that each student, in some way, should get to know a group of organisms. I'm not saying that they are going to be organismal biologists, which I am, a field organismal biologist, but if they don't know a group of organisms and learn it moderately well—it might be ants, in my case fishes, it might be algae, in your case corals—once you have that sort of systematic feeling for the diversity of a group of organisms, you have a handle on the diversity of the rest of the world, biologically. Without it, you don't. You just don't. Whenever a problem comes up in my mind, it may be a generalized problem, I can relate it to something I know. Let's ask, "What's the function of seeing in the ultraviolet?" I immediately switch back to fishes and ask why would it be important to this fish, but maybe not to another one, because I know the fish. I know their behavior somewhat; I know their ecology. If I didn't know that, I couldn't think in those terms.

Perhaps modern biology is just too diverse for one to be a generalist, so one must specialize. Yet, at the level of undergraduate training, I think we could do a better job of providing students with a broader biological perspective. For example, at Cornell, the undergraduate biology majors were all required to take a year of introductory biology, a semester of genetics, and one of biochemistry. Then they specialized in a concentration area: genetics, biochemistry, neurobiology, physiology, ecology, and so forth. For several years, as chairman of ecology, I suggested in the curriculum committee that all senior majors be required to enroll in an intense course in the history of biology, a course designed specifically to show the interrelationships among their specializations. It always baffled me that the other chairpersons of the other specializations did not think this at all important. So much for scholarship.

The lack of breadth also occurred at the graduate level. Often, when I sat on Ph.D. exams at Cornell, I would ask the student, "I'm going to name 20 biologists, and I want you to explain what they have contributed to biology." You'd be amazed how many students couldn't give me 50%. I'd ask people like [Theodor] Schwann. Even Louis Pasteur! They could say, he was a Frenchman, but they couldn't really tell me what he had done. Sure, I'd give them a couple of names in their field; sometimes they wouldn't even get the people in their field. There is something amiss. Students today don't get the historical perspective of what they are. That's too bad.

After all, what is science but a building of information? Why do we read papers, why do we publish if it isn't to communicate? Why go to scientific meetings and present papers? Why bother? There are many answers to those questions. But in the last analysis, I believe, it is that, as humans, we are inherently curious and social. And, how exciting is it for me to share my enthusiasm and knowledge with young graduate students. They have fresh ideas, different ways of looking at things that I don't have, can't have. I have my background, and they have different ones. They just have an energy. Sure, you see them fumble a little while they are learning. But then, we are not perfect. How many mistakes have I made as a biologist? Lots, I can assure you.

You are what you are, I guess. I am scientifically what people around me were. And they were wonderful and, God, do I like that. I revel in them, thinking back to those days.

And what about the future of field biology?

The future? Oh, there will always be field biologists, but they may be scarce and in much demand. Modern biology is an expensive undertaking, a business of sorts and a highly competitive one at that. If one is not funded, it becomes very difficult to pursue any research goals, let alone a goal which does not have peer approval. So, scientists turn to the "missions" of granting agencies. To succeed as an academician in the university of today requires a scientist to spend an inordinate amount of time writing grant proposals to compete for funding. Necessary? Yes. But, in my view, it removes the spontaneity from pursuing a new idea and realizing the exhilaration that comes from trying to create and do something new. So, Betsy, I guess I should just be classified as old-fashioned.

It took a minute . . . and two hours were over. I was fascinated.

Holger W. Jannasch (1927–1998)

Senior Scientist Emeritus, Department of Biology
Woods Hole Oceanographic Institution
Interviewed in his office 4 February 1998

Holger Jannasch was a wonderful raconteur, instantly engaging all those around him. I consider myself especially fortunate to have had several long conversations with him in the year preceding his premature death. His experiences with both the German and the U.S. way of doing science allows a consideration of the strengths and weaknesses of each system. He pioneered new techniques to expand his research on marine microbes in extreme environments, including the symbiotic bacteria fueling the deep-sea vent communities. Many of the leading marine microbiologists of today worked alongside Jannasch as postdocs or were taught by him in the MBL microbial ecology course.

You couldn't get into the university without having had some practical experience. Being on an island, by myself, being my own master, all that was very appealing. I did that after high school; I was 20 because I missed two years due to the war. It was a very important time in my life. I learned a lot. I had a lot of books to read, but I had work to do for the ornithological institute in Germany. I had to protect the eggs. OK. That was all right. But I had to count the eggs that were laid by a certain tern. When the eggs came, I had to go around all of the nests everyday. I recorded all of the shorebirds and the land birds that came to that little island in the fall migration. I wrote a long report with lots of interesting details. That settled it. I was not exposed to any more nonscience lectures that could divert me.

When I came back to Gottingen, it was too late for the winter semester. I worked for a Russian professor, a refugee who had a fantastic background knowledge of zoology and other things, too. He taught me systematics, which I never really appreciated before. The whole collection of deep-sea mollusks which came from the *Valdivia* expedition of 1898–1899 was in Gottingen. A bomb fell in that wing of the building. A lot of jars were broken and had to be relabeled. I learned how difficult that is. To look at a little slimeball, to find out what kind of gastropod that is. Again. The ability of a learned zoologist. He looked from a distance at that slimeball, and he said, "Look in the second volume of *The Taxonomy of Marine Mollusks,* I think it is page 270 or 280. That genus, I think is what we are seeing here." I couldn't make out anything. Then I prepared the radula. This is what you do in gastropods [to identify them], and sure enough it fell into this genus! But I couldn't decide which of the ten species of this genus it was. He came again, had one look at my drawing, one look at the micro-

Examining gear built in collaboration with WHOI engineers in the 1990s, Holger Hannasch contemplates the next frontier in examining microbes living in extreme environments. Photo by WHOI photographer Tom Kleindinst.

scope, and said, "Of course, it's this one." And he explained to me why. I learned tremendously how to observe, how to look carefully, and what the difference is between a superficial assessment and a careful assessment, which takes time.

I had decided to be a microbiologist. I was elected by the famous zoologist to be his graduate student. I didn't know that. He asked me into his office one day, and so I prepared something. I knew it was for a doctoral thesis. I had ideas about what I wanted to do. In those days, German professors were dictators. They were very powerful. He said, "You will be working on the biochemistry of the eye pigments of *Ephestia.*" *Ephestia* is a moth, which he worked on genetically. He also was the expert for taking pictures through insect eyes. He had also a Ph.D. in physics. A famous man. He showed what the compound eyes of insects would see. His lectures were marvelous, and so I wanted to work for him. It was an honor to be asked. Then he said, "You have ideas of your own, what you'd like to do?" Yes, I did. I got my piece of paper out. I read what I was interested in: to work on protozoa, how they were feeding in the natural environment on bacteria, and then repeat this in laboratory experiments. I particularly had laid out the ideas on how to do this in the laboratory, without changing the environment too much, keeping everything alive, and so on. He looked at me. "You really want to do that?" Oh, yes, I said. "Well then, I'm very busy. Get Mr. So-and-So." I was kicked out.

I missed a big chance or I got a big chance. It was a very decisive ten minutes in my life. Because if I had been his graduate student, in the old school system, you have a professorship guaranteed. Your career is laid out. You work hard, but as a student of that professor, he forms a school, and the school is important to him. He turns you out as a professor in some key university where he wants to be represented, and you better follow his line a little bit and so on.

I was back completely on my own, no professor. I still wanted to work on these protozoa and bacteria. The only department in microbiology in those days was at Gottingen. I went to that professor. He listened to me, and he said, "This is all very fine."

What was your Ph.D. on?

Aquatic microorganisms. Are they different from soil microorganisms? Can one make a distinction like that? I doubted that because wherever there is fresh water, it is in contact with soil. Constantly, bacteria are suspended in water. But it turned out to be different. In soil, microorganisms are usually exposed to high substrate concentrations on particles of soil. As soon as you get in water, the nutrients dilute out. The microorganisms have to have a very different system for picking up nutrients, much more efficient. So certain organisms just disappear.

So it's like oligotrophic or eutrophic?

Yes, oligotrophic in the water and eutrophic on the surface of soil particles. So that selected for certain organisms. These rich organisms, which we now call copiotrophs, require high substrate concentration when they do get into the water. I followed the fate of the spore formers, for instance. They get smaller and smaller because they divide but they cannot grow. But they still divide. Then they get below a certain size where they can't form a spore any more, and then they die out. If they dry up at the right time, they form a spore, and then they survive. All of these little details I worked on.

My thesis was sort of an accumulation of all of these facts. I was not satisfied with my thesis. It lacked the big idea. I complained to my professor, and he just agreed. But he said, "What you did was fine; you learned all of the techniques. A thesis doesn't have to be a tremendous piece of research." That was very true. It was a very good piece of advice. "A thesis is to learn your field. You have done so many things in the years you have been here, your thesis is acceptable to me. For publication, you abstract something," which I did. I still think he is really right. Often a grad student nowadays gets a bright idea, writes it up, and gets very famous, but he doesn't know anything. He just had a great idea, which is good, and he has to be credited for it. But learning details, spreading out, you'll never have a chance again in your professional life . . .

To really grow.

Yes. This is what you do as a student.

During my graduate studies I had been in Naples, Italy, to work at the Stazione Zoologica. There I met an American microbiologist, Claude ZoBell, who told me, "Whenever you can, call me and work in my laboratory in La Jolla, California, at Scripps Institution of Oceanography." So I wrote to him, and he invited me to come for a year. By that time I was married, and we had a baby. We went to California. That was a tremendous change of lifestyle, especially because I didn't speak much English.

La Jolla was very good, but I was floundering around. I didn't have much help. ZoBell was at the end of his career. He worked on everything in marine microbiology.

People are quite critical about his work, but he created the term "marine microbiology." He put it on the map. He wrote a book on marine microbiology, the first who would dare to do something like that. I was not really sure whether what I did was good or not. I used new techniques. One of them [Moore's techniques] employed paper chromatography for the study of metabolites and the chain of breakdown products in tissue. I used that for looking at metabolites in sea water. With ZoBell, everything I tried, he'd say, "It is wonderful." I wish he had been more critical.

Then I took a vacation with two or three friends. We went north along the California coast. On the way back we came to Monterey, to Pacific Grove. I remembered that when I told my old professor that I had an invitation to the United States, he wasn't that enthusiastic. But then he said, "Well, if you happen to get to the western side of the country, there is a man who is my American editor in *Archives of Microbiology*, and his letters and reviews are fantastic. If you meet him, you give him my regards."

Was it [Cornelius] van Niel?

Yes. When I saw we were in Pacific Grove, I said, "I have to stop and give regards to a professor who lives here." I went to Hopkins Marine Station and I asked for van Niel. The secretary said, "He went home." It was just after 5 o'clock. I said, "I could call him?" She said that by now he should be getting to his house in Carmel. I phoned, and there he was. Hearing my strong accent, he switched right into German and said, "Where are you from?" I just said, "I wanted to give you greetings from my professor." He said, "That's very nice, thank you very much. Where are you?" Well, I said, I'm sitting on your desk here at Hopkins. "I'll be there in 20 minutes," and then he put the phone down. Twenty minutes later, he came in. That was another change in my life. I didn't know it at that point. He sat down across from me, took the greetings again, and said, "What are you doing?" I said, "Well, you know." "No," he said. "What are you doing? No 'you know.' I want to know what you are doing, and you had better tell me quickly and in good short sentences."

Oh, he had me at the edge of my chair within half a minute. I told him that I was trying to extract nitrogen compounds from desalted sea water with Moore's techniques, using the chromatograph. He looked at me and said, "You want to do that, but that's impossible. This is ridiculous. That technique isn't good enough for sea water. First of all, free amino acids in the presence of sea salt are complexes. I don't think you can separate that. If you get sea salt out, you get amino acids out. You lose most of it. Let's look at" this book, that book, this paper, that paper. Immediately, the table was covered with reprints and books. We went over it. He was so interested just to make sure what he was thinking was right. So we worked on it.

It took a minute . . . and two hours were over. I was fascinated. He was so straightforward. That I wished from ZoBell. I thought, why didn't I do this work here? This was wonderful. Then, he suddenly looked at his watch and said, "Oh, I'm so sorry, I've got to go home. It was nice meeting you." I had heard that he was giving a summer course, but I didn't know any details of it. So, very naively, I asked, "Is it possible to take your summer course?" "No," he said. "Why is that?" I asked. "Because it is a basic course in microbiology, and you have a Ph.D. in microbiology, and it's not for you." "Oh," I said, "Dr. van Niel, the discussions of the past two hours may have told you that I'm far from being a microbiologist. I haven't even started yet." He grinned and said, "You've got a point there. All right, you are accepted to the summer course." With that, he departed. The secretary said, "You are accepted to the summer course? Did he say that? He takes only ten people, maybe one or two auditors, and he has hundreds of requests from all over the world. These are accomplished people, chemists, biochemists, all sorts of people he selects because they are very interesting, because they will make a good group." So I realized that this was a tremendous honor.

I went back to Gottingen with no job. I got a stipend. I started on the bottom again, but the atmosphere was fantastic. It was a very small villa again, stuffed full of people. I was in a room like this, not bigger, with another fellow named Norbert Pfennig; he became very well known later on. We were talking all day about what we had in mind. I learned a lot. Together, we figured out the chemostat kinetics, and I needed that for working in the aquatic environment at a high dilution of substrates. So I owe this to that colleague. Any question I had, I bounced it off him, he bounced it off me, and so it went back and forth, so together we understood all of the details of chemostat kinetics, which is not easy. That was a wonderful time.

The stipend that I had required habilitation, a feature that is only known at German-speaking universities. It is a procedure to prepare yourself for professorship. You have to write another thesis. You have to show to the faculty of science, about 50 professors, that you have done original work, that you are weaned from your professor, that you can lead graduate students, that you can lecture to the public and, specifically, to scientists. So, you get a stipend to work on a subject, and you deliver a new thesis to a committee, and then the whole faculty has to see it; they all have to agree. This goes back hundreds of years when the social situation was very strong, and they elected a new professor into their ranks. They were careful about it. In those days, it wasn't just the field. It was your family, where did you come from—all those things were evaluated. You may have some strong antigovernment feelings, and you wouldn't be accepted, even if you were a good scientist. Nowadays, it is much better of course; it is focusing on the science.

I did a new thesis, and this habilitation came around. It was a big experience. During this time, I got a letter from Woods Hole. I came back in 1961 for a meeting in Chicago that I was invited to, and I passed through Woods Hole and gave a seminar. After the seminar, I was invited to lunch by two elderly gentlemen. Usually, you are invited by the students, who get permission to take the speaker out. The students get to go to a restaurant and can eat a lot. They love it. It was very strange; these two were not microbiologists. One seemed to be close, but the other one wasn't. They asked all sorts of questions that really didn't have anything to do with my lecture. We went to the Flying Bridge, and I got martinis. I didn't really know them, very strong.

First time?

Oh yes, and I loved them. I was very happy. So I answered these questions, crazy questions—had nothing to do with microbiology really. That was it, and I flew back to Germany.

A few weeks later, I got the letter from Woods Hole: "We offer you a job." That was an interview! Ah, gee, I said, Woods Hole. I've been there twice. I love the area. I preferred it to Scripps. Right away. Even though Scripps has this beautiful weather. In those days, it was also much more picturesque than it is today. It is all overbuilt now. The human climate at Scripps was different. There were two classes, the hard-money class and the soft-money class. The soft-money people were the interesting ones. The hard-money people didn't show up all the time. They were not really doing that much. The soft-money people were a little angry about that. They worked very hard; they got credits but never a job. They hired people from the outside for hard-money jobs, who were famous. They came in and didn't do a thing any more. There were all these young people, 35 years old, long lists of good publications. They were never hired, and finally they left. It was not a healthy thing. Here [at WHOI] they were all on soft money. In those days, it was easy to be on soft money.

This was the 1960s?

Yes. Here, everyone was the same. The human climate was good. So I liked Woods Hole. I told them I would like to accept it, but I can't because I have accepted this stipend in Germany that would go a year, maybe two more years. The end is only when I get habilitation. So, I am sorry. And I was really sorry. I got a letter back from the director, Paul Fye, who said, "We put some time into looking into your work, so even if it takes a year, even if it takes two years, we will wait for you." So, I accepted.

I've never had microbiology, and when I read your autobiography, I didn't understand the principle of the chemostat.

Ah, yes. The chemostat is a wonderful tool. Some ecologists think the chemostat is a reproduction of nature, which it is not. It is another extreme. Batch culture is an extreme because it is a closed system. The chemostat is an open system, but it discriminates because substrate is flushing through, and everything that grows slower than the dilution rate gets flushed out. The organism that has the best growth rate with the available substrate grows faster, so you end up with a pure culture. A chemostat selects for the most efficient organism. That's what I wanted, so this is ideal. It doesn't reproduce nature, because nature is not constant; it goes up and down all the time.

And that keeps the diversity in the system?

Yes. Everyone has a chance, and the diversity is maintained. In the chemostat, it is not. But I used it particularly for this reason.

Low-temperature work, we did. We worked on psychrophiles [microbes that thrive in cold temperatures]. When you pick up psychrophiles, they die at 20°. You have to insulate, get them out, and then work in the cold room. I couldn't do any work on the effects of pressure until I had the tools. I was doubtful about these barophiles [microbes that thrive in high pressures]. ZoBell had never really isolated one. Then, in 1968, *Alvin* [a research sub] sank. They were lucky. They immediately put a bottom buoy there, with a pinger, so they could tell where it was. They retrieved it the next summer, and someone came in with the lunch bag from the *Alvin*. This rotting stuff. What should we do with it? Well, it wasn't very rotten. There was a sandwich. It wasn't eaten, and that was great. They just had gotten steel thermos bottles on *Alvin*, because often they fell and the glass would break. The jacket was compressed, the lid was bent in, and the plastic lid was cracked. So a certain amount of sea water got in, to equalize the pressure. It was like an inoculum. We poured this stuff out. It was seawater soup; we tasted it. We put it in Erlenmeyer flasks and incubated it at different temperatures. It went away at 30, 35°; it stank in a few days. So, hmm, that's interesting but very inconclusive. But, what to do about it? Then I thought, gee, this is interesting. We can work on the deep sea floor, and we avoid the decompression problems.

First, we worked together with other WHOI people and would put things down which came back up whenever they brought them up, one month, three months, six months, and we would compare what was decomposed. Each time, we put a sample in the refrigerator, at the same temperature as down there, but at one atmosphere, and there were differences. Clear-cut, the refrigerator cultures were faster. So, pressure has an effect, but they were endpoint measurements only—after three weeks, a month, a year. We had no idea of what happened in between. That wasn't very satisfactory. Then *Alvin* was rebuilt. It precipitated down to the very simple idea that you could incubate things down there, so that's what we started to do. This could get us so far, but endpoints again. Interesting, but not conclusive.

I talked here to the engineers, to Cliff Winget and Ken Doherty, about how to get things under pressure. That, at first, sounded much too expensive. This, then, became a delightful exchange of ideas. We had a very good idea but no money. I had just gotten an NSF grant, and in those days you couldn't ask again right away; you had to wait two years. We had a meeting of WHOI associates, and Fye told every one of us to be host of one table up at the golf club. I was at one table, and there were elderly ladies and gentlemen. They asked, "What are you doing?" I had just come from a discussion with the engineers, and I said, "Oh, I have something fantastic in mind!" There was a paper tablecloth, so I got a pen out, and I made a big drawing of it. They all looked politely. I couldn't care less; I was just in the middle of it. I explained how the pistons worked and how we would slow down the pistons to avoid the French press effect, so we wouldn't blow up the bacteria as they came in. We had a gas chamber underneath, so we could put a manometer on it, because we did not know the depth entirely of the cable—all of this. Oh yes, very interesting. So we left the table.

There was one man who hadn't said very much; he just watched. He came up and said, "How much will it cost?" I said, "Oh the steel would be so much and about $6,000." He said, "Go to your director tomorrow, and there will be a check for $6,000.

Go and do it." That is how we got the first instrument. It was done in two months' time. To go out here to 2,000 meters, there were all these cruises to test it. That started it. We developed on that. First, we brought the whole samples up, and we added the acetate to it, and we measured under pressure and deep-sea temperatures, the turnover.

Here in the lab?

Yes, because we needed time-course experiments. We could do that, and there was an end to that. We tested many substrates and so forth.

While that was going on, I never felt that I needed more projects. Then, the discovery of the vents came. I heard it the first day. When they came up with *Alvin,* a fellow named John Corliss saw them. The captain called that night, and if the scientist wants to say something, he can. I was over there [at WHOI marine operations], near the radio operator, and he said, "Come in, this is fantastic, you'll like this." I listened to it, and I must admit I didn't believe a word he said. Corliss said there were big clams down there and worms and all that. I thought maybe he was overdoing it a little bit. Yes, there are probably worms but not very many. He was all over the place, and he was a geologist. I got on the phone, and I said, "Do you have those in your hands?" "Sure," he said, "I have a clam right in front of me." "How long are they?" "About 30 centimeters." "How heavy are they?" "Oh, a pound or two." "And how many do you see of those?" "Hundreds." "And, about the worms, how big are they?" "Oh, about two meters long and so thick, and this reddish stuff." I thought, this must be something! It would never be supported by photosynthesis, 2,000 meters deep.

They knew the water was hot?

It wasn't hot, it was 25°, the hottest they measured.

Well, that was hotter than the surrounding water.

Oh yes, that's why they found it, because of the temperature. Dick Von Herzen found the thermal peaks. They decided, from his data, where to dive, so he shared the discovery very much. I thought, there must be something reduced down there. I said, "It could be chemosynthesis." I went back to my desk and wrote a proposal. Everything was granted.

Another thing you brought up in your autobiography was the microbial ecology course at MBL.

I am so delighted to be able to teach at the MBL. You have the students from early in the morning until late at night. You talk with them rather than formal teaching. They ask you questions, you ask them questions, and then you work it out. What I like so much in dealing with students is that they come up with questions that you don't anticipate. Sometimes you don't do them a service by saying, "This is all done, this has been tried ten times, with always the same result." Don't ever say that. Don't discourage them to do something, even if you believe you know.

Sometimes you could be wrong.

Yes, because they do something different. If they were to follow the "correct" protocol, yes, they would get the same results.

I remember that, as a student, I couldn't isolate certain organisms. I tried everything. Then Norbert Pfennig said, "You put in not 0.6, but 0.8 micrograms of B12, it may be too much." I said, "B12 can't harm anything." "Yes, it can." I put in 0.6, not believing that it would change anything, and the organism grew. So, you have to learn it that way. You don't always want to grow one particular organism; you want to get something new. The best thing, of course, is if you have an idea. B12 may be in the way; this organism may make it itself. I'll put something else in. Often it is chance. Students have a higher chance occurrence than we have. So, they are, in a way, more creative, with their own merits very often. We are now preparing the cruise again for Guaymas [Jannasch's last scientific cruise in April–May 1998]. I want to force myself to put in substrates where I know I was never successful in the past, but in a new composition.

What do you see as future trends in microbial ecology?

Right now, it's how can molecular phylogeny become useful to ecology? A lot of the molecular phylogenists think it is already done. They think they can answer ecological questions. They cannot, for the time being. This is very much more complicated. They are still pigeonholing. You can say they are stepping back in time, by describing organisms now with sophisticated techniques, and now the phylogeny, which is wonderful. But they instinctively wanted to get out of this being cast as pigeonholers. So, they say, we can solve all problems in physiology and in ecology. So far, they cannot.

It's just fun to teach people stuff. That's all.

HENRY JOSEPH THOMPSON (b. 1921)

Emeritus Professor of Biology
University of California, Los Angeles
Interviewed at his home 24 July 1998, with Joyce Thompson and Chris Van Gieson

Harry Thompson taught me how to teach. He utilized his creativity and broad knowledge of the diversity, physiology, and morphology of organisms to encourage undergraduate students (from beginning to advanced levels) at a large state university, UCLA, to personally "study nature." In the process, he also demonstrated to professors and TAs that they too could contribute to the courses, whatever their expertise happened to be (the exception being those specialists who knew no biology). He was an invaluable resource in the biology department, not only to graduate students but also to fellow faculty, to whom he passed his knowledge of plants, plant communities, and the art of field science. In his retirement, he has written a history of the World War II combat seen by the units in which he served.

I went to Santa Monica Junior College, where there was the best field botanist in southern California, Dr. Harry Lloyd Bower. He published a very famous paper in the late 1930s, "Moisture Relations in the Chaparral." All that kind of ecology had been done in eastern deciduous forest, but nobody had ever worked in chaparral like that before Bower.

He actually did the field observations and did the experiments?

Right, he was a field ecologist. I don't know whether I knew about him ahead of time, but I certainly knew the minute I got there. I took general botany from him and was a good student in the class so I got a little special attention from him. In my sophomore year, I was his laboratory assistant. If I had gone to some other place, like UCLA, I would have still been a dumb flunky my sophomore year. I wouldn't have reached the level of touching a professor until maybe my senior year. I thought it was the greatest job in the world. I'd grade papers for him. I also worked on his research project and got a footnote in his published paper, the first time my name appeared in the published literature!

Was this your first formal fieldwork?

That's right. All of the outdoor biology with my mother was informal, and this was very formal. He ran a botany club; we had an overnight field trip every spring. Then I was admitted to forestry school in Eugene, but this was 1941. The war in Europe

Paying close attention as his professor Harry Lloyd Bower (*right*) lectures about giant sequoias, Harry Thompson (*third from left*) enjoys a Santa Monica Junior College botany field trip in 1939. Photo from the files of H. Thompson.

was going on, so I decided not to go away to college. There really wasn't enough money. My dad was out of work at the time. You know how studios are. They fire everybody once in a while. I was going to go in the navy, but I was waiting for another guy to finish his college, to go in with. Then I went in the navy: Guadalcanal and Munda and Bougainville in the Solomons and the Marianas, Tinian and Iwo Jima and Okinawa. I was flying around the empire when the war was over. The atomic bomb saved my life. That is four years in a couple of sentences.

I graduated from Whittier and visited at Stanford and at Berkeley. My mother wanted me to go to Cornell, but I couldn't see going all that way, and I knew enough to know that there was a hell of a lot to learn at either Stanford or Berkeley; they were both good places. Stanford looked like a nicer place; it was sort of woodsy, outdoors, a real nice place.

All of this was possible because I was on the G.I. Bill. If there hadn't been a G.I. Bill, I probably would have gone to work when I got out of the navy. When I got out of the navy, as a naval aviator on flight pay, I was making $6,000 a year. That was in 1945. I went back to school, and in summers, I worked at the studios. My dad got me into the electricians' union. Although I only worked three months in the summertime, my annual pay, if I had worked regularly in the studio, would have been $6,000 a year. So then I finished my undergraduate education, went to Stanford, got a master's and then a Ph.D. and came back to UCLA to teach, and now this is five years and two big degrees later, my salary was $4,090. And I was glad to get it!

One of the fortunate things at Stanford was that I was able to team up with Ira Wiggins, the outdoor biologist. He was an old-fashioned plant systematist. His Ph.D. was the flora of San Diego county. In his lifetime, he did a flora of Baja California, a flora of the Sonoran Desert, a flora of the Galapagos Islands, a flora of the Arctic slope of Alaska, so he wrote these big floras. I didn't want to be a florist. I worked on population biology studies of certain groups of plants, rather than try to do all the flora of a particular area. I did my thesis work with [Jens C.] Clausen, [David D.] Keck, and [Bill] Heisey, who were a team working on outdoor plant biology. I worked out in the Carnegie Institution building on the Stanford campus for my graduate work and taught general biology as a TA for two years. The last year, when I was writing my thesis, I was the "pusher" for biology, that is, I coordinated all activities, like Chris [Van Gieson] used to do while you were at UCLA. I did everything for that course: gave lectures, ran the labs, typed occasionally when we couldn't get Miss Zilsky to type. That's where I learned to teach biology.

I took all of the zoology courses while a grad student at Stanford because I had seen one case and heard of others where guys who were botanists would substitute fields in the written exam. You know, Stanford had very strict written Ph.D.-qualifying exams. Six written exams, three hours or more long, beginning on Thursday morning and lasting through Saturday afternoon. Those exams were general botany, invertebrate zoology, comparative vertebrate embryology, comparative vertebrate anatomy, cellular physiology, and genetics. With the exception of the general botany, all of those were zoology. So I just took those zoology courses, and when I went back to Whittier that's what I taught: comparative vertebrate anatomy, comparative vertebrate embryology. There was no reason for me to go around and do special courses in botany. I already knew that stuff. So, that made me a good biologist.

I did outdoor population biology of the genus *Dodecatheon*: when they germinated, how they grew, how the population worked. Population biology. The lab work was growing them and making hybrids between them, the old Ernst Mayer, "a species is a group of organisms that is reproductively isolated" and all that. Unfortunately, botanists could test that, and bird people couldn't. I made a series of appropriate hybrids and grew them so you could really see how they worked. I did all the chromosomal pairing in the hybrid; the chromosomal numbers was the lab part of it. *Dodecatheon* is a big polyploid complex with the lowest number as 22 pairs. There are 22s, 44s, and 88s [different species].

In the summer of 1950, Ira Wiggins got a research grant to go to Point Barrow, Alaska, and study the vegetation both taxonomically, what these things were, and ecologically, where these things grew. We didn't know it at the time, but I've since learned that the purpose of the study was so that people could use our information and predict, by knowing what the vegetation was like, what the roadability of that land would be for tanks and trucks. This was all permafrost country. If you knew something about the vegetation, then you knew what the soil and subsoil types were. I guess they were still afraid of those Russians coming across in tanks. The DEW [Distant Early Warning] line was active. I went there as Wiggins's field assistant. We lived out on the tundra quite a bit of the time. I learned a lot from Ira; he was a very good field man. A lot of people don't realize that going out in the field, living in the tundra, is not quite like going out camping. When you go out camping, the purpose is to enjoy camping. Well, the purpose there, like any other field trip, was to go out and do this fieldwork, and you have to eat and sleep. There is nothing fancy. You try to make things as simple as possible. Ira always pitched in. The hard job was to scrub the pot. We'd throw all our stuff in there, stir it up, and eat it. He always took his turn. But, while he was washing the pot that I had dirtied, he was Dr. Wiggins! I've only referred to him as Ira since he died. It was a very formal relationship. That was a good summer.

I spent a lot of time in the Sierras at the Carnegie stations. The main base was on the Stanford campus, and then they had the Mather Station at 4,000 feet and then the Timberline Station at 10,000 feet. Bill Heisey's job was to run the Sierra field stations. The three of them—Clausen, Keck, and Heisey—divided up the administrative responsibilities. Bill's job was the field stations, so I always went with him to open the field stations each spring. He needed a strong body to take down the shutters and stuff. Those were all good things, because you got the undivided attention of a good scientist for a four- or five-hour drive over there, and then you lived there for a day. While you were lifting the shutters, he'd talk. There is probably less of that today. So that's how I got to be a biologist, although from the days with my mom I was always into plants.

The year that I was in the job market was the spring of 1952. I got the job at UCLA primarily because I was a botanist. They didn't want to hire another zoologist. They already had Waldo Furgason; he was a protozoologist. But they wanted a botanist who didn't know just botany.

They wanted a botanist who could teach other things?

Yes, general biology. That was the reason I got the job, that and my good personality, good looks, wealth, charm, and all of that.

I started teaching in the life science course. But Waldo could not team teach, like JIM MORIN and I did.

Team teaching, when people work together like you and Jim did, can be a wonderful thing. But if team teaching is just this person lectures this day, this person another day, with no coordination, it can be terrible.

Jim and I pledged to listen to each other's lectures, never give more than about four or five and then you'd shift back, you'd say, "This is an interesting topic, but Dr. Morin will cover that in a few days." We did that all along. I did that with Bart [George Bartholomew] too; he was good at that.

To me it was such a great educational experience in the introductory biology course team taught by you and Jim, even though for me, I was in the middle level as a TA.

But a good one.

Well, thanks. Why I had such fun and what impressed me so much about the way you approached the class was that you sought out specific graduate students (ones who could cover the diversity of life forms) to be the TAs. There were ten of us.

That's right, sure. We always had a couple of botanists, a couple of invertebrate people, a couple of vertebrate people, and no molecular biologists. They'd sneak one in at times. Usually the ones they were shipping to us had never had a course in biology. It was good for the TAs to have this mix of TAs from other areas. The other TAs were students on Monday afternoons, when they did the lab that they would teach later in the week. That was a good course to teach, when Jim and I taught it.

Chris: *It had a somewhat spotty career the rest of the year.*

There was one interesting thing that Bart and I did, when he and I started teaching that together, because we realized that we didn't want to bury our future in teaching this course. We wanted to get other people involved. Everett Olsen, "Olie," had come from the University of Chicago, where he was teaching advanced paleontology courses, dealing with only therapsids [fossil reptiles]. Bart and I went to him and said, "How'd you like to come and teach general biology with us?" "Oh well, you know," he said, "I've never done this." We said, "You come in and we'll set you up. You don't have to go to all the lectures but attend for about a week and a half. I'll give some of the lectures, and Bart will give some of the lectures, and we'll set you up to do the paleontology of this stuff." So I started and did some plant stuff, and Bart did some animal stuff, and all the way we'd say, "Some of the plants or animals that are interesting are extinct. We know them only from fossils, but we'll have this guy Dr. Olsen tell you about this." Of course, it turned him on, and he did a good job.

While you were at UCLA, new professors came into the department. There are at least two that I know, very bright guys, who had specialties in certain areas of biology, but they didn't have expertise with plants. Both told me that they were interested in switching to ask botanical questions, in PARK NOBEL*'s case, ecological physiology, and in Martin Cody's case, ecology, because they learned so much from you.*

Martin was a fabulous field population biologist/ecologist with birds. We started teaching FBQ [the UCLA field biology quarter] together, even before we had a biology department. I would offer a course in botany, TBA, time to be arranged, and they'd meet in this room. Martin would offer a course in zoology, TBA, and he'd have them meet in the same room. Then, we'd tell the students, this is the course, and we're going to teach it together, even though some of you are in botany and some in zoology. We started teaching the field biology quarter in that way. We would go out into the field; we'd have them full time for a third to a half of a quarter.

So we'd go out into the field after a few lectures on techniques, for instance, how to look through binoculars. We'd practice that in the botanical garden. There is no reason to take them to Brazil to say, here are the binoculars, there are the birds. I had so much trouble with students who couldn't noose lizards that once I bought some *Anolis*. I had the poles and nooses, and I let the lizards go in the room and I said, "Catch them!" They learned, so if they did want to work on lizards, they didn't spend all their time in the field learning how to noose a lizard.

So then we'd go out into the field. We often went to the Granite Mountains. The first afternoon, Martin and I would lead them on an orientation walk. I'd do the plant stuff, and he'd do the animal stuff. Then we'd have four set projects. We had two TAs. I'd choose a research project, each TA would choose a project, and Martin would have something. We'd divide the class into fourths, a fourth would spend the morning with me, another fourth the afternoon and the two other groups the next morning and afternoon, so I'd have two days' worth of data on the project. At night, I'd give a report on that, the TAs would give their reports, and Martin would give his report. That was a device for infiltrating the group with botany, don't you see! The students worked with me in botany, and the next day they would work with Cody on birds, but they'd be talking about the plant stuff that they did. Then we'd always have the student who would want to work on sheep. I'd say, I first came out here to the Granite Mountains when I was 14 years old, so I've been out here quite a few times and once, only once . . .

I saw one.

No! Never seen a bighorn sheep, but I saw the droppings while they were still fairly moist, I'd tell them. So we'd get them off that, and then they'd get onto a botany problem. I'd say, "What did you want to do with sheep?" They'd say, "Well, how many are there? What is their density?" You see, they'd taken an ecology course, so they knew all about density and area density. I'd say, "Look at these plants. They're not going to move, and you can take a little piece of toilet paper and twist it around, and you'll know you've done that one." I remember one guy, Dick Hutto, who is professionally very active now as a senior professor at Montana. He wanted to work on territories and spacing. That was a big thing with Cody. Dick wanted to work on the little lizard *Uta*. We got four or five cold, windy days, and you never saw *Uta*. So after about two days, I got him working on packrat nests. He never saw a packrat, but you could see the nests and see if they were big nests, where they live, or stopping, hunting, way-station nests and things like that. A lot of people went into spacing in plants. Before you knew it, Martin was working on plants, and in the past ten years probably 50% of his research has been on plants.

I took several plant seminars that you and Martin gave, one on cactuses.

And one on the Proteaceae. Because everyone worked on northern, boreal plants, I thought it would be interesting to look at a group in the Southern Hemisphere: Africa, Australia, South America.

It's obvious what these people got from you, and the students, too, whom you took to the field. Most of them probably didn't even want to do plants. They wanted to work with things that move, and so you opened up a whole new world for them. But what did you get out of being with these people in the field, the undergraduates, Park, Martin?

It's just fun to teach people stuff. That's all. The same thing that makes you want to teach courses. What could be more beautiful than teaching Martin Cody about plants? Or Park Nobel about plants? He is a hell of a student, I can tell you. In a year, he knew a lot of stuff about cactus. He started from zero. He knew a tremendous amount. The same reason I taught introductory biology for 34 years, give or take a couple of years' sabbatical, was the same reason I was interested in going into the field with Park. Martin is a little bit different deal because Martin is a hell of a field biologist, the best field biologist I have ever been in the field with.

Will you comment on trends in people who call themselves biologists and the kind of background they get today, as opposed to what you had, or even what I had?

I don't know what happened the past ten or so years; I know it more as an observer than as a participant. But in the 40 years or more, counting all my time at Stanford and 34 years at UCLA, across the whole department, all the biologists, the percentage of people doing fieldwork decreased, and the absolute number of people doing fieldwork decreased. The people that were doing fieldwork were people doing a little fieldwork, but they weren't field people. There was a lot more laboratory work, because that seemed to be what was selling.

Anyone who is doing fieldwork needs some sort of reference collection so they can go back and identify the things that they are working on. It doesn't mean they are going to be taxonomists. They need to have someone who knows the organisms, can identify them, and give them a name that will put them into the literature about that. Although everybody would agree that that is not a superscientific, dramatic part of fieldwork, that is an essential underlying part. And yet at UCLA, the fish collection went, the insect collection went. At Stanford, the whole herbarium left; it went to Cal Academy. There are lots of examples. It's important to realize that it's not just that those collections were necessary for some taxonomist to work on, like [John] Belkin, who did taxonomy of mosquitoes. The general insect collection at UCLA was important for people who were going to work on ecological questions.

Say, pollination.

Sure. After Stanford sent its herbarium to Cal Academy—it took a while, 10 or 15 years, to move—the modern problem of biodiversity came up. John Thomas told me that, at a Stanford biology department faculty meeting one day, one of the younger faculty members was expounding about biodiversity. He had discovered that the fish collections and the herbarium collections at Stanford had been given away. He couldn't understand this. Why, these dopey professors of 15 years ago had given away all of these collections! So, the interest in them isn't dead, but damn near dead because they are all dispersed and given away.

Nobody is going to open an herbarium at Stanford or get fish collections back at UCLA. It's easy to see that these collections were done away with. That's a physical thing. You can know the date when it happened, how many truckloads it took to haul the plant specimens from the Stanford campus to Cal Academy, where the UCLA fish went. But the subtle thing you can't see is: where did the students go who worked on those collections? They disappeared. There is no longer a field [plant] taxonomy course given at UCLA. There is none at Stanford. There is still a course at Berkeley. But [Robert] Ornduff retired, so I don't know if there is a course where people can learn what I taught Park and Martin. So, there are no students coming out like that.

Locally, the guys tell me they can't get people now to identify all of the plants for an environmental impact species list. The influx of molecular biology into biology sort of displaced the natural history. So you ended up with a bigger department but fewer people doing field biology, natural history.

The molecular biologists in zoology. We hired those guys. It's like saying, "Come and infiltrate our department." And the guy in botany, who never had a botany course in his life, he was a virologist by training.

He was the one who wanted to put the herbarium on microfilm, so the space could be used for another lab?

Another wet lab, which of course, by the time we built the lab was out of date. Somebody should have realized what was happening. Some of us. We were hiring all of these molecular biologists, and they were infiltrating the department and changing the department. They weren't an add-on; they were like a virus. I went around and asked one of the molecular biologists what he could contribute to the undergraduate teaching. I went around to everyone and asked that. He answered, "I can't contribute." He taught maybe a graduate course once a year. No interest. It wasn't just so much they wouldn't teach. They didn't know any biology.

Much of what I taught, however, I had experienced: mitosis, meiosis, the morphology of common invertebrates, common vertebrates. I had a personal connection with these topics.

Eugene Kozloff (b. 1920)

Professor Emeritus of Zoology
University of Washington
Interviewed in his FHL lab 14 July 1998

Eugene Kozloff's knowledge of both marine and terrestrial fauna of the Pacific Coast is encyclopedic; he is equally at home with a discussion of the evolutionary history of the protists and animals. He points out, however, that he is uncomfortable teaching about subjects of which he has no personal experience, such as DNA, ATP, and proton pumps. When I went to interview him, I discovered him in his small lab. The walls were lined with jars of specimens, and he was engaged in a lively discussion with two young, captivated students. In retirement, he continues to share his wealth of natural history experience with students and scientists at Friday Harbor Laboratories, as well as teaching adult natural history classes.

An important extracurricular experience came my way when I was in the fifth grade. My brother was taking high school biology and had volunteered to bring in some live crayfish. Well, I had never seen a crayfish before and had only a vague idea of what one looked like from eating lobster, which we bought occasionally from a peripatetic fish market that served our neighborhood on Tuesdays and Fridays. I was ten years old then, and that field trip to the Santa Ana River was a terrific revelation. Besides crayfish, there were minnows, water bugs, water beetles. It was a great experience, and I remember it almost as if it were yesterday.

It was in junior high school that I encountered the first formal instruction in general science and natural history. Then, when I got to high school, I was lucky to study biology and botany with Josiah Meryl Harper, for whom my brother had collected crayfish a few years earlier. Well, I had known J. M. for a year or two already, because I started hanging around his classroom a bit prematurely. This man was not just another "What-are-my-fringe-benefits-going-to-be?" type of teacher. If he decided that we were really interested, he took us on field trips, and we were welcome to pester him anytime during the summer. I remember an especially wonderful late spring excursion deep into San Diego county, where I saw, for the first time, many unusual components of the flora down there.

When I enrolled in Riverside Junior College, I was encouraged by Edmund Jaeger, a really fine naturalist, widely known for his studies of desert natural history. His background in zoology was very classical and a bit out of date. But he knew the Colorado and Mojave deserts, and weekend camping trips with him were exceptional experiences. Ruth Cooper taught me botany. Her encouragement and scientific attitude were very important in my development. Furthermore, sensing my interest in music, she and her family made it possible for me to attend concerts and opera out of town. I owe much to the Coopers, especially Ruth, for promoting my involvement in an avocation that has remained a lifetime interest.

When you went to Berkeley, you realized that you had had some old-fashioned science?

Yes, I discovered that my background in first-year zoology wasn't what it should have been. I wasn't aware of the most recent developments in genetics, embryology, and physiology. But in most respects, I was pleased to have been taught by a man whose knowledge of natural history, and especially of life in the desert, was so remarkable. As I've already intimated, Edmund Jaeger wasn't on the cutting edge of biology in 1940, yet it was he who filled up the gas tank, drove out into the desert, stuck a thermometer into the rectum of a poor-will, and showed that this bird hibernated.

Early in my career as a graduate student at Berkeley, I started working on the biology of saltmarsh gastropod mollusks. When I discovered some ciliates swimming in the mantle cavity of a little pulmonate snail, I applied some of the techniques of staining and silver impregnation I had been taught to use by Harold Kirby, who employed me a few hours each week. Well, I saw great beauty in those ciliates, and they were the subject of an early paper. Professor Kirby encouraged me to continue my work in

In 2000, Eugene Kozloff still relished working up flatworm samples in his Friday Harbor lab, work begun 40 years earlier. Photo by William Calvin.

protozoology, and I finished my M.A. and my Ph.D. degrees under his direction. Although I concentrated on protozoology, the research I did kept me in touch with invertebrates, and fieldwork was always a high priority.

Do I remember correctly that you worked on parasites in the kidneys of octopuses?

No, I never did. You may be thinking of Bayard McConnaughey down at the University of Oregon. He was a little ahead of me in graduate school, and he finished a marvelous thesis on the reproductive biology of dicyemids, sometimes called mesozoans, which live in octopus kidneys. No, I did not. But I worked later, long afterward . . . in fact, today is a landmark day, Bastille Day. I got interested in orthonectids, another group also called mesozoans, but I don't use the term. There are two separate groups: dicyemids and orthonectids. I've been the main investigator of this rather obscure group of parasites that lived in almost everything except octopus kidneys. I found them on Bastille Day in 1962, after looking for them for years, mostly because of my interest in ciliates.

I wondered if orthonectids and also these dicyemids could be evolutionarily linked to ciliates. I also became interested in acoel turbellarians, which I still work with, that's why there are so many jars in the other room. I wondered if ciliates, for example, were simply noncellular turbellarians of some sort. They differ from other protozoa in being diploid, and they transfer gamete nuclei by conjugation. Furthermore, they're very complex and have two kinds of nuclei. But, the main thing that links them to metazoans [multicellular animals] is diploidy. All other protozoans are haploid; the only diploid stage in the life cycle is the zygote, which then undergoes meiosis. Anyway, that's why I got interested in acoel turbellarians, and I have spent a lot of time hunting them, cultivating them, and studying various aspects of their biology.

And looking at them from a morphological viewpoint and at their evolutionary implications.

That's right. I unfortunately don't know more about this than I did 40 years ago! In general, however, acoels are now viewed as the simplest bilateral metazoans.

My niece, who took the invert course here at the Friday Harbor Labs last summer, is at Harvard as a grad student with Andy Knoll. Her interest is in metazoan evolution. Its time has come again.

There is a lot of interest in this now, and there are entire journals devoted to the subject. Much of the work is so molecular, however, that some of the biologists who specialize in it may not even know where the organisms they're grinding up live and may not understand how these organisms are constructed and how they develop. That worries me a little. Nevertheless, I'm glad to see that there's considerable enthusiasm now for trying to track the phylogeny of invertebrates through studies in molecular biology. I do regret not having any training in molecular biology. I suppose the closest I'll ever come is studying the ultrastructure of orthonectids, acoel flatworms, and a few odd ciliates. Knowing some of these organisms fairly well, it would be nice to apply the techniques of molecular biology. But, to be honest, I'm afraid that a lab full of centrifuges and scintillation counters isn't where I want to be. I like microscopy and consider much of what I do with a microscope to be natural history of sorts.

After graduate school, you taught at Lewis and Clark College, right?

Yes, and in terms of encouraging fieldwork, that was a great opportunity. Here was a small liberal arts college situated in a suburban and, surprisingly, still natural area near Portland, Oregon. I taught general biology and a course called field biology, which anyone could take; it had no prerequisites. It got a lot of young people interested in lifelong hobbies, like birdwatching. I had a letter the other day from someone down in Portland, who said, "Thank you for getting me interested in native plants; I'm growing them now." I had this student 40 years ago!

You could take field trips during a lab period.

Right. We usually didn't have to drive anyplace; we'd just walk out of the classroom into nature. I had a wonderful time teaching field biology. An outgrowth of this course, although I didn't know it at the time, I was just taking pictures to use in class, was my little book called *Plants and Animals of the Pacific Northwest,* a field guide. It has about 320 color pictures, of which I took all but a few mammals. When I came to Friday Harbor, that was in 1966, I had good opportunities to develop *Seashore Life,* which was the first of my field guides to come out.

In your experience with labs, starting from your own days at Hopkins Marine Station in the 1940s, teaching at Oregon and at Pacific Marine Station in the 1960s, until today at Friday Harbor Labs, where you've worked for over 30 years, have you noticed any changes in student interest in invertebrate zoology?

I don't know about the difference in students so far as their interest in invertebrates is concerned, but I notice a difference in their backgrounds. When I first started teaching invertebrate zoology, as a summer course for graduate students and advanced undergraduates, most of the students who came into it had had some sort of zoology or biology course in which types of invertebrates had been dealt with. They knew something about *Hydra* and *Obelia*, the earthworm and the grasshopper.

Then, we started getting more and more students whose backgrounds included plenty of biology courses, but whose experience with whole organisms was limited. They didn't understand the anatomy of the earthworm or the life history of an *Obelia*.

I wouldn't say interest in invertebrates has fallen off, because these animals are being used in research in subsciences that didn't exist when I was starting out. But, invertebrate zoology as a college course is suffering, and we've had to move what used to be called general zoology or general biology into upper division courses. There we are teaching students structure and function that they didn't get in the lower levels. But look what they are getting down there. They are getting material that wasn't in the textbooks when I was a student, or even after I started teaching.

Much of what I taught, however, I had experienced: mitosis, meiosis, the morphology of common invertebrates, common vertebrates. I had a personal connection with these topics. By the time I was teaching general biology at the University of Washington in the early 1980s, I was dealing with considerable material I didn't understand, because I had no personal contact with ATP [adenosine 5[1]-triphosphate] or the spiral helix. A good deal of my teaching had to be secondhand. I struggled through it because our mission was to make sure that the students got a fairly good idea of what DNA and messenger RNA were. But it never felt real, because I never had laboratory experience with the material. On the other hand, the general course, which dealt with diversity and morphology, was always satisfying. It was something of which I have firsthand knowledge. I enjoyed teaching general biology even though I realized I was in deep hot water much of the time during the stages of trying to explain DNA, RNA, ATP, the potassium pump, and how nerves transmit impulses. I have a lot of respect for the new recruits who have to teach this material, even though they have been trained to do the job.

I'd like you to comment on the sociology of a marine lab. I know you feel that a marine lab, particularly one like Friday Harbor, is a wonderful institution, there is a wonderful atmosphere generated here. What are the most important elements that contribute to that?

Well, thinking of Friday Harbor, I can think of one feature that contributes to its reputation: the ease with which one can work here. Just about everything is easily accessible after you get off the ferry and unload your electrodes; you can be at work in a few hours.

You get your own animals?

You have to get your own animals! It is a little different from MBL in Woods Hole, where it's a tradition to deliver material to the lab in which the investigator is working. Collecting your own specimens here at Friday Harbor gets you into the habitat of the animals or algae which you need. It's a good experience for a molecular biologist to see where this stuff comes from. I think Friday Harbor's success as a teaching institution and as a research institution is the ease of working here. It also has considerable housing, which is not so easy to find around some other marine labs. It also has been good to have a reasonable amount of subsidy for graduate students for their tuition and their transportation. Funding for individual scholarships has been a high priority for FHL. So far as I know, we have no NSF support for students coming in the summer. Another thing that has been good is that we usually have one or two postdocs who come for a year or two. These young scientists add a lot. They have limited duties, such as organizing a seminar program and introducing the speakers. But they are usually live wires and have a good influence on older scientists as well as on younger ones.

Dennis [Willows, the FHL director] and I talked about that. I think a real key to the success of an institution is to have the balance of neophyte with experience with all those stages in between.

That's right. This is a great place for that. And besides that, we've had good instructional programs and given some courses that could be hard to find in this day and age, such as invertebrate zoology and invertebrate embryology.

They got the money, and our time was spent doing science—dirty hand science.

ELIZABETH BUNCE (b. 1915)

Senior Scientist Emeritus, Department of Geology and Geophysics
Woods Hole Oceanographic Institution
Interviewed in her home 6 October 1998

Sharing a love of the outdoors and of sports proved an entrée for me into several discussions with Betty Bunce, an energetic and opinionated octogenarian. Her physical capabilities combined with her quick intelligence, work ethic, and sense of humor enabled her to work effectively at sea, in a "man's world." Data she and others gathered from the Puerto Rico Trench provided an important component of the sea floor–spreading story (see NAGLE). The first woman to serve as chief scientist at sea on an American oceanographic vessel, Bunce continued to be called upon in that capacity even after she retired. Her home hallway is a photographic gallery of the ships on which she has conducted science.

I wasn't interested in science. I shared a house at Smith College with a gal in the geology department. I was one of the first three graduate students in physical education. Dorothy Ainsworth started the master's of science program after we graduated. I was class of 1937. We enjoyed it, but gosh did we work, because we had to back up and take some courses, such as physical chemistry, plus run a full schedule of activities in the phys. ed. department. We were assisting, coaching, and I played field hockey and lacrosse competitively with the Boston Association. We were running what you might call a fairly active life.

What had you majored in, in college?

Government, history, and economics. Constitutional law, to be precise. I coached some of my housemate's boyfriends, who were Yale law students, because we had a far better grounding in constitutional law than they were exposed to. They would come up for the weekend. We would sit on the floor in the smoking room—I smoked like a chimney in those days, most of us did—and I'd guide them through Evans in constitutional law. What a tome!

So, through your roommate, you got into the geology program at WHOI?

There wasn't any geology. I had been teaching at prep schools, and I visited Betty at Woods Hole one summer during the war. To get into the institution, past the security guard, you had to fill in a job application. I was hired because they needed someone who was willing to work outdoors. I could use my hands, and I didn't mind getting them dirty. They couldn't get a man. The physicists were all being piped out to Los Alamos, and the others were drafted. I became a straight physicist, and I hadn't had a physics course. That's the way we did things then. We learned by doing. My phys. ed. stood me in good stead because I could work outdoors in all weather. I didn't mind being on the island, Nonomeset, when we were doing comparisons of high explosives under field conditions. We had safety precautions and, let me tell you, we minded them.

Basically, the group was looking at various sizes of explosives?

Yes, their relative effectiveness, which you measured by peak pressure and impulse, which is the area under the peak pressure curve. Your gauges are set up, piezoelectric, eight of them, all set up in a line. All the records were done photographically and were on cut film. Eight cameras. You had to bring all that stuff back to the lab, throw it in the water, and let it wash while you went up to the mess and got some supper. You came back, took them out, dried them, hung them up, looked at them carefully, made measurements, and then decided what you needed for explosives the next day, in order to get a comparison. If you were lucky, you were home by 11:30 or 12 and were up by 6 the next morning.

I got to thinking, sooner or later this war is going to end, and the need for this group is going to fold. What am I going to do? I decided to go back to Smith and get a master's in physics, never having had a physics course. Gladys Anslow, chairman of

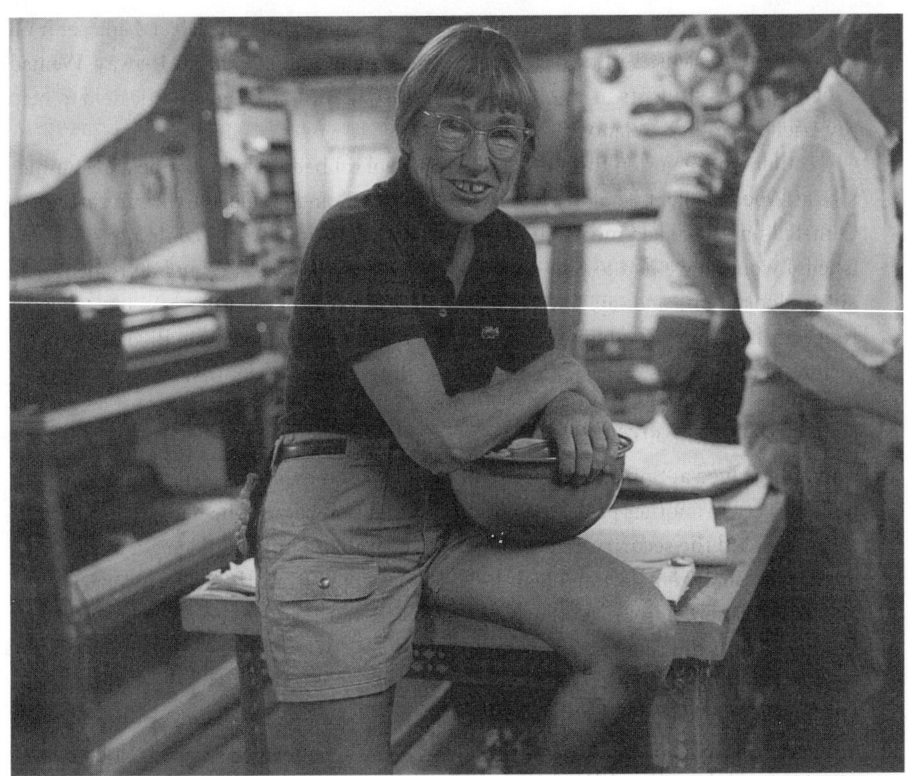

Betty Bunce pauses briefly before heading out to work on the deck of a WHOI research vessel in the 1970s. Photo courtesy of WHOI Media Relations.

the physics department, who headed up the graduate school and was also a very fine physicist, agreed to take me. I decided I would really like to find out what I had been doing for two years and find out what it was all about. I didn't know what F = ma was, to say the least. I was a government, history, and economics major; I was excellent in constitutional law. What good did that do? I was a female jock, and there were dog-goned few of us those days. They offered me a fellowship in the physics department with the understanding that I would do 20 hours a week of teaching and assisting in the lab. Mainly, I was assisting in the lab before Miss Billings retired, and the rest of the time I would take required courses to get myself up to speed. Calculus? I had to get a trig course under my belt first. I hadn't had anything since algebra in my sophomore year at Northfield. I was good at it then, but you lose your skills. I sweat blood over my homework that first semester. I couldn't get out of my own way for having to work.

We had five labs a week. Miss Billings taught me all I know. On-the-job training. I learned my basic physics from being her assistant in Physics 11. The first thing you learn is: don't try to fake it. The first time a student asks you a question and you go waffling around, and a student looks at you and says, "You know, Miss Bunce, you don't really know what you are talking about, do you?" You know, you are right, but

I will tomorrow. That was a lesson, believe me! I would finish the lab at 4 and set it up the next day for Miss Billings, clean up all the equipment, put the stuff away. We had a lot of physics students in those days. Then I would head for the athletic field to help my friend Ann Lee Delano in soccer.

I couldn't have gotten through graduate school if I hadn't had vacation and summer jobs at Woods Hole. I wouldn't have been able to eat, and I happen to like to eat! Betty Olmsted finally staked me to my last graduate year. Otherwise I could not have finished. I would have had to pull out and find a job without it. Try and do that. People always look at you and say, "Why did you drop out?" Money! Oh no, you could have found it. Umhum! She loaned me enough to get me through. She died a couple of years ago of Gehrig's disease. I got my first scotty from her family.

Did she go on to a career in geology?

Sure. She stayed at Smith as an assistant professor and then as professor. But, she gave her seniority to another professor, stepped behind him because he had a wife and family and needed the security and the extra salary. When Smith College got a new president, who shall remain nameless, he axed 17 people from the science departments "because women didn't need that kind of education, they were going to be wives and mothers." Boy, did that stir people up.

Let me guess. That was in the 1950s?

Yup. We had welcomed him as the president, because he seemed so good. Nobody ever asked him, What do you mean "women can't"? So, 17 slots shot down.

From a school that had had a strong tradition in science previous to that.

Yup! It was one of the tops in geology. Howard Meyerhoff made it that way.

When they offered me a slot here, at $3,600 a year in 1947, I came back to WHOI, back to my apartment looking out at Little Harbor and the coast guard dock. I belonged to the Navy 7 group again. I was taken back in. It had continued. It was mostly Harvard people, who had discovered that they could do a lot of good fieldwork, which they couldn't do at Harvard. You don't have salt water to test hydrophones and things like that. You can't do it in a swimming pool because you get echoes. And, you can't handle small explosives.

Tell me about early cruises. You worked out of WHOI.

We didn't work out of WHOI; we *were* WHOI. Our funding was raised through writing proposals, or at least our bosses did. For years, I didn't have to write proposals. Brackett [Hersey] raised the money. He and George Willard were the darlings of the Office of Naval Research because they knew how to run a program. They got the data. They got the money, and our time was spent doing science—dirty hand science.

I spent a fair amount of time on the RV [Research Vessel] *Bear*. We did all of the survey towing the big 20-element hydrophone array up and down Vineyard Sound, to Gay Head and back, then over to Buzzard's Bay, up and down. It is the only precise seismic study of the upper sediments of either area. It is done by producing a loud sound and recording the echoes returning from different layers. You can ask the guys at USGS. We did it. We published it. There were about five of us who got our hands dirty and did the work.

That was when seismic profiling first developed?

We were the people who did it: Brackett Hersey. We used an acoustic source rather than explosives all the time. Lamont [Geological Observatory of Columbia University] towed a hydrophone array around the world, throwing over a two-and-a-half-pound explosive block every ten minutes. They had the 'round-the-world thing. We did the fine structure, with the sparker and the boomer, which let you define your upper sediments.

It got boring, yes. It was interesting. It was also a nice way to spend a hot summer's day, out on the fantail, watching your gear. Not in the lab, watching your recorder, which you better do! I had a student, he was one of Brackett's less successful hirings; he fell asleep with his face in front of the PGR [precision graph recorder]. We lost two hours of records. We had to go back and do it over. He fell asleep! It will do this to you; you are standing in the lab watching the paper come out, you can hear the sounds, but you are watching the paper. It is hypnotic. I know because I was the recorder when we did the Narragansett Bay work for the bridge foundations. Ours are the data the army used.

Tell me a little about what you were doing in Puerto Rico.

We were studying the Puerto Rico Trench, with the sparker and the airgun. Nobody had ever done a pure survey of the trench from east to west. Everybody always did a transect [at right angles to the trench], and we did too. But the problem is, with the trench, you get these tremendous reflections. You have to be able to sort out the echoes from the sides. Yes, the acoustic reflections, holy mackerel, so you were not able to sort them out. You could hear them! You always wore headphones, one on one ear and the other ear cocked for something else, while you watched your records so you could mark them, if you heard things that were not part of your study. The odd whale saying, "Is that my mother?" Or the shark gnawing.

So this answers my question: Why does an experienced person need to be on site?

You have to be there while the data are being recorded because you have to sort out the sheep from the goats, and you cannot tell anybody who doesn't know what you are looking for and who has never done this. Oh sure, you may be a brand-new Ph.D. So what?! You have to be able to collate your ears and your eyes and also to have the experience to say, "That's just a multiple," and not say, "Oh look, we've got deep reflectance."

Previous to this, what work had been done in the trench?

Lamont had done a lot and so had we. We knew what the trench looked like. What I now wanted to know was: What did the north slope look like as it was going down? What kind of layering did it have? We wanted to know what the sediments looked like on the bottom. It is entirely different from looking at a picture of the trench. So you had to modify your attack. You had to use discrete sources, like the sparker or the boomer, rather than two-and-a-half-pound blocks of explosives, which gave you multiple multiples. You had to be on your toes.

You were trying to discriminate more finely what the layers in the trench were?

Yup. What the layers were and what the trench really looked like. It is historic, of course. The deepest hole in the Atlantic is at one end of the Puerto Rico Trench. What does it look like, sedimentologically? We worked all the way from east of the island of Anegada, toward the Midatlantic Ridge, and came in. Brackett, George Willard, Doc Ewing were early pioneers in the study of the Puerto Rico Trench and outer ridge. I was just lucky to have come along at a time when interest was still high enough to keep funding going.

Tell me about being a woman on a ship full of men.

The only problem was that you couldn't take your shirt off! It was hot. I had a whole bunch of really old cotton shirts. They were tattered by the time they had been through the washing machine on ship and out in the sun; they got very thin. You had to figure out how you were going to maintain your female dignity, if you didn't wear any underwear under the shirt. It didn't work because you wanted to be able to unbutton the shirt down and up, so you needed to keep one button closed, and you needed to wear a bra. They are hot!

Your views on opportunities for women in science?

I haven't the remotest idea. If you want it badly enough, go out and get it. You have to assert yourself. Just because you are a woman doesn't mean you can't take a shovel and dig some dirt. Come on, guys!

You said the other day something that I feel is true. You have to treat people, men and women, equally.

Yes. Equally. That is why I got along at sea. That is why I got along on the *Glomar Challenger,* on *Chain,* on *A II* [*Atlantis II*]. Actually, I was lucky, well not lucky. My stepfather briefed me years ago when I was going through the physical education program. That was the way I got my cornflakes, because they would pay me to be a teacher. He would say, "I don't want you to look like those Irish women hockey players with those bulging muscles below your knees!" They wore brown cotton stockings and brown tunics, without doubt the most unattractive combination, and they had a big, blow-up picture of them in the *New York Times.* He said, "Good God!" But, you know, that has stayed with me. You are still a woman. If you want to be a boy—you're not, you ain't—you never can be. I'm sure you can be one of the group, but don't try to be that kind of one of the group. You can use the bad language, but you use it not loudly in the lab. Unless, of course, it is warranted, like the day the shark took our magnetometer. The jerko colleague, just graduated with a Ph.D., who was on watch in the main lab, didn't even notice that the maggie was a straight line and nothing coming out on it. Wow! That is one time when a couple of the guys stood there and said, "Oh, boy!" Because I blew up, I boiled over.

With justification!

That's what the skipper said. Because this guy complained. He came from a moneyed family, and he wasn't used to being treated like a "slave." The skipper said to him, "Then, what are you doing here? It was explained to you. You had a job to do and you didn't do it. And personally, when Betty said that you ought to be keelhauled, I agreed with her!" He got absolutely no sympathy from anybody.

One must be enough of a generalist to have some knowledge of the system as a whole—its physics, chemistry, geology, and biology—to have any real understanding of how it works.

JOHN H. RYTHER (b. 1922)

Senior Scientist Emeritus, Biology Department
Woods Hole Oceanographic Institution
Interviewed in my office 27 October 1997

My first introduction to John Ryther was through the literature, reading his papers on primary productivity in the oceans when I was a graduate student. However, it was only while interviewing him that I discovered the real reason behind all of his hard work through the years: his passion is fishing, and fish are ultimately dependent on primary production. His research has been so comprehensive and so focused on such key issues, both basic and applied, that he is considered by many to be Mr. Primary Productivity. When I meet him these days on the streets of Woods Hole, he is usually headed out to a good fishing spot.

When I was discharged from the U.S. Army Air Corps in 1945, I mistakenly accepted a reserve commission. I entered Harvard as an undergraduate, and I continued there as a graduate student. As I was about to receive my doctorate in biology, the Korean War erupted, and I was called to active duty. I asked for a brief deferment, which was granted until September, so that I could finish my degree. I had the summer to kill with a pregnant wife and no more G.I. Bill checks coming in. I obviously needed a summer job.

While in graduate school, I had a small, sideline job trying to restore the run of

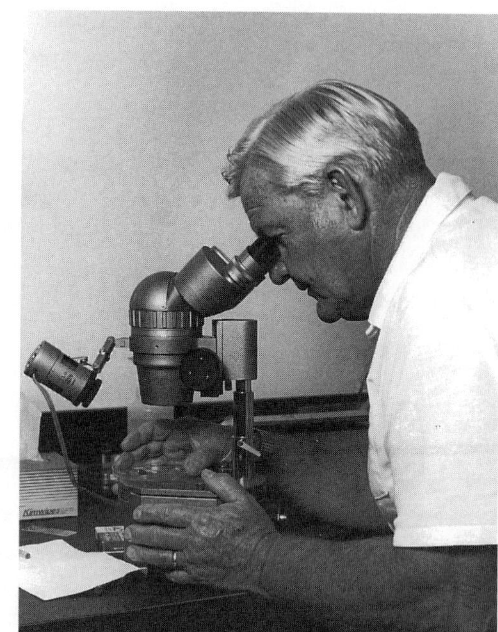

John Ryther (*left*) identifies phytoplankton for the Great South Bay Project in 1951; (*right*) doing the same basic species identifications for a project more than 40 years later. Photos from the files of J. Ryther.

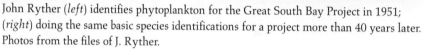

anadromous brook trout to the Mashpee River. This is a small estuarine stream on Cape Cod, leased at that time by a private club in Boston, renowned for its salter trout fishing back to colonial times. The salter trout population had virtually disappeared. The concept was to heavily stock the river with hatchery-reared brook trout, with the resulting overcrowding forcing the fish to go to sea and thereby resume anadromous behavior. At that time, the Harvard biology department provided a small summer scholarship for a student wishing to work at the Woods Hole Oceanographic Institution, for which I successfully applied. I used the scholarship to continue work on the salter trout project. My wife, Jean, and I spent a memorable summer of 1951 at the little fishing camp on the Mashpee River.

When September came, I said to the air force, "Here I am. Where do I go?" They replied, "We don't need you any more." So I was left without a job in September. I had assumed, while in graduate school, that I would end up at a university teaching freshman biology. I threw myself at the mercy of Alfred Redfield, then associate director of WHOI in charge of biology and chemistry. He allowed me to stay on in some sort of fellowship capacity, at least for the fall, while I looked around for a job.

At about that time, the townships of Islip and Brookhaven in New York approached WHOI. They requested that the institution investigate the condition of Great South Bay and the contiguous embayments on the south side of Long Island to determine the cause of the decline and collapse of what had once been a very prosperous oyster industry. The two towns each contributed the munificent sum of $1,000 for the study.

Grants of any size or kind in our field were so rare in those days that Dr. Redfield jumped at the opportunity. A group of six or maybe more WHOI personnel spent several months carefully studying the chemistry and biology of the bays, documenting for the record what everyone already knew. The tributaries to the bays were virtually lined with duck farms [the burgeoning Long Island duckling industry], and the bays were badly polluted from their wastes.

They were just being eutrophied?

I'll say. The bay water looked like pea soup from the blooms of tiny green and blue-green algae that had displaced the normal phytoplankton flora. These blooms could not be utilized as food by the oysters, and periodically, during periods of calm water, they produced anoxic conditions. Since I had worked with phytoplankton for my doctoral research, and since there was no one else at WHOI working on algae at that time, I was elected the "phytoplankton expert" of the team.

Also, at about the same period, the early fall of 1951, I read a small announcement in the newspaper that a new organization had been created in Washington—the National Science Foundation—and it was accepting proposals for grants. I quickly applied for and received what I believe to have been the first NSF grant to WHOI, to study the "etiology of the phytoplankton blooms of Great South Bay." It was a pitifully small proposal, my meager postdoctoral stipend and a few other odds and ends, but in making the award, NSF informed us that they did not pay the salaries of scientists, which was the whole point of the exercise. But Redfield didn't want to turn down the institution's first NSF grant, so he was stuck with me for a year, a term that eventually stretched out to 30 years and my retirement.

What was your graduate degree in?

Biology. One can't do much other than that at Harvard, which doesn't, or at least didn't then, offer degrees in specialized subjects. I became a strong advocate of that policy. I did my dissertation on interactions between phytoplankton and zooplankton.

So you were interested in plankton at that time?

Yes, and I continued working in that area when I came to WHOI and worked on the Great South Bay project. During that project, other WHOI scientists worked on water circulation, nutrient chemistry, and so forth while I worked on the phytoplankton. I got the causative bloom organisms in culture, and I studied their environmental physiology. Why were they able to persist and dominate to the exclusion of the normal plankton flora? That, of course, is a problem that is still very much with us and more critical than ever with the spread of toxic or otherwise harmful algal blooms in coastal waters around the world.

By the end of the Long Island project, I became involved in another problem. In the early 1950s, the Danish oceanographer Eirner Steeman Nielsen introduced the use of radioactive carbon in measuring the photosynthetic rate of natural phytoplankton communities, the so-called C^{14} method of measuring primary productivity of the ocean. He then used the technique on a 'round-the-world cruise of the Danish

Galathea and came up with an estimate of oceanic primary productivity an order of magnitude lower than Gordon Riley, then of Yale but formerly of WHOI, who had used much cruder and less sensitive methods. Since Riley was at Woods Hole when he made his estimates, Redfield felt that someone at WHOI should resolve the discrepancy. I was elected. Initially, I spent some time on methodology. I then made measurements of primary productivity in a number of different places around the world, usually as part of a team working on plankton ecology and dynamics. The work was primarily supported in those days by the U.S. Department of Energy and its precursor organization, a curious but warmly welcomed source of funding for the field of biological oceanography at a time when no one else was particularly interested.

By the 1960s, I began to get involved in administrative work. WHOI grew rapidly in the post–World War II era, from about 50 people when I first arrived in 1951 to several times that number a decade or so later, but it was still completely unorganized. When Dr. Paul M. Fye became director, he was largely responsible for the institution's growth in terms of buildings, the research fleet, and facilities in general as they exist today, but he had little contact or communication with the scientists, who still had to raise all of their own research funds through grants and contracts. Eventually the staff began to rebel about their lack of direction and general care and feeding. Quickly, Fye and the trustees appointed two staff committees to make recommendations concerning the status and organization of scientists and their research staffs. What resulted was a more or less university-type organization, which radically changed the structure and operation of the institution. As chairman of the Committee on Organization, I was rewarded for my efforts with the dubious distinction of being the first chairman of the biology department. In retrospect, my most rewarding accomplishment as department chairman was the securing of an NSF grant for a substantial number of postdoctoral fellowships each year for several years. Some remained as WHOI staff; a few are still in residence. Others are now directors of marine laboratories all over the country.

But I was never comfortable with the administrative duties of the chairmanship and stepped down as soon as I felt I had done my bit, only to find myself engulfed in a much larger administrative capacity. An international organization was initiating plans for an International Indian Ocean Expedition [IIOE]. The U.S. participation, to be funded by NSF, was to include a major effort in the field of biological oceanography. I was asked to develop the plans for and then serve as director of the U.S. Biological Program of the IIOE, running it out of WHOI with the help of my able deputy, Ed Chin, and my close colleague Dave Menzel. The IIOE is a separate story in itself and one I can't get into here except to say that it kept me rather fully occupied for several years in the mid-1960s, including some of the most memorable oceanographic cruises of my career.

Following the IIOE, and still with our Atomic Energy Commission sponsorship, we made several cruises on WHOI ships in highly productive coastal upwelling areas, Peru and southwest Africa, and in highly impoverished central-gyre oceanic regions. I was struck by the fact that much of the oceans could be characterized as extremely nutrient poor and unproductive. I then began to think about the relationship between primary productivity and fish production.

Aha! So your main love has always been fish?

I guess so. I figured, what the hell is the use of measuring primary production unless it tells you something useful, such as how it relates quantitatively to fish production in the sea? At that time, some prominent fisheries biologists, U.S. and Soviet, were predicting potential harvests of a billion tons per year. My simplistic model, based upon primary production and food-chain dynamics, predicted a sustainable yield of about 100 million tons. The figure was not original with me, though I was perhaps the first to claim that the 90% of the ocean beyond the continental shelves was capable of producing only a negligible fraction of the world's catch. At the time my paper was published, the total annual catch was 60 million tons. In the following decade it staggered up to about 100 million tons.

Which is the current level, right?

Yes, the annual catch has been hovering at about 100 million tons for the past ten or more years.

It surprises me when I talk to people who don't know much about marine science or fisheries. You say to people, "We've reached our limit, and most of the world's oceans are overfished," and they act surprised. Despite all of the articles, most people don't know that. Even people in fisheries don't want to hear that.

Yes, they hated me for saying that. My paper was published as a lead article in *Science.* It attracted a lot of attention at that time and some outright hostility from NMFS [National Marine Fisheries Service] biologists, but most now agree with my lucky guess, and some think it was greatly overestimated, since, as you say, many if not most of the world's stocks are overfished. That was my brief excursion into world fish production.

Somewhere along the line, I became interested in nutrient-phytoplankton relationships, eutrophication, and coastal pollution, perhaps as a reawakening of the beginnings of my career at Great South Bay 20 years before. On a coastal Atlantic cruise in 1967, Bill Dustan and I found that nitrogen was usually the critical limiting factor to phytoplankton growth. This was in contrast to the situation in freshwater lakes, where W. T. Edmonson and others had shown conclusively that phosphorous was the bad guy that controlled algal growth and eutrophication. Our publication, another lead article in *Science,* also caused a bit of commotion, as the phosphorous in detergents was, at the time, widely blamed for eutrophication; manufacturers were frantically replacing phosphate in their products with other substances containing, of all things, nitrogen! This exacerbated the problem at least as far as the marine environment was concerned. I was amused to see a recent editorial in *Science* on coastal eutrophication with the headline "The Nitrogen Problem Really Sneaked Up on Us." A 30-year sneak, by my reckoning.

I then hit upon the harebrained idea of "controlled eutrophication," growing phytoplankton in wastewater under controlled conditions, feeding it to commercially valuable filter feeders, different kinds of mollusks, and then growing commercially valuable seaweeds [e.g., for agar] in the residual nutrients regenerated by the mollusks. Waste recycling, was of course, not a new idea, but ours may have been one of the first such marine systems. After some small exploratory research, I raised enough money for a moderately sized pilot-scale system of ponds and raceways: the WHOI Environmental Systems Laboratory [ESL]. NSF had just inaugurated a new program of support for applied research called RANN [Research Applied to National Needs]. We were able to get funding from them to test and evaluate our system Initial trials were encouraging, but while we were coping with the inevitable start-up problems during the first couple of years, the short-lived RANN program was discontinued, and no other support appeared to be available. It was an idea whose time had not come—and still has not come.

Meanwhile, the directors of Harbor Branch Foundation in Florida, J. Seward Johnson and Ed Link, became interested in our ESL project and its possible application in treating the wastes of their facility. So, I moved our group to Florida and set up a small system there, with HBOI [Harbor Branch Oceanographic Institution] support. At the same time, the DOE reentered the picture. During one of our periodic oil crises, alternative energy sources became the fad, of which fuel from biomass [i.e., fermented to methane] was a prominent contender. It turned out that the seaweed used in our waste recycling system, *Gracilaria*, grown under optimum conditions, has one of the highest rates of organic production of any plant on earth. So we forgot about waste recycling, growing oysters, and so forth and concentrated on growing *Gracilaria* and other seaweeds, maximizing their yields and fermenting them to methane, both at WHOI and HBOI and again with DOE support. That ended in 1980 with the end of the [Jimmy] Carter administration. The Republicans weren't interested in alternative energy, DOE dropped its support of biomass research, and we were broke again.

By 1981, I was still interested in aquaculture, but there was little or no federal funding available, and no one at WHOI was interested in the subject. I had by then put in 30 years at the institution, was eligible for full retirement, and saw little prospect of continuing to raise my own salary for the indefinite future by doing something I didn't much want to do anyway. Besides, the people at Harbor Branch were definitely interested in aquaculture. The then-director of HBOI, Bob Jones, enticed me to join them and set up a new Division of Applied Biology, to include aquaculture, and that's how I spent the last several years of my active career. I then fully retired, returned to Falmouth and accepted a scientist emeritus position at WHOI, where I can work at my own pace, doing what I want to do, without ever having to write another proposal.

You mentioned your oceanographic cruises and your Great South Bay work earlier. How important do you feel field experience is?

Extremely important for the kind of work I have always done. Without it, an environmental scientist has no firsthand feeling for, or knowledge of, his or her subject. I get very impatient with modelers who spend their whole careers looking at a computer screen and never getting their feet wet, because I just don't think they can understand how the environment works unless they see it firsthand. When I first came to Woods Hole from graduate school, I got involved in the Great South Bay project. Some of the old-timers referred to it as "hip boot oceanography," but it was fieldwork. I wouldn't have known what laboratory culture studies to do if I hadn't seen the nature of the problem with my own eyes and been able to measure the existing environmental factors that produced it.

Do you feel that people in the past 20 years have lost that viewpoint?

Certainly a lot of people have. In the area that I was last involved in, aquaculture, I saw countless examples of people trying to grow things without any knowledge or understanding of the environment they were trying to grow them in. I quickly learned that a successful aquaculturist must first be a knowledgeable ecologist.

How important is scientific communication among people from different generations, and is the opportunity for that diminished today?

I think so. WHOI has gotten so big now that people don't communicate. They are in different buildings, different campuses. When I first joined the institution, we all worked in one small building, everyone had coffee together in one room. Physical and chemical oceanographers, geologists, meteorologists, and biologists would all talk to one another.

Something is lost without that.

Yes, but it seems to be the inevitable consequence of growth and specialization. I still like to think that in the areas I've always been interested in—biological oceanography, environmental biology, in general—one must be enough of a generalist to have some knowledge of the system as a whole—its physics, chemistry, geology, and biology—to have any real understanding of how it works.

Finally, just recently, I was able to complete the circle. *Trout Unlimited* asked me to supervise a summer student in an up-to-date and state-of-the-knowledge survey of the behavior of anadromous trout in salt water. I chose to do the job myself and spent a fascinating summer in the library and visiting trout streams from Long Island to the Canadian Maritimes, finding out what had been done in the past 47 years since my days on the Mashpee River.

That's great. So it really does complete your circle here at Woods Hole?

Yes, it does.

A summer at the seashore is an even better alternative to a weekend field trip to the shore, but as the former became more difficult, the latter was disappearing altogether.

John M. Kingsbury (b. 1928)

Professor Emeritus of Plant Biology
Cornell University
Former Director, Shoals Marine Laboratory
Interviewed in his office 13 November 1997

Having had personal experience in administering a marine lab located far from the main campus, I read with great interest Jack Kingsbury's account of building Shoals Marine Laboratory in his delightful book, *Here's How We'll Do It*, before I interviewed him. The title itself embodies the tenacious determination required to build and then sustain a field facility, so characteristic of Kingsbury. Driven by his passion for teaching and for the natural world, almost single-handedly he created SML to provide undergraduates with the opportunity to "get their feet wet" and to study nature with experts. In retirement, he runs a publishing company and continues to teach natural history in the Cornell Adult University.

For 30 years in the classroom at Cornell, I taught about algae as organisms living in an environment. I used slides from my own experiences at marine laboratories on various coasts to portray the variety of marine environments. However, no amount of photographic material can compare with being there yourself. Until the courses got too large, I included a field trip to the coast in each marine algae course. We would leave Ithaca on a spring weekend, departing the earliest I could get them off campus Friday afternoon, and we'd drive half the night—the roads weren't what they are like now—until we got to somewhere in eastern Massachusetts, where we could spend the night. Early the next morning, we would head for low tide at good spots for observing and collecting on Cape Cod. On Sunday, we'd do the same at rocky habitats on Nahant and Cape Ann and then head back for Ithaca, driving most of the night, arriving in time for classes on Monday morning. Two full days on the shore, carefully chosen with regard to the tides but not the weather, unfortunately. We hit bad weather some years, but basically, it worked out well, and the students got a lot out of it.

The first time I taught marine algae, it was a new course at Cornell. I had five students, of whom three dropped the course before the end of the semester. The last two or three times I taught it, the enrollment was over 50. When it was small, we could take the field trip to the Cape in state cars. When 25 to 30 students were enrolled, we used a school bus. Luckily, one of the students was a licensed bus driver. Effective field trips for larger groups reached a point of diminishing returns. I encouraged students to take the marine botany course offered at MBL in Woods Hole. For some years in the 1960s, Cornell's invertebrate zoologist, John Anderson, and I had been teaching our respective subjects in the summer courses at the MBL. As interest in the marine sciences increased nationally, admission to the courses at MBL became much more competitive, and we had trouble finding slots for our own undergraduates. In fact, the courses at MBL were being taught more and more at the graduate and even the post-graduate level. A summer at the seashore is an even better alternative to a weekend field trip to the shore, but as the former became more difficult, the latter was disappearing altogether. The problem of the landlocked undergraduate interested in the marine sciences got worse and worse, locally and nationally.

It was this problem that led me and some colleagues at Cornell and the University of New Hampshire to start teaching an introductory course in field marine science at the Isle of Shoals in 1966. That course led to the creation of the Shoals Marine Laboratory [SML] in its own facilities on Appledore Island, which opened in 1973. In the later years at Cornell, when the marine algae course grew too large for a weekend field trip, I recommended to those students who were particularly marine oriented that they should take the course at Shoals, and most of them did.

It was obviously a tremendous amount of effort by you and, even if passively, by the students to do this field trip. Why bother? Two questions: what did the students get out of it and what did you get out of it?

Well, I think I'll answer the last one first. If a person is a teacher, and I like to think that I am a teacher, successful teaching is its own reward. I think field experiences are a part of good teaching in a subject like mine, and conducting them made me feel that I had added an important dimension, which needed to be added to those courses. The students appreciated it and came back from the field glowing.

One of the reasons for getting the Shoals Lab started was because I couldn't do field trips with 50 students very well. They needed to be able to get their feet wet. I've answered your "Why bother?" question in great detail in my book *Here's How We'll Do It: An Informal History of the Construction of the Shoals Marine Laboratory, Appledore Island, Isles of Shoals, Maine*, published by the Bullbrier Press. Certainly, it is a disaster if you get undergraduates all excited about marine things at an inland campus like Cornell, and they never get a chance to get their feet wet, to get on a boat, to get seasick. They don't have any idea what the realities of marine fieldwork are. If they decide to go to graduate school somewhere in marine sciences, then comes the rude awakening. I think it's a tremendous waste of the campuses' resources and the national marine resources for them to commit to a graduate program and then drop out, to say nothing of the effect on the student's own life. It isn't right to wait for the graduate years to give students who need it a field experience in marine science.

That's what I felt with the WIL kids. Many went on, but some of them said, after the research, this was a wonderful experience, but this isn't me.

Exactly. Our successes at the Shoals Lab are just as great when we give students the chance to find out this isn't for them as when a student starts down the path of a major productive career in marine science. Even the ones who decide not to go on take away much from their marine field experience that is useful to them for the rest of their lives, whatever they decide to do. Typically, participation at the Shoals Lab or in the Sea Education Association programs (which I have been associated with since their beginning) is genuinely a life-changing experience.

They are useful personally and, I think, also useful to society to have that kind of an appreciation both of how people like this work and also how organisms work.

And how you work, smelling diesel fumes on a boat in a rough sea!

My connection with SEA resulted in 1990 with my being shanghaied for two and a half weeks aboard SEA's 125-foot-tall ship *Westward* while we traced the track of Christopher Columbus in the New World with a small class and a filming crew from WGBH, Boston. For days on end, we bucked into steady 25–30-knot tradewinds and heavy seas under power, regularly burying the bowsprit and flooding the decks with ankle-deep water. Everyone aboard, even pretty seasoned people, got sick at one point or another. I think the most difficult kind of lecture to give is the one when you're on deck with the wind whipping your notes, you can't be below or everyone would be sick, and your glasses are fogging over. You keep one hand on the rail at all times, and you try to keep track of the lee rail, in case of sudden need. Your inflection loses something in consequence!

But it's a reality.

It's good for kids to see that, and that's why the Shoals Lab got started, for the undergraduate. I hope JIM MORIN keeps it that way, and I think he will.

The Shoals Lab is primarily for the undergraduate. The marine science course attempts to give a broad overview of the marine scene, not studying just the algae, just the invertebrates, just the fish, or whatever, but seeing some of all of that, from several people who can connect to one another well enough to bring a unity of perspective. We also include practical aspects, such as how various types of commercial fishing are done, with demonstrations at sea of bottom dragging, gill netting, and the like. Nowadays, many of our best faculty members are our own former undergraduates.

How important that is. That is one of the problems with some of the field programs that have lingered along, but the continuity of instruction hasn't always been of the quality that you would like.

I like to think that out at Shoals, in the early years that I can speak personally to, that we had a great array of good lecturers who were free to do their thing. I would fill in the holes between one expert and the next expert when that was needed, so it could develop in the student's mind as a coherent picture.

Another important point is that the students always had full access to the experts. On a small, isolated island, you have the advantage that the students and the faculty can't get away from each another; most of the faculty appreciated the full-immersion situation. They became the only show in town, and it goes 24 hours a day. A tremendous amount of career information gets passed back and forth that way. The students get to see these people from a variety of backgrounds and talk with them one on one about what it's like to be whatever that person is. What do you worry about? How much do you get paid? You know, that sort of thing.

Wasn't it difficult running a marine facility in a remote location at a great distance from the main campus?

Actually, that's one of the big problems the Shoals Lab has continually. Many of the administrative things done on an otherwise uninhabited island ten miles offshore dock to dock are, to put it mildly, irregular when the paperwork comes back and hits the midlevel administrators in Ithaca and Durham. They don't know how to accommodate it. So you have to search out the person at the right desk and cajole them to come out to the island, so they can see what the reality is out there. You spend a lot of effort doing that, and you finally get somebody who is helpful, genuinely helpful. Not to say the other people don't try to be helpful, but it doesn't fit their situation. The helpful person lasts some five years or maybe ten, if you are really lucky, and then, all of a sudden, they're gone, and you've got to start all over again. And, don't forget, we have two universities to which we must report.

It's difficult to get the top-level administrators to take *any* interest. Of the four presidents that Cornell has had since the lab was founded, only one has ever gone out to Appledore. And that one went out partly because he had a summer home in Maine, and we were sort of on the way.

Well, I'm here, so I'll go out?

That's a little harsh. He was genuinely more interested in our program than the others have been. They do all kinds of lip service to the excellence of Cornell's undergraduate teaching in their annual reports and in presentations to trustees and alumni, and they emphasize the importance of all those year-abroad programs and the like. Why wouldn't they go to see one of their own best examples in action? It was not for lack of invitation.

One of the recruiting efforts that was most successful [at West Indies Lab in St. Croix] began because I said to this supportive dean, "If only the faculty who might recommend the lab to their students could see the lab." He said, "Do it!" and he gave me the money to run that kind of program. It paid off. My philosophy was, if people only knew what we had to offer, we'd have to turn people away.

Exactly. I agree with that 100%! It's obvious when you think about it that the overhead runs the same whether your programs are full or not. When they are full, you have a little extra money to do good things with. It is one of Jim's most urgent jobs, to recruit good students and fill the spaces all summer. He needs all the help he can get from faculty members who have actually gone out to Shoals, seen the program in action, and how it changed lives.

The Shoals Lab differs from an academic department on campus in important ways. At Cornell, it has been officially designated as "an enterprise operation." That puts it in the same financial boat as the university press, the art museum, the arboretum, and even the campus store. Unlike an academic department, it does not go to a dean in supplication for its budget. It has to raise its own money and construct its own budget. In its budget review, the administration is concerned primarily that the bottom line not show a deficit and very little, if any, with the academic program. Several faculty committees go over it from time to time. Cornell and the University of New Hampshire contribute little money to the program, some 10–15% at most of the annual budget. The lab has little endowment. Most of the income comes from, and must come from every year, the students via tuition. It would be a much easier situation if the two universities would find some way to put some of their vast undesignated endowment incomes back into the teaching effort at Shoals. It would be very nice to have a stream of income coming in from that.

How did you end up at Cornell?

They needed someone who could teach veterinary students about poisonous plants as well as develop an unrepresented area of botany at the graduate level, either bryology or phycology. I was hired right out of Harvard. The two years of animal husbandry helped a lot. I was ultimately professor of botany in the College of Agriculture and Life Sciences and also professor of clinical sciences in the Veterinary College.

All the time?

Most of it, until the Shoals Lab got started full summers, and then I had to resign from the Veterinary College at that point, due to lack of time.

So your interests there were primarily what the poisonous plants were and how they were distributed?

No, no, it was the toxicology of the plants. What they did in animals and in human beings. I wrote the standard text of its time on that.

You know what I do for a living now—publish books. I'm working now with Phil Sze on a revision of my book *Seaweeds of Cape Cod and the Islands.* After 30 years in the classrooms of Cornell, as I came near the possibility of early retirement, Cornell decided it didn't want algae any more and it didn't need poisonous plants taught by a botanist, either, so it didn't pay me very well. It became clear that a pension based on my pay wouldn't suffice, so I retired while I could still earn money elsewhere. In a way it's sad to stop practicing what you were educated to practice, but the situation at Cornell was not conducive to continuing.

Are you glad you made that choice?

Yes, very glad, even apart from matching my pension from 30 years of teaching in just ten years of publishing.

Do you miss the students?

People said that I would. But I still lecture to eager students in the Aquavet-R programs and, with my wife, have been leading Cornell Adult University [CAU] programs in various parts of the world for 20 years or more. Undergraduates will absorb almost anything you'll say to them. Lecturing to accomplished adults is another matter because you get so much immediate response and from lifetimes of various experiences.

We found that too. Some of our summer classes at WIL would have high school seniors up to retired judges. We'd have wonderful interactions.

That's what we get in the CAU program where you have 40 people listening to a lecture and reacting to it. It can produce a very stimulating situation, and that's what I like now. I suppose I do miss the undergraduates somewhat, but I don't miss having to be in a particular place at a particular time, properly prepared to do a good job, several times a week, without fail. That is somewhat like having to milk a cow, twice a day, 365 days a year. And I certainly don't miss committee meetings. Perhaps even more important, I don't have to feel at all responsible for the things that are going wrong on the university campus. I can say with good conscience, "If they want to make their mistakes, go ahead and let them." But they better not make any serious mistakes about the Shoals Lab!

Being involved with so many individuals is one of the important things in life.

John Vernberg (b. 1925)

Emeritus Professor of Biology
Director, Belle Baruch Institute for Marine Biology and Coastal Research, retired
University of South Carolina
Interviewed in his office 5 November 1997

I first met John Vernberg in the mid-1980s at a meeting of marine lab directors. Although he was an experienced veteran and I was a new kid on the block, he quickly made me feel comfortable in this group. In the late 1960s, Vernberg left a secure position at Duke University to go the University of South Carolina and build from scratch the Belle Baruch Institute for Marine Biology and Coastal Research. His humor and gentle manner belie the determination and persistence that were a necessity for him to build and sustain (and then rebuild a hurricane-ravaged) a coastal marine lab through the years. When I interviewed him, he had recently retired from being director but was still actively engaged in research.

I got my Ph.D. in 1951. I went to Duke and was there for 18 years. Then I came here to USC to start a program, a field station, and so I have had the chance to observe some marked changes in that time. You asked about major milestones. That's a hard one to answer because it's cumulative. Rather than being macroevolution, it's been more microevolution in development. To a certain extent, I guess we as scientists are an opportunistic species. As you do your research, some new questions open up, and so if you have ten new questions to answer, you obviously take the one that's most appealing to you and in some respects the one that is capable of being funded, because research increasingly costs money. To get around to this, I think one of the major things that happened was the development of the National Science Foundation. I think I had my first NSF grant in 1954, and it was for the grand sum of $7,800 a year. Gosh, I thought this was great. Oh my gosh, what am I going to do with all this to get our research done? Of course, now the value of the dollar has changed, even getting back to what a research grant today will fund: equipment, supplies, travel, support of technicians and graduate students. So the cost of doing research has gone up.

I can ask, "What kinds of congressional actions really influenced my life?" One was the G.I. Bill. Without the G.I. Bill, I probably would have gotten through college, but it would have been very difficult. My wife, Winona, and I both went on the G.I. Bill. She had been in the service. I like to think that it was a fantastic investment on the part of the federal government and a big return to them, in terms of our paying it back in taxes. It was great. Of course, the other was the creation of NSF.

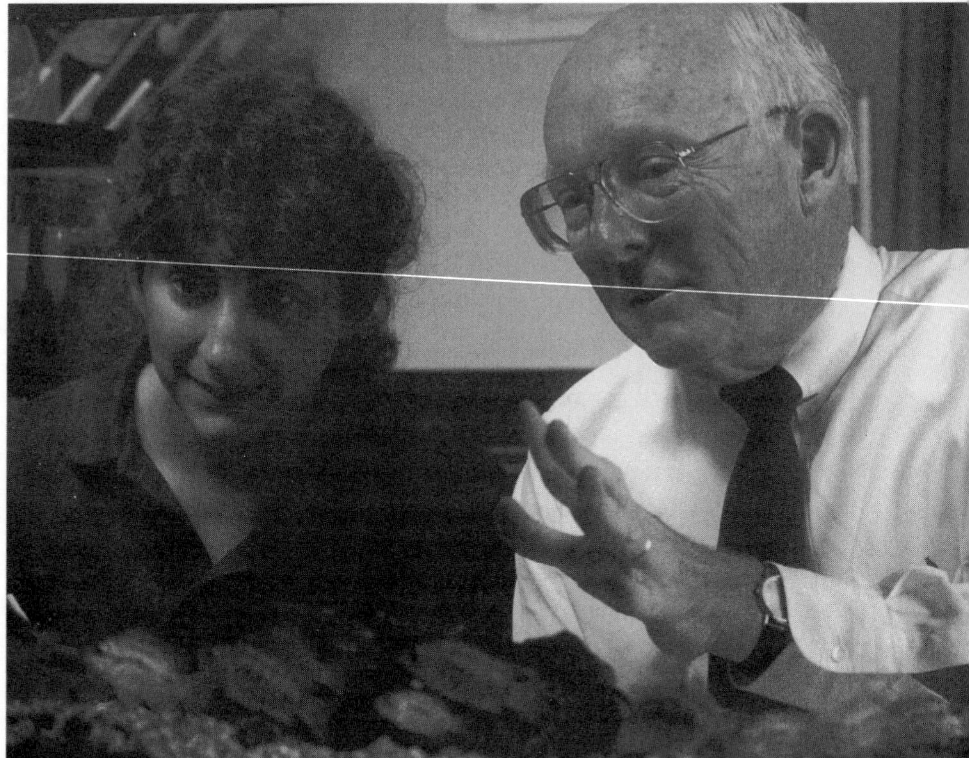

John Vernberg and a young student share the fascination of watching fish behavior in the mid-1990s. Photo from the files of J. Vernberg.

The National Science Foundation has done so much for science in research productivity. I think it has help set the United States off from other countries, if you want to look at it, a competitive advantage for the United States. I've been funded by the National Science Foundation continuously from 1954 until just two or three years ago. That has been a wonderful way of broadening the kind of research you can do.

And it's also played a role in your life in terms of funding facilities that you've used and probably courses that you've taught.

Oh yes. During the development of NSF and my particular career, much more than just research was affected. Through the years, NSF also supported teaching programs and, as you well know, programs to enhance field facilities for teaching and research. I taught courses for high school teachers in marine science. I particularly remember teaching an ecology course for college teachers in Colorado. JOHN RYTHER would lecture in the morning and I in the afternoon. Of course, the facility grants from NSF for enhancing marine labs, not for us exclusively but for visitors to come, to provide better sea water, for example, have had a wonderful effect on generating good science. So the tentacles of NSF have reached out and influenced really all areas of science.

In a very positive way.

Yes. Oh, I'd get turned down for a proposal sometimes and I'd cuss them, but nevertheless the total feeling, the total impact has been very positive. Of course, having NSF stimulating basic research, I think that's had a strong impact on the research priorities of some of the regulatory agencies, such as EPA and NOAA [National Oceanic and Atmospheric Administration]. Even though we've had budget cuts and a shifting of emphasis, comparing the growth of science from when I started in 1951 to today and then looking at what had occurred before, there is no comparison.

Before the National Science Foundation started, there were usually very limited amounts of research money available, which in turn restricted the kind of research. The National Marine Fisheries Service labs were sort of havens for academics to go and work; there was some space available. With the rise of more university facilities, I think a lot of researchers shifted to there to work, although NMFS and other labs, like that of the Forest Service, still provided some facilities for visiting scientists. But most academic research was geared to university-affiliated facilities or places like Woods Hole Oceanographic or the Marine Biological Laboratory, which are really like university affiliated. They are both independent, but they are not federal facilities in that sense. It is interesting to have watched, with the increase in the number of people going to college after the Second World War, the increased demand for classes in every area, including more classes at field stations, for example. That has mushroomed. To try and get back and weave this into me. There was such a growth of different kinds of research going on, it greatly influenced the kinds of things that I was thinking of doing and was able to do.

What I think I hear you saying is that intellectually it also opened up new ideas.

Very much so, just as the Second World War brought people together who never would have interacted before the war. I was thinking of my navy experience. I met guys from different parts of the country; we were stationed in different parts of the world. It expanded your mind, it broadened your horizons to realize there was this diversity. I think the same thing happened with the rise of the National Science Foundation, the expansion of scientific organizations. Field stations became increased in number and capability; you met more and more people and so you weren't as insular, you weren't as provincial.

So field stations were magnets to bring people from different scientific disciplines as well as different academic settings and different generations.

Yes, and so for the undergraduates, as you know, field stations are the best way to get to know people, much better than in a campus setting. In a field station, your dignity is not harmed. The interactions between people are less formal. I think it is stimulating to people to make them think in terms of how one can use the experimental approach that, say, an embryologist might have, to look at something that you might want to do. Or dealing with physical oceanographers. Some of the data that we would like to have are, to them, so mundane, so easily done. We had a symposium a few years ago, bringing together those individuals instrumental in new techniques, new technology in remote sensing, and on the other side were ecologists, who were stymied by their lack of training as to what was available from the new instruments.

Ah, but they had the questions, and they knew the data they needed to answer their questions.

Right. That's why I think it is so important to get these people together. It is important because the answer to these questions is sometimes limited by technology. On the other hand, people are great at developing technology, but they don't always have the questions. Their questions are different from a field ecologist. You get the two together, it's . . .

Exciting?

Yes!

How important do you think it is that there be experienced people in the field, people who have been places, seen how things are in different places, that sort of intuition and wisdom?

I think, perhaps it's because I'm older now, I like to feel there is a niche for older people. But seriously, I think a blend of having some experienced people blending in with kids who have really got some questioning minds is very good. Now I think the experience depends on the individual. It can also stifle young people from following some crazy ideas that they may have because an older person will say, "Ah, that doesn't fit into my concept of what's going on, and I'm experienced." I think there is a trade-off here.

But doesn't that force the younger person to take risks and to prove something to the more experienced people?

Oh yes, sometimes it is wonderful to tell students, "That can't be done," and then it will be a stimulus to them to go out and show you that you are incorrect!

You know that that happens, right?

Oh definitely! Sometimes you can do this purposely to stimulate students, but sometimes they show you, so it's a feedback to you, as long as you've got an open mind on things. Gosh, I didn't know that! Sometimes a young person looks and says, "Well, the emperor doesn't have any clothes on," and will ask what might appear to be a very stupid question because he is not prejudiced by what the dogma is, and I think that is very illuminating.

There is a problem with the progression of people with time. They go from being a student, young faculty members, if you are lucky you get a number of grants, and the first thing you know, you become separate. You become somewhat removed from what is going on in the field. That happens and, to safeguard against that, you need people to manage, you need people who are good in the field. The reward system is such that people who remain active in their research exclusively don't always rise as high in an organization as people who rise in management. The financial returns are not always as great.

You and Winona seem to have done a balancing act with continuing to do research and being involved with the creation, management, and facilitation of science by other people. I suppose part of that is personality.

It is a conscious effort, but it is due to one's personality and outlook on life. You know, it was an interesting thing you said about what causes changes in your life. Sometimes, a casual remark . . . When we were considering whether to leave Duke, where I was a professor—I had a very nice program, a nice lab—I was considering whether to come to South Carolina where the program, well, there wasn't one. Art Hasler from the University of Wisconsin was visiting the Duke lab at Beaufort at the time. I got talking with him about it. He said, "Well, you know, John, sometimes in your life, you take from the system. When you're a student, you take ideas, you take whatever you can to develop your own ideas. You are a taker. Sometimes, you've got to put something back into the system, and in certain cases putting back into the system is to become director of a lab and for you to be enmeshed in starting a new program." I've always remembered him mentioning this. When you do get to the point where you can be supportive of other people in developing a program, you do it.

Despite its frustrations, it has its own reward.

It does. I guess some rewards from this you don't really appreciate until a period of time has gone by, and you look back and say, "Oh gosh this has been done," but it probably took 15 or 20 years to do it. During that period, there were probably some times when you said, "Oh God, I'd like to throw in the sponge." When you think how many letters of recommendation you write for people, it's not trivial, the amount of time it takes, but you do it. Or refereeing papers. You evaluate your life sometimes and you say, "Well, what's important?" Looking back at things, I think being involved with so many individuals is one of the important things in life. Whether it's been favorable interactions with people or abrasive ones, it is still part of your molding of yourself.

I think one of the low points was after Hurricane Hugo [both Vernberg's lab in South Carolina and our lab in St. Croix were heavily affected by the direct hit of this category 4 storm in 1989]. You know this very well, but to go down and see the lab, the condition it was in after working for so many years to get some semblance of a laboratory facility, it was heartbreaking. But at the same time, it was interesting because Wendy Allen had a video camera and recorded the scene. She and I talked, and I told her we would return, we would recover. Tomorrow, tomorrow is going to be a better day. It took a while, but we did it. The lab did recover, and the lab is much better. It is extremely nice. These are the milestones that are important to people.

SUMMARY: CHANGING TIDES

What is dramatically evident is that the biologists of this generation were broadly trained in organismal biology as undergraduates and had to demonstrate their breadth of knowledge as graduate students. Not only were they expected to become expert in some specialized discipline, but they were also to have a general knowledge of the biology (taxonomy, morphology, physiology) of one or more taxa. For instance, Holger Jannasch, a microbiologist, was first trained in zoology; John Vernberg and Eugene Kozloff knew vertebrate as well as invertebrate zoology; Harry Thompson, John Ryther, and Jack Kingsbury knew animal as well as plant biology. These biologists stress the need to know the organism with which you are working.

By the 1940s, the field biological sciences were clearly becoming more than simply descriptive, and questions of evolutionary, ecological, and physiological natures were pursued as the graduate research and then as the careers of each of these scientists developed. McFarland even today asks of animals he observes in nature, "What are those animals doing?" Vernberg wants to know, "How do the beasts in this world get by?" Thompson asks, "How did these plants speciate?" And Jannasch asked initially, "How are aquatic microorganisms different from soil microorganisms?" a question that was eventually to lead him to the ocean's depths and other extreme environments.

Funding for the natural sciences, particularly for the oceanographic sciences, increased dramatically during and after the war years because the military (especially the navy) needed to know more about the environment in which it operated. The initial goals of research had immediate military applications. How could explosives most effectively be deployed? How does sound propagate in the ocean? But, as Betty Bunce insightfully stated, the clever scientist could use the same data to better understand how the ocean and the Earth worked. Oceanography blossomed as a discipline in those years, supported by the Office of Naval Research. We will see in subsequent chapters that that source of funding continued to grow during the Cold War years, as national defense needs dictated a much greater understanding of many aspects of ocean science, including the topography of the ocean floor, currents, weather patterns, and continental shelf processes. In contrast to the earth sciences, Ryther and Thompson recounted that the funding for the biological sciences was rather limited and often quite applied. Vernberg explained (and others illustrated in their life accounts) that two acts of Congress helped define the science of this generation: the G.I. Bill and the National Science Foundation.

In the 1960s, as most of this World War II generation were well established in their careers, the explosion of scientific research, funded by ONR, NSF, NIH, and other sources, led to an enormous growth of academic departments and research institutions. The interest in the ocean sciences also resulted in the established marine laboratories being flooded with students, and many chose to limit their classes to graduate students, denying undergraduates the opportunity of a field experience. The positive aspect, of course, is that much more research and education in the natural sciences was supported (which will be explored in more detail in chapter 4), and a financial climate developed in which universities were willing to allow new field institutions to be built (e.g., the Shoals Marine Laboratory, the Belle Baruch Institute for Marine Biology and Coastal Research, and the West Indies Laboratory) to support the demand for field teaching and research space.

The negative aspect of rapid growth and expansion is that a bigger institution (or department) cannot operate the way a smaller one can. For example, no longer was WHOI a one-building facility where physical oceanographers, chemists, biologists, geological oceanographers, meterologists, and engineers all gathered for morning coffee and discussions of the newest research results. Ryther explained how the creation of a university-type administration (with separate departments for the various scientific disciplines) in the early 1960s may have been an organizational necessity, but dynamic face-to-face discussions among colleagues (a crucial element in the production of new and innovative science) were greatly diminished.

There were not many women in this generation who became scientists, but a few did. Betty Bunce, in addition to serving as chief scientist at sea, also served for a period as chairman of the geology and geophysics department at WHOI. Dr. Winona Vernberg (John Vernberg's wife) became dean of the School of Public Health at the University of South Carolina and served as provost of the university for a period. Dr. Isabel Abbott (Don Abbott's wife and my first field instructor), after years of second-class academic rank due to nepotism rules, was finally awarded a professorship at Stanford University and given the recognition she deserved as a teacher and researcher. These women succeeded in science, on their own extensive merits, of course, but also because they were afforded the opportunity to obtain enough practical experience during the 1940s "doing men's jobs." Given this chance to prove themselves, they gained recognition for their expertise. No one doubted the contributions they could and did make to the advancement of science.

What has changed the most about science and the training of scientists during the lifetimes of this generation? Probably the most oft-mentioned lament is the fact that students aren't taught about organisms—morphology, physiology, and systematics—nor about their relationship to the environment in which they live. The systematic collections in universities have largely disappeared. The appreciation that science is a historical process, built upon the past work of many others, is lacking. While each of the scientists I interviewed acknowledged the vast potential of the powerful new technologies available to investigate questions, many expressed concern about the increasing lack of communication among scientists. Students are becoming specialized in subdisciplines at an earlier stage in their educations; breadth is sacrificed for narrowly focused depth. Languages in subdisciplines are becoming so specialized that a Tower of Babel is being created. Finally, for these World War II adventurers, the business aspect of doing science today has taken some of the fun out of it.

3

The Silent Generation of the 1950s

> That's also another beautiful aspect of science,
> you never know what's new around the corner.
>
> *Peter Glynn*

The 1950s are remembered as a period of stability in natural sciences as well as in society. Former general Dwight Eisenhower was president for most of the decade, and the military was respected by almost everyone. This generation of scientists is characterized by a seriousness of purpose and a devotion to duty, which for many included military service. Nevertheless, there is a clear hint in each interview of not wanting to be trapped in the wrong career. Most of them, as children, were budding natural historians, curious about facets of nature. Interestingly, each became an explorer, visiting remote areas of the world in a period when travel was still an exotic activity. These trips, undertaken on their own initiative (unlike the organized groups of the present day) were memorable journeys during which adventures and new discoveries contributed to the development of self-confidence. While students, each worked alongside experienced older scientists as field assistants or at marine laboratories. Scientifically, these researchers have often embraced a global or integrative approach to questions about natural phenomena. This is due partially to the geographically diversified field experiences each has had. It is also the result of their maturation as scientists when the theory of plate tectonics revolutionized the way natural science was done; it forced scientists to look at others people's data to help solve their own questions.

The pendulum has swung so far from descriptive biology that we're into theory without description.

Stephen A. Wainwright (b. 1931)

James B. Duke Professor Emeritus of Zoology
Duke University
Interviewed in his office 6 November 1997

Steve Wainwright's intense Socratic approach to education and science has influenced many people (including me) well beyond his own distinguished academic progeny. His own story, recounted in his direct style, weaves together a love of organisms, a curiosity to figure out how they work, an imaginative intellectual leap to adapt a mechanical engineering approach to biological structure, and a love for his students, to whom he has taught the art of fieldwork. Illustrative of his broad educational philosophy is the nonmajors course called "structure," which he taught for many years at Duke University.

One of the papers I always had my students read in their first week at West Indies Lab, in fact, the first paper I would discuss is 1969 Wainwright and [John] Dillon sea fan orientation.

That was maybe the best paper I ever wrote.

In that paper, you were completely clear, in terms of the logical steps, about what went next in the study. That is a wonderful introduction to students who have never read primary literature before. How do people discover things? How do people do research? One goes out and observes a pattern. If the pattern is real, one can quantify the pattern and then one can begin to ask, "What are the mechanisms that result in that pattern?" Your paper was a perfect example of that process in action.

You're right!

I grew up in Indianapolis, Indiana, with parents who encouraged me and my brothers to do whatever we wanted to do. I wanted to do nature study and I did. I was told by my second-grade teacher, 20 years after the fact, that on the first day of second grade, she said, "Well, now, here we can study anything we want to this year, what would you like to study?" Our class sat there totally dumb, no one wanting to say anything. After the class, she felt somebody tugging at her skirt, and she turned around and there was Stevie Wainwright saying, "I'm interested in lions and tigers and bears," and that was the beginning of it.

I was lucky that in public high school in Indianapolis, I could take a full year of botany and a full year of zoology, and that was based on fieldwork. We had lectures first, but in botany, we collected leaves in the fall and had to classify them, preserve

Steve Wainwright conferring with colleagues at the opening of the new interdisciplinary Whitely Center at Friday Harbor Laboratories in 2000. Photo by William Calvin.

them, and make botanical specimens. In the spring, we had to sketch wildflowers as they came in bloom and talk about the habitats they bloom in and identify them. In zoology, we collected insects and formed a labeled collection in the fall and watched and recorded birds in their migration in the spring. Some summers I went to nature-like summer camps, and my father would take my two brothers and me fishing in southern Florida. We'd go down to the coral reefs on the Florida Keys, and we learned how to snorkel there, when that was a very new activity right after the war.

I came to Duke University. I knew by that time that I wanted to be a marine biologist. I knew what invertebrates were. Duke, of course, had a wonderful marine lab that I went to in the summertime. I first took a course in invertebrate zoology from C. G. Bookhout and then took another in marine ecology from I. E. Gray. After that, I knew that's what I wanted to be: a marine invertebrate zoologist with an interest in ecology. In the summer of 1953, Don Pritchard, director of the Chesapeake Bay Lab and a blue water oceanographer, taught a graduate course in oceanography at Duke Lab. During the course, Jacques-Yves Cousteau, who had just invented the aqualung, sent two of them to his friend Dr. Pritchard so that the aqualung might be introduced to American marine scientists. Pritchard was a spirited, macho individual but had never dived at all. I was the only person at DUML who had diving experience, and that was just snorkeling. So a group went out on a really exciting expedition to be the first Americans to use aqualungs. It was a thrill for me, a lowly undergrad, to be able to give pointers on how to use and clear a mask and how to be comfortable underwater with mask and fins. Visibility in the estuary wasn't great, but I showed them around my study site on Shackleford jetties.

After graduation, I spent a period of time in the army and then I went to the University of Hawaii for a year as a graduate student. I went to Robert Hiatt, who was professor of marine ecology. I said that I wanted to talk to him about doing research. He said, "Fine. Read this paper before you come see me on Monday." This was the

[Howard] Odum and [Eugene] Odum paper ["Trophic Structure and Productivity of a Windward Coral Reef Community on Eniwetok Atoll," *Ecological Monographs* 25 (1955): 291–320].

From 1955 on Enewetak atoll.

That's right, on the metabolism of a coral reef. That fascinated me, and I spent the next 15–20 years being interested in reef-building corals, as opposed to coral reefs, because I discerned that to be too much of a chemical problem, and I didn't want to do chemistry. I was really interested in the corals that built reefs, their intracellular algae, and their boring algae. So I guess I've always been interested in the structure in biology, not just as dead structure, but structure that did things.

At that point, I got married to a marine botanist, and she is still my wife and mother of our four children. Then I went to Cambridge University and became an undergraduate again. I'm sorry to say, that was my first real education. It was my first challenging intellectual education where I was forced to learn to think, but not only to think about the logic of the science I was reading, but also, did I like it? Did I find it to be good science or bad science, both of those concepts. I was asked to have an opinion about what I was doing, which I had never been asked to do in America.

I went to graduate school at the University of California at Berkeley. My major professor was Cadet Hand, and I was a fellow student with Len Muscatine, Chuck Stasek, Bill Neuman.

That was a great part of the Smith tree.

Yes, it really was. While in England, at Cambridge, I had learned the rubric "functional morphology," and I knew that's what I wanted to be. I mean, I cared about the structure, but I cared about it doing things. So, my Ph.D. was on the functional morphology of coral skeletons. The kind of function I was interested at that time was: How is the skeleton made?

The second year as a postdoc, I went to Woods Hole Oceanographic to study the question: How can epipelagic organisms, such as chaetognaths, larval fish, and so forth, be transparent? Because, if you take any one of those things and drop it into alcohol or boiling water, it becomes opaque. It's full of protein. A chaetognath and a fish larva are mostly muscle, and I can't see through my muscle, so how come I can see through their muscle? So I studied that phenomenon, and I also did my first mechanical experiments. By that time I really had the idea that there must be simple principles of mechanical engineering that could be used, if one knew them, to approach biological support, mechanical systems, skeletal systems. You could learn a heck of a lot about the mechanical function of the systems.

I came to Duke as a faculty member in 1964 and was encouraged to pursue this engineering-biology interface.

Encouraged by whom?

By the department, by my colleagues in a very genuine and warm way.

They thought this was new and different and interesting?

That's right. For my job interview, I gave my seminar in which I said that. As part of the interview process, I went around to speak to individual faculty members, including the great [Knut] Schmidt-Nielsen. I was ushered into his presence. He was very shy and, therefore, a formal person. One sits up very tall and straight and at the edge of one's chair. He doesn't make small talk. He asked a few questions and said, "This thing you said you want to do [connect mechanical engineering and biology]. I hope very much that you will come here and do that. It would please me very much." He stood up; the interview was over. I didn't hit the ground for the rest of the day! It was wonderful.

Twelve years later, three other people and I published a book called *Mechanical Design in Organisms,* which is still the only North American textbook in solid mechanics of organisms. It is being replaced now by a book by one of my former students, Mike LaBarbera, another good field biologist. I've had wonderful luck. I didn't necessarily attract good graduate students. Good graduate students tended to come to this department: people interested in marine biology, marine invertebrates, particularly. MIMI KOEHL had never heard of me when she came here.

But the Duke zoology department had a good reputation?

In those fields connected with the Duke Marine Lab, that's right. So I've had wonderful students, who were wonderful people, and many of them went on to make me famous. And I love it. So that's it.

Steve Vogel was hired two years later, not because he was an expert in biomechanics, but he was an expert in fluid mechanics. He and I have been the closest of colleagues ever since.

What a wonderful synergism.

That's right, exactly, a synergism, like with Chuck Pell, who we'll get to later on. Steve and I taught the course biomechanics, he teaching half of it in fluids and me teaching half in solids for 15 years. He continues to teach the whole thing now that I'm retired. In 1990, I met Chuck Pell, an artist who made animated dinosaurs. He was tired of making things that looked like life on the outside but had gears, wheels, and motors on the inside. At that time, in my research then on fish, I was trying to make a blueprint, an outline of a fish, and put into that blueprint only those structures that we know are in the fish, that have to do with its swimming, leaving everything else out. I'd been working on this for 15 years and was only learning that the fish is enormously complex inside. I wasn't getting any closer as to how this thing swims. The summer after Chuck Pell joined me, we formed the BioDesign Studio. It was his job to make three-dimensional models of biological systems that I and other people were doing research on. So we all went to the Duke Marine Lab and started to take fish apart. Chuck was there, making models of what we would see every day—of this part, of that part. From this we wrote a paper with Mark Westneat, a student who was the dissector. I was the computer notebook for the dissection, and Chuck Pell was the designer of the models. We came to understand the morphological basis of fish swimming. It was the first time anyone had ever done it.

Where did you publish that?

Journal of Morphology. From then on, the BioDesign Studio began to produce really clever things. One of them was the little toy fish we call the Twiddlefish.™ We formed a for-profit company off campus entitled Nekton Technologies, Incorporated. We found a factory to produce these Twiddlefish commercially. They are now for sale at science museums around the country. We then wanted to do more things and we are. We now have three navy grants to turn this Twiddlefish into a propeller to drive submersible vehicles for the navy, and we have other plans too.

So I'm in retirement. I'm still director of the BioDesign Studio in the department. Any student can come and learn to make models and get some instructions about the models, as can any professor. I ask professors to pay modest lab fees. I'm on the board of directors of Nekton Technologies. I'm not technically adept enough to get into the design business, but I'm fascinated with how we use the same physical and biological principles that we use in scientific research to analyze organisms and see how they work, acting as design engineers. We use the same principles, but we are using them in the other direction. This is now a big thing in this country. It is a subject with the new name of biomimetics.

Engineers use inspiration from biological materials, biological systems, and things in nature that we don't have in humanmade things. They find interesting properties and interesting performances of various organisms and their parts. They think, we don't have that. It's been up to them to design it. So biologists are requested to come and talk to engineers. When we published our book in 1976 on mechanical design, our last statement in the book was "In the future, this subject will depend on us getting engineers interested." Well, that was apparently very easy because now there are many engineers and groups of engineers in this country and in Japan and Europe doing biomimetics. The bottleneck now is that there are not many biologists to talk to, biologists who are willing to sit down and tell them what they know about these systems. If you care, there is funding.

My feeling—and I somehow get the notion as I talk to young faculty and graduate students that the feeling I'm about to express to you is an antediluvian one, it's out of date—is that the essential reality in any biological study is the organism or the organisms being studied. Everything else is created in the mind of a human being. We are so hooked in this country and certainly in the Duke University zoology department and other departments I've visited throughout the country, we're so obsessed with the theoretical aspect. We got there because in the middle of the twentieth century we tried to flee from purely descriptive biology. We've gone all the way. The pendulum has swung so far from descriptive biology that we're into theory without description. Concerning theory without structure: it's easy in many departments of biology or zoology in the country, now, to graduate with a bachelor's degree without ever having taken a course where you studied a whole organism. Many schools will grant biological degrees where a student is not required ever to go out into the field, but many programs also allow one to graduate without ever having seen a whole organism in a lab. I think this is too bad, and I'd like to see it change in the other direction. In a sense, I've given up on the university, although it has been wonderful to have been at Duke for 33 years.

I've discovered that the greatest learning machines in the world are children. We as educators, all of us, somehow don't appreciate that enough. We don't realize the enormous challenge to keep a child's mind inquisitive, inquiring, receptive, synthesizing, to keep up to that challenge. No grownup can keep up with it. It is really tough.

For the last 15 years I taught a course called structure. It was not a biology course; it was an interdisciplinary course, actually taught in an art studio, but with no prerequisites. Every year about 15 people would take it, never more than two students in the same major, students from every major all over the College of Arts and Sciences. Every assignment I gave them was something they had to make with their hands. I told them, "When you make this, you are not making it for me, you are not going to hand it in, you're going to show it to your classmates and tell them how you made it and how it expresses something that you've learned in the part of the course that we are studying." I found in doing that all those years—and it happened every year the same—that within three weeks every student in the class was experiencing an increase in self-confidence and collegiality. They started caring about each other and motivation. If I can now codify that in the educational project that I'm involved in now with Norman Budnitz and Kiisa Nishikawa, so I can give these simple instructions to teachers of any subject of any age children, somebody should care. Because isn't lack of self-confidence one of the major reasons Johnny can't read?

None of the ideas I'm telling you is original with me, but maybe the collection is. So it's the hands-on; it's allowing the kinesthetic sense to add to the visual and the auditory sense. That is what we've left out of our education. Johnny and Susie can play, but they have to come out of recess sometime. In the first grade, you must sit very still, put your hands in your lap, listen, and watch your teacher read and do numbers. I want to put hands-on back in the curriculum. This business about having them do their assignments for each other, oh boy, does that fire them! We will be producing operational exercises, which we will feed through teachers in local schools to try them out and then publish them and put them on the Internet.

But to come back to your original question. What's being lost is the tangible part of biology in the curriculum around the country. With that is starting to be lost even the visual. Yes, you may watch videos of crawling caterpillars, flying birds, and swimming fish. It is not the same as going out to the stream to see the swimming fish or to the woods to see the flying birds. So field trips are really important.

I had a student once named Gigi Roark, an undergrad. She got her degree, married a medical student, and stayed in Durham teaching lower- and middle-school science. She was a gifted teacher, one of the three most gifted teachers of anything whom I've met in my whole life. She then came back to graduate school after seven years of teaching. She said, "In order for me to be a better teacher, I need to experience research." So she wanted to get a master's, which she did with me, on the functional morphology of sea anemone tentacles. During that time, she took a course here at Duke in southeastern flora and fauna taught by a botanist and zoologist here. They would take long three-day field trips. On the Monday returning from the field trip, she would come and sit in that very chair, she could just hardly keep from exploding. It was revealed to her, the total epiphany, how important "the field trip" is.

That's the thing about the human mind. We synthesize stuff from many different, diverse experiences. That is why it's so important. She used to give me the most wonderful lectures on the importance of "the field trip." So that's another part of this. How can we do a field trip in math class, in history class, and I'm sure we can do it as part of the educational process.

In every field.

Every field, absolutely! To take you out of your routine and make you do this. It opens your mind, makes you receptive, in a way you wouldn't always be.

The tactile thing is illustrated potently by the experience you just had in the BioDesign Studio when you experienced the swimming of the fish. The minute you put it in the water, your posture straightened, your pupils dilated, your eyes opened wider, and you did a vocalization, "Wow! Ooh!" and everybody does it. We've watched thousands of people do this now, and even the most cynical can't prevent themselves.

Well, I thought it was going to swim but actually seeing it.

No! It wasn't seeing it. It was feeling it! That's what did it. You can watch someone else do it.

It takes off from you.

That's right. It'll pull itself out of your hand. One out of ten people doesn't hold it tight enough, and it gets away.

The human brain that evolved in eastern Africa around three million years ago evolved in an organism, a species, a creature that used all of its senses everyday, from when it was born 'til the time it died. If it didn't get the right feel of things, it would be naturally selected out. All right, we've taken that creature and educated it through only two senses, sight and hearing. Those tactile connections are still there; they are just asleep. We're not using them, that's all.

It makes for a much richer world all around, if you do.

That's right. Those are some of the things that if you don't use the tactile sense, you are missing all its connections. Everything in the brain is connected, and we want to get this back into the educational curriculum. So it's total biology that we're after.

Another aspect of biological education at the university level interests me. People who got their Ph.D.s in the 1940s and 1950s and then joined faculties, some of them were very productive as far as publications, and others were not necessarily so productive in terms of publications. But many of those people played important parts of their careers as teachers and as real resources in departments, for instance, Don Abbott or Bob Page at Stanford or HARRY THOMPSON at UCLA. By the late 1960s, "publish or perish" had taken over. What do you feel about the diversity [of academic roles] among faculty?

I certainly agree with the notion that to teach biology, particularly at the university level and particularly at the graduate level, without doing research at all, that would be a hard thing for me to defend. I think you need to do research.

The people I'm talking about actually did research, but not lots.

I think marine labs have been full of people like that. Marine labs have been wonderful places that really inspire people. I'll never forget my first courses at Duke Marine Lab, but Bookhout always published at least once every year or two.

I don't regret that we are all encouraged to do research and that the research is publishable, but I certainly do not like the idea that you need to publish more than once a year. The idea of having to have a grant—and now, the idea of having to have a large grant—that's the part I hate. It's the overhead that counts.

Of course, there is a lot of research that can be done without a whole lot of money.

Of course! My whole life has been that way. Steve Vogel, my colleague, is the world's best at designing highly sophisticated, highly precise equipment to measure things in an experimental situation and designing and building the gadgets on a shoestring, literally. He once designed an instrument to measure the pressure difference in the burrow of a trapdoor spider from the bottom of the burrow to the top. This was going to be in some number of millionths of an atmosphere. There was no instrument anywhere to measure this. So he designed and built a gadget that did it for $13.75.

What is going on in the late twentieth century that couldn't have gone on before is that we can now approach an ecosystem and know enough about the climate, enough about soil, enough about the plants and animals to begin, if we have the right experts, to get at some of the broader questions. So collaborative research is there. It will be rare to have a researcher have more than one-quarter of the publications be single authored. In any profession that is a selection of a selection of a selection, there are still only going to be about 2% of people who can think. Most of the people can do the work, can do things honestly and well, and can contribute, so why shouldn't they?

I guess ultimate success will be dependent on someone who can think and synthesize.

That's right—and how to keep charlatans out of it. This is the history of science; it's a very uneven and an undirected process. I'll mention another trend. Today, a second- or third- or even fourth-year graduate student may give a talk in a research group and never even mention the name of another researcher, as though science isn't a social process. Which it is, dammit! There's good science and there's bad science, and these things matter. And it matters who does it.

He spent a lot of time doing that . . . not giving us the answer, but helping to train the eyes.

FREDERICK NAGLE (b. 1937)

Professor of Geological Sciences
Chairman, Department of Geological Sciences
University of Miami
Interviewed at his home 27 February 1998

I used to look forward with great anticipation to Fred Nagle's visits to WIL to teach in the geology field course because he is an exceptional raconteur. At Princeton, he fell under the influence of Harry Hess, one of the dominant figures in the plate tectonics story. The Caribbean Project based at Princeton University in the 1950s and 1960s was the brainchild of Hess. His brilliant and integrative mind was behind the science; his ability to read people correctly was behind the more than 35 hardy, self-reliant, and bright grad students he recruited to map serpentine zones rimming the Caribbean; and his political and social skills were behind the money he raised from commercial interests and from regional governments eager to be apprised of their natural resources. Hess hoped that his students would continue to focus efforts on the advancement of knowledge of Caribbean geology, as Nagle has done throughout his career.

The idea of being able to convert things in rocks for useful products, say wire or automobiles, that seemed to me to be fascinating. We lived in an area of Long Island that you would now say is terribly overdeveloped, but in those days we had five acres, and our neighbors had 50 acres, much of it wild woods. I spent a lot of time alone, winter and summer, playing, walking, and I'd be very comfortable with that. I also spent a lot of time building radios, everything from crystal sets to things with lots of tubes. I learned a lot of physics this way and electricity. My father let me set up all kinds of antennas on our property. I could set up long wires, and there would be Pittsburgh, coming in on the earphones! My friend started building TVs. I didn't go that far.

But it connected you to a much larger world.

Yes, it did. It was a very interesting world. I also read books. In our family, it was expected you would do your schoolwork, keep up passing grades, take up something you were interested in, and if you could play a sport, you were to play a sport. That is what was expected.

At Lafayette, I started with a major in mining engineering, but I switched to English. I had been a writer on the prep school paper. I was writing some of the sports

Harry Hess and his student Fred Nagle (*left*) in the mountains of the Dominican Republic in 1959. Photo by Carl Bowin.

news for the Lafayette paper. I thought, well, hell, I think I'll become a newspaper person, a comment that did not please my father. He said, "You are already involved with a fraternity. I know you guys drink too much because we drank too much. Now you are involved with the newspaper people, they tend to be drinkers, too. I just can't see you making a career out of this." But I loved it! At that time, Jim Dyson (who was chairman of the geology department and an old college friend of my father) approached me and said, "Come over for dinner." In the first year and a half, he didn't bother me. Yet he had been often at our house before I went off to prep school. At dinner he said, "Why don't you take a course in geology? You might enjoy it." I said, "Sure, OK." That is how I got started. I took that course, and it was fascinating. We had field trips every week, and then we'd go back and talk about it. I thought, this is great stuff! Again, I was surprised people would get paid to do this. I started getting very serious about minerals.

Lafayette had one of the best mineral collections in Pennsylvania, due to a professor, Arthur Montgomery, who, in fact, had been a Princeton student. Unfortunately, when it came time for me to take some of his courses, he was on leave. I was very disappointed. But, what happened was that Montgomery asked Princeton for a replacement: Leslie Coleman, who got his Ph.D. in 1955, worked at Lafayette for four or five years, and then went to the University of Saskatchewan, where he is now an emeritus professor. So, as a recent Princeton graduate who knew Harry Hess, he picked up on me as time went along and said, "You ought to go to graduate school." I said, "No way, I'm not going to graduate school." He said, "What are you going to do?" I said, "Either I'm going to work for a mining company or I'm going to try and play ball again." My father had told me that after I graduated, "If you want to take a few years and make it in the minor leagues, why don't you try that, and then figure out what you want to do." I asked Dyson, "Do you really think it would be worthwhile to go to graduate school?" He said, "Yes."

I went to Princeton to visit, and I was impressed. The faculty was terrific, and I met Harry Hess, who was chair. I didn't know it, but he was looking for people who he thought could survive working alone on theses in the Caribbean, in politically unstable conditions, where one had to make fast decisions about this, that, and the other thing, and those who were reasonably athletic and also had very good grades. He and his wife would go over the applications and say, "Do you think this one has the personality, from letters of recommendation, to be in the Caribbean Project?" which was his highly funded project. We had about 35 Ph.D.s over 20 years. I guess the decision was yes, because that was one of the acceptances I got.

I had done a senior thesis with Dyson, a field thesis, which was also sent on as part of my application. I mapped some things, and I also worked with a certain type of serpentine that—I didn't know at the time—was one of the rocks that Hess was very interested in. Serpentines often represent old oceanic crust. The ones I worked on, as a senior, did not. At that time, 1957, the Woodrow Wilson Fellowship Program was picking people for their teaching potential. They wanted people to go through a very strong Ph.D. program, do well, and become teachers. I didn't really want to be a big-time researcher. I either wanted to work and make a lot of money or, if I was going to be in the academic world, I wanted to be a teacher.

When I got to Princeton, they immediately made me a lab instructor, despite the fact that I had the fellowship money, and I didn't need to do that. They said it might be good to get the experience. I was a big hit in my laboratory section; they were mostly hockey players. The department liked my teaching and kept me on in a number of courses for a number of years. Of course, I loved being at Princeton. Everything was happening there. I really didn't want to leave. The military finally pulled me out to fulfill my ROTC obligation.

What about the Princeton experience was so good?

I think it was because the faculty who were there were topnotch, on top of their fields. Plate tectonics had not yet come in. In fact, Princeton at that time was anti–continental drift. We didn't really believe in it, but there were a lot of features that suggested that at one time these things must have been together. One of the things that bothered Hess, for instance—and he later became a great supporter of plate tectonics—was that to make continents, which were less dense material, move through the mantle, which is more dense, was about the density equivalents of trying to push spaghetti through steel. I remember this analogy very well.

So this was the geophysical argument against continental drift?

Yes. There were some very strong arguments against it, although the surface geology, the ore deposits, the fossils provided an awfully good match between Europe and North America and between Africa and South America, and we all knew that.

Hess's major research project was to zero in on serpentines in the Caribbean. He wanted to study mountain building. He had studied serpentines himself in the Appalachians with this theory in mind, and he had also worked in the Caledonians and the Urals in Russia. His idea was: let's send out a bunch of students to the Caribbean, which is a young, active volcanic arc, much like some of the Appalachians must have been 300–400 million years ago. These might be nascent Rockies, Appalachian systems, these island arcs, like the Caribbean and northern Venezuela, which has similar kinds of mountains and serpentines exposed. So, we'll follow the serpentine belts. He plunked us down on the different islands and in northern Venezuela and Colombia to track out the serpentines and then to unravel the geology around them. We had no idea that they represented oceanic crust. The first inkling of that came from one of Hess's students working in Puerto Rico. Pete Matson, later to become chairman at Queens College in New York, said, "Look, these are the oldest rocks we know on the island, they are in anticlinal folds, they must be old oceanic floor." Then, the same kinds of rocks were dredged. Right after the Second World War, they started to be dredged more and more off the Midatlantic Ridge, off fault zones in the Midatlantic, and later from the Puerto Rico Trench. This rock type turned out to be a very common thing on the sea floor. Other than basalts, serpentines were abundant.

By the time I was finishing my thesis in the mid-1960s, everything that Hess believed about the importance of studying serpentines had been changed, mainly by his own students. I said to him once, "Harry, aren't you disappointed? All those ideas you had and published on in the old days don't seem to be right. Aren't you discouraged?" He said, "On the contrary. All the questions now are new. You guys have opened up my eyes. All the questions have changed. I have a reason to go on!"

What a broad view he took to put together a program like that in the first place.

Yes, he did, and the camaraderie among those students was such that we all talked with each other about our different areas. We all had offices together. We'd come back in the fall and compare our adventures in the tropics. We'd get out our rock samples, compare, and see if we had the same kinds of things. Sometimes we did, and sometimes we didn't. I didn't. I was up in the north coast of Hispaniola in the Dominican Republic. The stratigraphy I was coming up with matched nothing of the other students who had worked there. We'd work our first year in the field with a more experienced graduate student before we went off to our own area.

That was part of Hess's program, too?

Yes, to make sure we didn't go berserk, for one thing, but also so we had some experience in how to handle things, both in the field and also in personal situations. How do you behave if you come across a funeral? and those sorts of things. Politically, who do you talk to if you go to a town? Who is in charge? I was with CARL BOWIN and Curry

Palmer for parts of the summer in Hispaniola, but, for my own work, I had crossed over the mountain ranges and gotten up to the serpentines on the north coast. Two years went by, and I realized "my" geology was different from those areas to the south. There was a younger mountain range in between. My stratigraphy didn't match theirs at all.

Sometime in the early 1960s Brackett Hersey and ELIZABETH BUNCE, famous names from the geological world at Woods Hole, had been involved in dredging the north wall of the Puerto Rico Trench. These results were discussed in a 1962 conference. I remember that very well because I was there, Carl Bowin was there, Harry Hess was there. The Woods Hole group was not, and Harry presented the Woods Hole data. He said, "I've just come from Woods Hole," and Carl Bowin, who had already graduated and was now up at Woods Hole said, "Oh my God, he's going to tell them what we found." I said, "What the hell, it's Harry Hess." So, he says, "The stratigraphy on the north wall of the Puerto Rico Trench, they've done the dredging, it is serpentine on the bottom, basalts in the middle, limestones on top, and it looks very exciting."

Afterward, I went up to Harry and said, "Maybe that's it," meaning that a trench sequence might be what I was mapping. He said, "I was hoping you'd notice." "I should be talking about that, right on the coast there, and if you look at the gravity anomolies, it looks like an upthrown section of the Puerto Rico Trench." Hess said, "You're doing good, Fred. You've got it!" I did a paper on that, one of the first papers from my thesis. There is a series called Benchmark Papers in Geology, which was for seminal ideas. Two papers came out of that work, and they were the only two papers I ever had in the Benchmark series. I presented that first paper at a conference and thought, this is crazy, no one will believe me. In fact, all the geologists there congratulated me on that. It was the Woods Hole dredging; they were careful and had good stratigraphy. We knew it was layer cake because we had seen the geophysical data.

But you didn't know what the composition was?

No. Nobody did until the dredging was done.

Harry would come down to visit each one of us each year. The maximum number of people he had at any one time was 17 people in the field. He'd bring money to all of us. Some of us were broke; some were not.

Where did the money for these projects come from?

It went through a fund at Princeton, which at any given time was between $500,000 and $1 million a year. It came from the Venezuelan government; it came from all of the oil companies operating in Venezuela. Mining companies, like Falconbridge Nickel operating in Hispaniola, put in money, and they also gave us vehicles and parts. Falconbridge was interested in the serpentines because on top of them were the laterites, which concentrated the nickel. They were interested in serpentines anywhere, especially in Hispaniola.

So the economic interests from both the government and industry supported this basic research.

Yes. All of these theses were published in the same format, and the copies were distributed back to the governments and all of the contributors to the project, the Caribbean Project.

It was the biggest U.S. project in the Caribbean in geology, ever. The only thing comparable was that Oxford University had a number of students for a number of years going in and out of the Caribbean, and there were others from the University of the West Indies in Jamaica and the seismic research station in Trinidad. The French had a few people, too. We all knew each other. It was a very cooperative effort. In those days there were no maps, no nothing. It was a great exploring effort.

That 12th Caribbean Geological Congress in St. Croix in 1989—those were the people who came?

Those were the ancestors; mainly their students and the next generation now attend. Hess asked us to continue the work, and many of us did that.

Carl Bowin and I were going to do a book on Hess. We became very close to him personally. I knew him very well, and I understood his dry humor. There were times he'd be so mad at me and my jokes that he wouldn't talk to me for days. For example, he ran the USS *Cape Johnston* in the Second World War, a troop ship. He'd talk about combat and being on the bridge. He finally was made a rear admiral. He had stayed in the reserves. That group was very important in the funding of research after the war. They were part of the Office of Naval Research. There wasn't any National Science Foundation then. Almost all of the money for research in ocean science came from the navy. All of the major oceanographic institutions had navy connections, one way or the other. Hess would serve in the reserves every summer after the war. He would work on maps and other things in oceanography. The first paper he wrote on plate tectonics was actually a U.S. Navy report before it was ever published. He showed it to some of his graduate students, of whom I was one. I said, "Harry, you are risking your entire reputation on that paper." We were very frank with him. He said, "I know. Do you think I ought to publish it?"

This is when there was still that controversy: is there or is there not continental drift?

Yes, and he was going back and forth. He matched all these data across the oceans and said, "It looks pretty good, doesn't it?" I said, "I couldn't publish that sort of paper." He said, "I know, but I can take the chance." It turned out to be the most quoted paper in geologic literature!

The kind of things I would do with him . . . I'll tell you one story. In the Dominican Republic way up there on the north coast, in those days it was pretty isolated. With me was a Princeton undergraduate who was about to go into the navy, George Hunter Ware. He turned out to be a good geologist. After dinner, we were sitting out on this hotel veranda enjoying the sunset in Sosua, very romantic, a coral reef out in front, quiet, and we are talking navy and military. Hess was talking about how in combat you'd give the general quarters warning and how he could be on the bridge in two minutes and so forth. I don't know what got into me—we were drinking rum—and here is Hunter, who is about to go into the navy, listening with great respect. I said to Hunter, "I'm going to wake him up tomorrow morning and give him a general quarters." So I manufactured a megaphone, and I got a bell from the kitchen and got myself up at quarter of six in the morning. He was in one of these rooms—no screens, it just had curtains and a veranda all around—in the bed, with mosquito netting tucked all around. So I go, "General quarters!" and rang the bell. The bed turned upside down, and he got tangled up in the mosquito netting! I took two steps back, put my shoulder into it, and popped it free. He said, "Dammit, Fred, that's it! This is the last of your pranks. This one, I don't like. I'm not going to talk to you for two days!" And he didn't.

Another story with him in the field. One of the things was refrigeration. I would mark on my maps with red dots.

Where you could get cold beer?

Exactly! I didn't realize it, but I had given him that set of maps to look at my work. He said, "What are these red dots? Do I want to know?" I said, "Yes, those red dots represent all the gas refrigerators that I know about in my field area where you can get something cold. I will bet you $10 that I can get you to a cold beer within 15 minutes from anywhere in my field area should you call upon me to do that." He never did that until after the two days when he didn't talk to me. We'd eat meals together, and he'd talk to Hunter, but not to me. We'd go along in the field and he'd say to Hunter, "Would you tell Fred that I think this means . . ." So we went through two days like this. At the end of two days, the first thing he said to me was, "I need a cold beer. You've got 15 minutes." So I said, "Get in." We were in the middle of nowhere. He must have realized this was a most difficult location. But he didn't know about the shortcuts I knew. We went over the mountain. He said, "I don't want to tip over." We drove over the mountain on a donkey path and got to a place with a cold beer—not only cold, but it was solid ice—which they had a habit of doing. They would turn on the refrigerator and put it on max. So the caps popped off, and we were eating beer ice. At that point he said, "You know, I think you are going to get a Ph.D." So, he forgave me.

He made me realize that by studying the books only, you just can't do it. There was another time we were going along and he said, "What have you mapped in here?" We had stopped at a bridge and gone down to eat our lunch under the bridge. He says, "You've mapped an unconformity here." I said, "Yes, but Harry, I don't know what it looks like. I've never seen it." An unconformity is a time gap. I said, "It can't be a fault, it must be an unconformity. You've looked around here for a couple of days, do you have any idea where it might be?" He said, "Yes, you're sitting on it." And I was. The point was, I had never seen this kind of feature except in textbooks. Here it was, a slight difference in the bedding angle, and here was this plane, and this represented a change. It was a time gap.

An erosional surface and then the next layer was laid down?

Sure. I knew it had to be there. I had been under this bridge before—it made a good place to sit down for a picnic—and not noticed it. You are totally blind until someone has actually shown you something, even if you've seen it in textbooks. So he spent a lot of time doing that . . . not giving us the answer, but helping to train the eyes. This is perhaps the most important thing that I learned from Hess.

We used to say that it took us about two or three weeks to get our eye in, meaning, look at a feature and then translate it into something useful geologically and then be able to make sense out of it from there. When you first come into the field after a year of looking at textbooks, you've lost that. If you don't do it frequently, the eye doesn't register.

When I was at Princeton, there was a course taught every year called advanced general geology, which all of the senior majors took and all of the first-year graduate students. The faculty would come together, and each would give a lecture on the latest in his field and how it related to continental drift, as it was called in those days. As I said, the faculty didn't believe in it; the paleontologists did, but Hess and the geophysicists did not. Other people came through the department. For instance, one thing that was coming in from Woods Hole was heat flow measurements. They would put a probe down on the floor in the sediments, and they'd have a thermocouple, and they'd get some measurements of heat flow. You could make a calculation as to how much heat was escaping through the sediments and from the crust below. Well, the high heat flow values were all out in the middle of the ocean. The sediments were thin out there; we all knew that. But they'd find a place to plop one in, you are talking maybe 3,000–4,000 meters, to drop this thing down. Another thing that was coming in was bottom data, which the navy wanted. The bottom topography wasn't too well known until the Second World War, and after the war everyone had those bottom sounders. In particular, Lamont [Columbia University] was doing that. A lot of data were funneled up to Bruce Heezen in Lamont, which was OK with the navy. He produced the bottom maps with a good draftsperson, Marie Tharp. It turns out we could see in those map compilations what looked like a rift zone out in the middle of the ocean. [Lynn] Sykes and some other people at Columbia were plotting earthquake locations, because they were doing it for the military. The military wanted to track nuclear explosions, so a whole bunch of seismic networks had been set up. The networks were also turned loose on earthquakes, even small ones, and they could be pinpointed pretty well.

So we had heat flow data, we had earthquake data, we had the topography. Everything was pointing out toward the center of ocean basins. It looked like there was a narrow zone which from the topography looked like a fault zone. It looked like what is called a basin-range type faulting where you'd be pulling crust apart. The high heat flow was there. Interesting, around the island arcs, there would be very low heat flow by the trenches. We studied the earthquake patterns from those regions, and the earthquake patterns defined a descending slab. They would plot in a zone. We didn't make any connection between that and the crust right away, but we said, "Hmm, that's interesting. There is a piece out here in the middle of the ocean with these earthquakes plotting along this line, with high heat flow. It looks like the crust is spreading apart in that region." Pretty soon Hess gets the idea.

Another thing that interested Hess, one more piece of evidence, was that, from the seismic evidence, the ocean sediments were very thin. Even the thickest sediments in the ocean are not very thick. We assumed slower sedimentation rates because you are farther from land. Hess made several calculations; others did too. If, in fact, the age of the ocean basins was Precambrian, there ought to be not just 300 or 400 meters of sediments, there ought to be a couple of miles. Why is it so thin everywhere and even thinner, almost nonexistent, in the middle third of the ocean, which happened to be these ridges?

We had all that information. All of this stuff was coming into Princeton before it was published. So this all comes together, and Hess gets the picture and puts it all together. He says, "It must be that these continents are attached to the mantle, and the damn thing moves like conveyer belts." He finally publishes this paper in the Sears Foundation, which showed the continents in a cartoon bolted to the mantle with a great big lug nut, and it showed these currents moving underneath. So the actual plates are not just continental crust; some contain oceanic crust as well. The continents were carried passively. Once he got that concept, he didn't have to push spaghetti though steel. He was a total believer from that point onward and a great supporter. The whole faculty swung with him, at which point he hired [Walter M.] Elsasser from Maryland, one of the great geophysicists, who had been looking at the problems of mechanics in moving things. For the next few years, at least half of the geology department graduate students were physics students, which we considered very strange. They were interested, though, in earth problems, and we soon began to interact with them. It was an exciting time, the late 1950s and early 1960s, to be at Princeton. If there were something new, the class, whatever it was in, was interrupted. It was that quick.

Most of the things with Hess were done by telephone with his colleagues or with us. He was a member of the Cosmos Club in Washington. He would go down there occasionally. Later, he was also a member of the National Academy. A spur of railroad came into Princeton, called Princeton Junction. He'd say, "Go with me." Five or six of his students would go down to Washington with him. That's how he kept track of what we were doing. He'd say, "Bring your maps, bring any rocks you want to show us." Commuting time was used to work on the project. He also did a lot of his research on the phone.

I must tell you. This is the true story of how I got to be a Caribbean thesis worker. I had no interest in the Caribbean when I first came to Princeton. I didn't know he was interested in me doing this. What you had to do in the first six months, in addition to courses and the language exams (incidentally, I flunked the German reading exam nine times, which is not the Princeton record, which at the time was 14, held by Gordie Taylor, a Canadian), you were to sit down and get all of the current material on the subject you were interested in, and you were to present a thesis problem. What I was interested in was to follow up on some of the things I had done for my senior thesis on pegmatites, which are these very coarse igneous rocks with big crystals that develop all around the world. There is one mine in Russia that is said to have one crystal which is 70 feet long. The pegmatites in the Black Hills of South Dakota have crystals that are four or five meters long. Some have valuable minerals—beryllium, lithium—so they've been studied.

So, this is what I wanted to continue to work on. I was to come up with a presentation to be given in December or January to the faculty member who would be most likely to take me on in that particular project. They would appreciate it if there were some faculty member who could handle this topic. I found out later, if you had funding, that was of even more interest. They forced you to do that. They wanted to see your presentation, how you would handle the literature, what you would get out of it, what you thought was important to do. So I did this.

You didn't know that you had already been preselected for a different project?

No. I had this pegmatite thing. I worked all fall. I put my presentation on index cards. What you were to do was to come before a faculty member and give your presentation in about a half hour and place on the board the important phase diagrams or whatever it was you wanted to put up there.

So, I was ready, and Hess was my listener. What I wanted to do was to grow big crystals in the laboratory, to make them artificially, to see what kinds of conditions were necessary. It was said aqueous environments would do this. So, OK, I'd get these high-pressure titanium test tubes, which were available after the war, and apply heat and pressure.

Recreate the conditions, to understand the conditions under which they may have formed.

Exactly. I had it all planned out. Here was Harry. We were all alone. I think all the time he had no intention of letting me do this, but he wanted to see what I would come up with. He even directed me to some of the literature, and he'd check with me to see the progress I was making, first-year student and all that. We set the date.

I'm very nervous. I come in. Harry is sitting at his desk. Harry had lots of desks. They were all crowded with crap, so he'd just get another desk. He had several offices. This thing had stuff piled everywhere. I could see him, but I had to keep my head in one position where I had a clear view. He was smoking like hell, and I was smoking like hell, so the room was full of smoke. And it's hot, because it's January, and the heat is on full and the radiators are banging. That was distracting me a little, too. I was very nervous. He said, "Why don't you put some of your data on the board." I said, "Yes, I have some." He said, "Fine, take a few minutes to do that." I turn around, I'm putting my data on the board, and I look over my shoulder and I think, jeez, he is asleep. I continued, and you could hear the chalk on the blackboard. Pretty soon I hear him snoring. There are just the two of us in this room. I didn't know what to do. I finished, and I sat down. I went through my cards and practiced in silence. He was still asleep. Now, I was a little panicky. I started looking at my watch. I thought, I'm going to try and get away with it. I made a loud noise and said, ". . . and that's what I'd like to do for my thesis on pegmatites!" I wondered, well what is he going to say? He looked at his watch. "Very nice presentation, Fred. I like the stuff you put on the board." He glanced at it quickly. "I know you want to grow these things, and we are getting that equipment so it would be possible for you to do that, but to tell you the truth, Leon Silver is going to do this at Cal Tech. He is well along. He has several students on this. You've done a nice job." To this day, I don't know whether or not he was really asleep, and I never had the nerve to ask him.

He said, "I don't think it is a good idea for your thesis." I was totally crushed. "What am I going to do?" "Why," he said, "you are going to go to the Caribbean." "Where?" He said, "I need somebody in the Dominican Republic." "Where is that?" I asked. He said, "Just relax, take a couple of weeks, talk to [Curry] Palmer and Bowin, and see if you are interested." I said, "Is there support for that?" "Oh yes," he said. So that is how I got interested in the Caribbean, and I stayed interested in the region for the rest of my career. For the next 30 years, I was in the Caribbean every summer and sometimes in the winter, in the field, sometimes consulting work, sometimes field research funded by the National Science Foundation. Almost all of it involved mapping, and now that money is much harder to get.

Curriculums are not set up really for students so much to teach them integrative science. If you say, "Oh, hell, let's do something cooperative with another department," then someone isn't going to get paid, and if they don't get paid long enough than you lose a faculty slot. I don't know how to do this, but this is one of the major problems we are running into in both research and teaching an integrative pattern. We have to change the cost-accounting system so popular now on many campuses, which has been brought into the universities with "the suits." It was easier in the kind of institution you used to run, the West Indies Laboratory, and you can do it with field projects.

The other problem is that you had a classical background—a lot of physics, math, chemistry.

Now people are more specialized. This is one thing Hess did with all of his students. Since he was putting us out there to do a mapping project, where maps had not been made before, we might run into anything. He picked undergraduates who had a broad background. Not all of us made it. He put some of the students with other people. He made us increase our breadth of knowledge as graduate students, also. He made me take a course in paleontology and a course in sediments. In fact, he had me at the very least sit in on all of the geology graduate courses offered my first two years. A lot of graduate students in other departments, even in the 1950s, weren't doing that. We'd go out there and make these maps for the first time in the middle of nowhere; oh, yeah, that's a fossil. I think I'll collect it because so-and-so can identify it. If we were primarily trained in say, ore deposits, we wouldn't be able to do these general mapping projects. As a consequence, I became well known for reconnaissance geology. You could drop me off in the middle of nowhere, I would drive around, and I could come back and say, "OK, I think there is a problem here." This is something that could be done. As an outcome, what I did in my work and my thesis and areas I got to see, probably created another six or seven Ph.D. theses. I was very fast, I could say, "OK, there is this kind of rock," and I could remember what rocks were where. I had that kind of a memory. I got very good at that kind of reconnaissance, broad geology that is sort of dying out as NSF-fundable projects.

Right now our Department of Geological Sciences is going through a curriculum revision, and we are trying to integrate what we in geology do with our colleagues at Rosenstiel [School of Marine and Atmospheric Sciences]. I've stayed out of it. I'm over 60. The younger faculty are going to have to teach this curriculum. It is interesting what they are coming up with. The emphasis is changing away from the things I learned. I can hand-specimen identify probably 250–300 minerals by sight or with very simple tests. That took a lot of work and memorization of things like crystallography, what crystals look like, luster, hardness, all these kinds of things you use in the field. Physical chemistry has become ever more important. Computers are terribly important now in all of the commercial aspects of geology. A lot of the descriptive stuff is being played down in the curriculum. There are some problems with that approach in my mind. Students are not going to be able to do the fieldwork, if you send them out on reconnaissance-type missions. The general feeling now is that that kind of training takes up too much time. They want me to integrate three of the courses that I've been doing into one and then do advanced courses on mineral deposits and Caribbean geology, instead of teaching the basics. Hal Wanless is going to try and combine stratigraphy and sedimentation. I don't know what kind of a geologist we will produce when we come out of this. This will free up slots to take more mathematics and more computer courses. I don't know. I'm not sure who, in the end, integrates this stuff for the students. Maybe, people who are old, like me. No one is now getting prepared to do that for future generations.

All my motivation has been to understand the Earth, not to find the gold deposits or to strike it rich with oil.

Carl Bowin (b. 1934)

Senior Scientist Emeritus, Department of Geology and Geophysics
Woods Hole Oceanographic Institution
Interviewed in his office 15 April 1998

Carl Bowin, another Hess product, was recruited fresh out of Princeton to run a newly conceived marine gravity anomaly program at WHOI. His expertise for the position consisted of a few measurements taken while doing fieldwork in the Dominican Republic. He taught himself how to program computers, a brand-new tool, through trial and error. Bowin is the first to have used computers in research on an oceanographic vessel, which allows for real-time interpretation of data and avoids the lengthy delays between making observations, interpreting the data, and asking the next question, which had characterized this type of research previously. Although Bowin switched his scientific inquiries from geology to geophysics when he began his career at WHOI, he explains that his whole approach to a scientific question is framed through his personal experience with geological phenomena in the field.

It began when I was four. Our goat died. This was in Los Angeles. Somehow I knew about dinosaurs. I told my parents I wanted to wait a few days and then dig up Rosalie, the goat, to put her bones together. I never did get back to it. As I got older, I was eligible for the Museum Students, a group which Gretchen Sibley ran. Throughout junior high school and high school, I joined kids from all over [Los Angeles] county who came to the museum, where we were comfortable because we all liked paleontology, entomology, and science. It was normal, for all of us had the same feelings, while we were freaks in our own schools. It was just a great environment, a pleasure. Science became a great interest for me.

I did my undergraduate work in geology at Cal Tech. As a class project in a field camp, we mapped a swath across the Owl Creek Mountains. There were some very nice structures, some overturned beds. As you were mapping it out, you could see the structure develop, drawing maps of the surface geology across this small range and then seeing the cross sections that you could develop from those strikes and dips, measured, of course, by using a Brunton compass; there was no GPS [global positioning system] at that time. That was real geology.

As an undergraduate, I had summer jobs with Texaco Oil in Wyoming, being an assistant for an oil geologist in the field.

I think the oil companies played an important role in the development of the current generation of geologists.

Of course, and there was a practical nature to the geology. You could see it being used. I came along, not interested so much in the economic aspects of it, but for the science, for knowledge. All my motivation has been to understand the Earth, not to find the gold deposits or to strike it rich with oil. I wouldn't mind having done that, to fund the other part, but it was for the enjoyment of understanding the Earth.

I completed my master's degree on the geology of a portion of northern Maine. I worked on a very interesting contact between an intrusive gabbro and adjacent metamorphic rock. As one approached the igneous contact, the andalusite crystals in the rock became longer and longer until they went to the next phase, chiastolite, and changed to smaller crystals again. I was fascinated by this increase in length. I went to Princeton University and there met Harry Hess. Harry was known for studying the serpentinites of Venezuela and Colombia. In those days it was known that there were serpentized bodies in the Dominican Republic, and they were mining nickel from the laterites above it. They didn't really need to know the geology, but because the cost of students was so minimal, it was a nice investment for them to make and maybe we would find some more for them. I happened to be the next student who came along.

When I started at Princeton, I took classes, but I didn't do much fieldwork, because I had already had lots of field experience. Another student in my year, Charles Helsley, told me that the Fortran manual had arrived, and I got a copy. He also told me there was a Defense Department–funded research company off campus with a computer, a 650, the first commercial computer with 8K memory that they allowed Princeton graduate students to use at night. I went there that noon and was shown how to get in, where the key was kept over the door, where the fuses were to turn on the lights, and how to bring the power up on the computer. I typed in my program. It was about those andalusite crystals and how they changed size and eventually changed phase. I had a plain vertical axis giving off the temperature of gabbro and looked at the data to see how far back the smallest andalusite crystals were. I took that as a starting temperature and then used it to examine the increased temperature where I found the largest crystals. If I then calculated how long it would take the heat to migrate, I would learn how long it would take the crystals to grow. So I got involved in a heat flow question, how to make the calculations for the heat flow, and then how to model it.

I wrote up this program and took it over there that evening. The place was all dark. I got the key, opened the door, put in the fuses to bring on the lights, and turned on the power to the computer. I put my cards in the hopper and pushed the load button, but it didn't read the last card. So I tried again. But again, it didn't read it. So, I turned off the power to the computer, turned off the lights, shut the door, put the key back, and went home. The next day I went back to the Quonset hut and asked them, "What do I do?" They said, "If it didn't read all the cards, push load again. Your stack may be too large." Often the deck was larger than the hopper. When you would get to the last card, you would load the next deck and push load, until all your cards were read. So, that's how I learned computers. Trial and error, a step at a time. I never got that

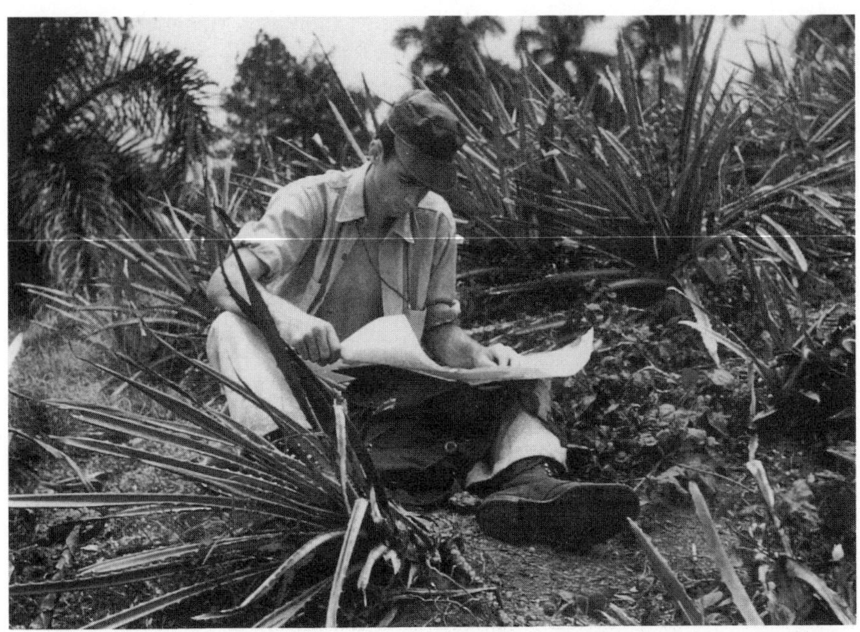

Carl Bowin (*top*) checks his map in the Cordillera Central in the Dominican Republic in the late 1950s; (*bottom*) on the RV *Chain,* operating the first computer onboard an oceanographic ship in the early 1960s. Photos from the files of C. Bowin.

program worked out; the computer wasn't fast enough to cycle enough times. I just got involved in other things.

Harry [Hess] was interested in this heat flow project. He came to me one afternoon and asked me about the heat conduction. I told him what I was just telling you, and he thanked me and went away. Shortly after that he came out with the ocean age solution paper, in which he stated that the oceans were young and not ancient features as had formerly been thought. I'm the only person acknowledged in that paper. To us students, the idea was out of the blue. What had clicked in his mind? I never did really see how he came to that conclusion, but he was right! I thank him for the acknowledgment. I hope I helped a little bit by talking about the heat flow conduction. Maybe the curve that I drew?

He never talked to you directly about the ocean floor?

I don't think he talked to anybody until he had that manuscript, and then it was published in GSA [Geological Society of America].

Maybe the conversation with you was the last piece he needed for the puzzle?

I wouldn't take that much credit, but I think it helped solidify something. He came to me with a question about heat flow. He had the germ of the idea; he was looking either for confirmation or further things to help glue it together to see if it were viable or not. That was the role I apparently played.

Here is a picture of me in the field climbing the mountain. I did my thesis in the Cordillera Central. On one of these climbs going up, I was standing there dripping wet and looking at the rocks on the surface, but trying to look down into the mountain. I wondered whether gravity measurements would help, or at least be a big aid, in trying to understand what was going on down deep within the mountain. I talked to Bill Bonelly when I got back to Princeton. He loaned me his Worden Student Gravity Meter, the only one Princeton University had. I had finished writing up my thesis, my jeep was still down there, so I went back, midyear, to make gravity measurements. I made measurements over much of the Dominican Republic.

I came here to Woods Hole because of Harry. He used to go to the Cosmos Club in Washington a lot, and Brackett Hersey, the man who hired me, used to go there also. I graduated in 1960. I stayed a year at Princeton as an instructor while I waited for a job to open. In the conversation between Hersey and Hess at the Cosmos Club, Brackett mentioned he wanted to start a marine gravity program. Harry told him he knew just the person—because of my experience with gravity—my 31 measurements in the Dominican Republic! I came to Woods Hole, I had the interview, and Brackett gave me the job. Two things happened. I submitted a proposal to NSF to make measurements in Haiti, the other half of Hispaniola. Those were the days when proposals were funded very readily. There weren't very many people, and NSF had lots of money. That's a bit of a simplification. It was a good proposal, of course. A lot of good proposals are coming in now, but the agencies only fund a third (or fewer) of them. I liken it to waves that are coming in, waves of science, waves of interest. If you happen to be riding one of the crests, then you have proposals funded regularly, and you'll have substantial funding. If you happen to be out of phase with that interest or are doing something that is not in the popular view of that time, or perhaps (even worse) contrary with what is known to be the view held by the community, then you'll get axed.

Those are often the new directions.

Yes, a few years later . . . drifting continents!

I came here to start the marine gravity program in 1961. To do that, I didn't know much about gravity, except having made the measurements in the DR.

You could take that instrument on ships?

No, so I went to [Lucien] LaCoste and [Arnold] Romberg, who made a land gravity instrument. Then they made one that was able to swing; they had it on a pendulum. I went down to Washington to see how the Coast and Geodetic Survey made its measurements, because it had started making sea measurements along the coasts. I remember seeing this room with five tables and women at these tables reading the records which came off their LaCoste and Romberg instrument, which had a strip chart. It took them three days to process one day's data! With this system, you wouldn't have the results until months after you got back to shore.

It took me three milliseconds to think, I've got to put a computer together with the instrument, at sea, so we can see the results in real time. At that time Brackett Hersey had lots of navy money, so he said, "Yes, we'll do it." In tackling this problem, I interviewed all of the major computer companies, and the one that seemed to have the most interest was IBM. But the engineer who designed the computer said it wouldn't last 45 minutes at sea. It was the 1620, an office computer, to which was added two other units. It worked!

He said it wouldn't work at sea because . . .

Of the environment. It was Brackett—I had never been to sea before—who said, "Sure. It will work!" We rented the IBM machine (it was too expensive to buy) plus we paid monthly fees for an IBM maintenance guy to go to sea with us. It was very costly. Also the research vessel *Chain* did not have air conditioning at that time. We had to build a special room within the main lab which was air conditioned for the computers. It was the very favorite room for people to visit at sea in the tropics, like off Puerto Rico.

On the ship, we had the computer, which could compute the gravity signals. Because it knew the latitude and longitude (every once in a while we'd get a fix from the bridge, and it could dead reckon knowing the direction and velocity of the ship until the next fix), we would have offsets that we could go back and smooth. But at least we had preliminary numbers. We had the water depth from the fathometer, we had the magnetometer, we had the gravity anomalies (and it could compute both free air and Bouguer anomalies because we had the water depth). We had these all displayed on a CalComp plotter in real time, so we could make judgments, if we passed something interesting: what was the response of the magnetic field, the gravity field, and the topography to that feature? and we could decide whether we passed it completely. If we saw something interesting, we could then turn the ship, go back, explore it some more, sample it by dredging, and then go on to whatever our main destination might have been. This is unlike the Coast and Geodetic Survey. They would get their results months later and perhaps say, "Oh, there is something interesting." They'd have to plan a whole new cruise, a few years later, to get back to it. So, this was very effective in making good use of ship time, good use of your time and investigations.

One of the interesting sidelights of the gravity data: I was trying to make a gravity map of the world. Fairly early, I became interested in the global aspects of gravity anomalies, thinking that comparing features in response from one place to another would give information about what was happening inside the Earth. I had a goal of trying to look at the gravity field over the Earth, both on land and at sea. I began collecting data. The information from the Dominican Republic and Haiti were my first two collections of measurements. When I would visit countries, I would try to obtain information. In all of those years, the gravity data for the Soviet Union and China were considered classified, a military secret. These were complete blanks on my map, which was published in 1972. The gravity data have military significance through what is called the "deflection of the vertical." As the gravity is pulling down toward the center of the mass of the Earth, if there is a large mass anomaly pulling toward the side from shallow in the crust, it pulls the vector for gravity over a little bit. It is deflected. It is offset a little bit, maybe up to a minute or a minute and a half of arc. That minute and a half is very important if you are launching an ICBM and it is traveling 6,000 miles to hit a target. That minute and a half *at* the launch site means it might not hit the target that one is after. However, when a missile is coming at you and it has but a few seconds to integrate the error as it approaches you, that minute and a half deflection means it will land a meter and a half away from where it is targeted. It is not significant.

Local differences in the Earth's gravity (due to mass anomalies) result in large-scale but low profile (about 100 meters) irregularities in the geoid [the Earth's equipotential surface, in particular, the average sea surface in a region]. This effect extends above the Earth's surface as well. Picture an imaginary spherical surface, above the Earth's surface, which is composed of all points of equal gravitational pull. This gravitational equipotential surface also has topographic highs and lows; where the Earth's mass is anomalously high, the equipotential surface has a topographic high, while if you have a lack of mass, the equipotential surface goes down. The greatest topographic high [of gravitational equipotential] is over New Guinea, while the greatest low is over Sri Lanka. If you look at the whole pattern, it does not show a coincidence with plate tectonics.

You now have a pattern over the whole Earth, but the pattern isn't consistent with what is happening at the level of the crust. Do you think this is reflecting something about the mantle?

No.

The core?

The core-mantle boundary. I came to my conclusion by modeling how these highs and lows were composed in the harmonic coefficients. I've been pursuing that evidence, which struck me as a geologist.

You got into this field—fundamental questions about the mantle-core boundary—through geology?

Right, but I haven't done any geology since I've been here, only geophysics. The way that I approach the solution to the problem of the core-mantle boundary is through the empirical geologic way of thinking. In contrast, most geophysicists approach the solution by modeling a fluid equation formula, which they solve by trial and error until they get the viscosities to work right, to give them a model that matches. We are having trouble getting details of the core-mantle boundary. We think it is bulged under New Guinea and depressed down under the south of India. The geophysicists have not tried to make detailed mass models, because every mass point affects every harmonic degree. That is what I've been pursuing for ten years now, fighting against the tide to do it.

What are all those species doing there?

Alan Kohn (b. 1931)

Professor Emeritus of Zoology
University of Washington
Interviewed in his office 12 May 1998

Alan Kohn sought work at a different marine lab each summer as an undergrad. When he first visited Friday Harbor Labs, he realized that his quest for the perfect lab was fulfilled. It was there in 1966 that I first met him, although I soon learned that 25 years before, he had been my aunt's third-grade pupil in Connecticut. A Hutchinson student, Kohn worked on aspects of the ecology and evolution of the snail species *Conus* throughout his career. His baseline research laid the foundation for the use of *Conus* toxins in practical medical applications.

I went to Princeton, which is kind of far from the sea. The next summer I wrote around to all the marine labs I could think of and asked if they had any jobs for the summer. I got one positive reply. It was from Hopkins Marine Station, from Lawrence Blinks. He said, "You can come and wash the dishes for C. B. van Niel's microbiology class for $300." I said, "OK, I'll do it," never having been to the West. My boundary of the West was the Pennsylvania-Ohio border.

There were some remarkable students in van Niel's course. It was the premier course for microbiology. At Hopkins they told me, "If you work on Saturday, you can get in your 40 hours, and we'll pay you $350!" They knew I wanted to sit in on Don Abbott's invertebrate course. I sat in on the lectures, but I couldn't do the labs because I was washing dishes. I was also able to go on the field trips, all of which were in the early morning, because that is when the low tides were. You know, Don Abbott taught me how to teach an invertebrate zoology class. I still have my notes.

The following summer, again I wrote around to various labs, and I spent it at Narragansett Marine Lab being a flunky. I drove the truck every day to get the mail in Kingston; I built lab benches.

Just meeting people as you could?

Yes, especially David Pratt. I did the research for my senior thesis there. I studied the feeding biology of *Venus mercenaria*. When I got back to Princeton, I did a similar thing with a freshwater clam. It lived in a pond nearby, and I would borrow a friend's car, a 1940 Chrysler, every week to go out and collect. I spent the final summer between senior year and graduate school at Woods Hole, working at MBL.

In the winter of my senior year, I applied to Scripps and Yale. I thought I wanted to do oceanography. I went to talk to the director of the Bingham Oceanographic Lab, a man named Dan Merriman. He probably gave me the best advice that anyone has ever given me, after hearing me spout off about what I wanted to do. He said, "I think you should go across the street and talk to Mr. Hutchinson."

So I went across the street to talk with Mr. Hutchinson. Evelyn wasn't there. His door was open and inside his door there was a chair, and on the chair was a stack of reprints of a paper he had recently published in *Ecology*, which you may know, called "Copepodology for Ornithologists." So, I picked up the top one and thought, oh, that's a curious title, and I started looking at it. It was really interesting. It was a time when people were starting to study interspecific competition. But the only people really studying this were ornithologists, birdwatchers who knew their birds very well. But, you know, Evelyn was interested in freshwater plankton. And he said, "You can learn something from copepods. Ask these questions of other animals too." Finally, he came in and we chatted. Larry Slobodkin was there; he was just finishing his Ph.D. He asked me what I had done in Rhode Island, and he kind of challenged me about those clams and how I could figure out what they were doing. I had also been interested in the tropics. Someone at Yale, I can't remember if it was Merriman or Hutchinson said, "Oh, we have a working agreement with the Bishop Museum, and people can go to Hawaii." Unfortunately for me, between the spring when I agreed to come and the fall when I arrived at Yale, they changed the rules, and the money was now only available to postdocs. But, they had told me all about this, so between Evelyn Hutchinson and Percy Morris, whom I had known for years, they scrambled enough money so that I could go to Hawaii. I think it was $350! So that was 1953–1954.

Then the question was what to do. There was a graduate student at Yale, a mol-

Asleep on the deck of the fishing/research vessel *Hydah* in October 1976, Alan Kohn (*center*) returns to dock at the Friday Harbor Labs after a field trip with students taking a University of Washington zoology course. Photo by E. Vermeij.

luskan physiologist, Sidney Townsley, who had done a master's in Hawaii. I said to him, "What I would really like to do is something in Hawaii that deals with marine invertebrates, preferably with mollusks." He handed me [Alan reaches for a book] *Reef and Shore Fauna of Hawaii* by C. H. Edmondson, autographed, no less! I looked through the molluskan section, and I got to the section on *Conus*, family Conidae. He talks about them and says more than 30 species are known throughout the islands. I said, "What are all those species doing there?" I was taking Evelyn Hutchinson's course, principles of ecology. This was a time, as I said, when people were beginning to look at interspecific competition, between pairs of species. Hutchinson had some neat stories about African freshwater flatworms, how they divide up habitat and so forth, and here were 30 species of the same genus in the same place. Oh, what are all those species doing there? And that was the question I asked in my dissertation. Completely unbeknownst to me [Theodosius G.] Dobzhansky, who was at Columbia at the time, had some of his students doing the same thing in South America with multiple species complexes of *Drosophila*. So I wasn't the only person trying to look at more than two species in an interspecific competition system, but I thought I was.

So, all right, that's what I did my dissertation on, essentially the comparative ecology of *Conus* in Hawaii. In 1954, I wrote to Robert Hiatt, who was the director of the

Hawaii Marine Laboratory, and said, "Here is my project. I have $350 from Yale"—which was enough—"can I come?" It turns out it was pretty expensive to rent a room there, but he said, "If you live in the lab and don't occupy one of the cottages, then you won't have to pay for housing." So I said, "That's fine." I was right next to the communal kitchen and dining room. I didn't know at the time that I had to share this room with a whole bunch of mongooses. In cages. Mongooses that people were using to test the effects of ciguatera toxin. But it was OK, as long as they were in cages. The price was right.

So that summer I started to get good data about the comparative ecology of *Conus*. But first, it took me two weeks before I found one! I'm sure I had walked on them in the meantime. The person who was really helpful, someone whom you probably knew, was Chuck Cutress. He was a perennial graduate student. He showed me where *Conus* were and how to find them.

At that time the U of H was running the Enewetak lab; they had plenty of money from the Atomic Energy Commission. The military supported people, but we could only go between operations; you couldn't go when they were shooting off bombs. But you could go anywhere you wanted on the atoll that wasn't hot. You could get an airplane to take you to an island if it had an airstrip; you could get a helicopter to take you if there wasn't an airstrip. They had all this hardware, and these guys had nothing else to do. It was very efficient for the kind of field stuff that I did, which has always been descriptive-correlative. I didn't even try any experiments until I came here. That was the summer of 1956. I went to Enewetak for a few weeks. You could also go to the Caroline Islands, so I went to Ponape as well as Majuro in the Marshalls. You could go anywhere; they had all these airplanes around. I did transects across the reef usually and collected specimens. I usually pickled them for gut contents. By this time I knew that feeding biology was important. But I recorded the habitat, population density where it was dense enough to do that, or by the number I saw per unit time. My first paper, in 1956 in *PNAS* [*Proceedings of the National Academy of Sciences*] was on the capture of fishes by *Conus*. So we did a little bit of morphology of the radular tooth and the proboscis. Marian [Alan's wife] did some nice drawings for it. She was the illustrator at the University of Hawaii zoology department at the time.

Enewetak was the start of going farther west, where diversity is higher. After I finished my degree at Yale in the spring of 1957, I had another opportunity to go west because, fortuitously, the Bingham Oceanographic Lab had an expedition that was sponsored by a wealthy Texas oilman, a natural gas person, actually, Alfred Glassell, who was a big-game fisherman. He wanted to combine this big-game fishing expedition with a tax dodge and contribute specimens to Yale. Jim Morrow, who was the ichthyologist, was the leader. Will Hartman did sponges, and I did mollusks, and Jim did fish. So we went all over the Indian Ocean. We flew to Colombo, Ceylon, which is now Sri Lanka, and met the ship in Trincomalee. We went to the Maldive Islands, the Chagos Islands, and the Seychelles. By this time Alfred Glassell wasn't catching very many fish, and he was getting a bit disgusted with that and he wanted to head back. So Will Hartman and I decided the Seychelles was really nice. We jumped ship. We took our stuff ashore, stayed in the Seychelles for three months, and worked on the reefs of the Seychelles. We took a British ship back to Mombasa. This was before there were any airplanes on the Seychelles, before there were any Americans, and before the satellite tracking station, which changed the Seychelles forever.

After the six months on the expedition, I spent the rest of the time in Europe working in museums, doing taxonomic work on *Conus*, looking at the original specimens described in the eighteenth and nineteenth centuries. This is the other side—the taxonomic side—of my life. It was the first time I had been to Europe, to Geneva, Paris. Geneva has a big collection of *Conus*-type specimens, including Lamarck's collection, and the British Museum has the largest number of types.

Was it through comparative ecology or through taxonomy that you got into evolutionary questions?

Both. I think it was mostly because the comparative ecology came to a dead end. I did as much as I could. As I said, I haven't done very many field experiments. In the 1970s, with Paul Leviten, who was a graduate student of mine who died of cancer at too young an age, we tried to do some experiments in Enewetak on factors affecting density and diversity. Caging *Conus* in, caging *Conus* out. They didn't work too well. All we got were data on cage mortality. But, influenced by Bob Paine ecology, we tried that. You would have to be at that isolated site for a long, continuous period of time to make it work. At that time, it was, and it is still, difficult to do that in the tropics. But the correlative-descriptive ecological approach had essentially told me what it was going to tell me, so I decided to look at something else. That is one way I got into asking evolutionary questions. The other way was, here is this gastropod with big, heavy shells that fossilize very well. There is a very good fossil record of them, but nobody had ever used it to learn about the evolution of the group. And the evolution of this group is spectacular. Doubling time is six million years. Much faster, an order of magnitude faster, than bivalve speciation. So that raises the question: where and when did all these speciation events occur?

So I thought, maybe you could learn something about this by looking directly at the fossil record, which I knew was good. But, you know, I didn't know much about paleontology. So I did that on sabbatical in 1981. I set out to do it, to play paleontologist. I went to two places where tectonic forces had pushed erstwhile marine environments up above sea level. I wanted to go to areas where the modern reef was rich too, for comparative purposes. So we went to Okinawa for several months. Okinawa rose quite rapidly, and the geologic history is known quite well, in fact, all the way back to the Upper Miocene. Then, after that, I went to Fiji, different tectonic history but some similar results. What I really wanted to learn about was *Conus* evolutionary history, changes in diversity, and could I associate this with independent evidence from the geology, from the matrix in which the fossils are preserved. The answer to this in a nutshell is, yes, you can do it but only on a crude scale, not as finely as I wanted.

But I got interested in all of the other gastropods that co-occurred with *Conus*. Gastropods are really dominant in the Pleistocene fossils of Fiji, lower Pleistocene, somewhere between 1.5 and 1.8 million years. There are a lot of outcrops that had been studied by other people, amateurs and professionals, especially Harry Ladd. He wrote a paper before World War II and another series in the 1960s and 1970s. The thing that impressed me most about the fossils of *Conus* and other gastropods (they really dominated over the bivalves) was evidence of predation. Lots of drilled holes and lots of repair scars. So I set out to do paleoecology and paleobiogeography as well as evolutionary history of *Conus*.

You were just saying, "I got into these other gastropods," and I was thinking, wow, that's a big job.

Just the taxonomy is difficult.

Then you were saying you got into this paleoecological question, and I thought, wow, all in a year's sabbatical?

Well, we collected the materials. The Fiji Pleistocene is done. Now I have Pliocene from Fiji, not as extensive material, and some Miocene. I keep hoping some student will come along and want to work on it. Oh, who knows, now that I am retiring, maybe I'll have more time.

You know, these guys [*Conus*] have venom. That is another whole story. I published my last paper on venoms in 1963. That is now a big industry, mostly through the efforts of Baldomero Olivera at Utah, who started where I left off when pronase was invented, the enzyme that breaks all peptide bonds and was the active ingredient of the enzyme detergents. It destroys all of the activity. Olivera went on from there and sequenced the peptides of a few important venom components. But there are hundreds of them, if you look at HPLC [high performance liquid chromatography] peaks, 50 peaks or so for every different species, and we only know what a few of them do. Some of them are becoming increasingly useful in medicine. There is a company in California called Neurex, which makes a pain killer, a calcium channel blocker. They use it in cancer patients with unbearable pain. It probably isn't very good for you, but these are terminal patients. There is one that overcomes side effects for some drug that is used for Alzheimer's disease, and there is one that is used as a diagnostic drug. Because it is very specific for a particular calcium channel, it will distinguish the peripheral from the central nervous system ion channels. There is a lung cancer, a small cell carcinoma, that has two versions. You inherit the tendency to get one. They differ slightly in amino acid composition of the calcium ion channel, and so this conotoxin can distinguish between the two. The taxonomy is becoming very practical. So is the conservation, because these things have all sorts of pharmacological applications.

Your particular group, the cone snails, are well known because you've worked on them for the last 35–40 years.

Almost 50 years. I started doing taxonomy back in Hawaii because you have to put names on these species. They don't come with names. They are like babies; somebody has to name them. I ran into taxonomic problems right at the beginning in Hawaii. There were arguments among shell collectors: what species is this? If you study an animal in nature in the field, you get a feel for what the animal is doing. You can tell whether species A and species B are really two species or whether they are variable members of a broad population.

To get back to your point. You have to know what the animals are doing in the field to have that intuition. You cannot just collect shells.

That's right. Our *Manual of the Living Conidae* says "volume 1" on the bottom, "Indo-Pacific," with about 60% of the species. I felt comfortable working on the Indo-Pacific because I've seen maybe 30% of the species alive. They wanted to do another volume on the rest of the world, but now Dieter Röckel is older than I, and Werner Korn is director of a museum, so I don't think it is going to be done.

The future? I don't know about the future. What we used to call invertebrate zoology or invertebrate biology is now a part of what people call biodiversity. That is frustrating. It gets a lot of lip service, but it doesn't get very much support. The problems of understanding biodiversity are not going to be solved unless you have people who are expert in a broad variety of organisms and comparative ecological and evolutionary approaches.

If you try to have someone become a pianist and never let him play the piano, you're obviously not going to succeed.

James M. McLelland (b. 1935)

Charles A. Dana Professor of Geology
Colgate University
Interviewed in his office 12 November 1997

Jim McLelland, whom I first met when I visited Colgate to recruit students for WIL programs, begins his interview by portraying a magical period in his life. As a freshman at Yale, he was taught by "ultimately three of the most outstanding professors in a faculty of outstanding professors." Perhaps this explains his own dedication to educating students and his interest in passing the tradition of scholarly excellence from one generation to the next. He helped create the Colgate field geology program three decades ago (1967) and has spent parts of many summers teaching it. This field experience, as well as the emphasis on student independent research, helps make Colgate one of the outstanding undergraduate programs in geology in the country.

When I went to college at Yale I knew nothing at all about geology; I didn't even know it existed. I was going to be a great young American novelist. I asked somebody what was a good course to take to get past my science requirement. The answer? "Ah, take geology and astronomy, a semester of geology and a semester of astronomy. 'Rocks and Stars.' You'll like that and you don't need a lot of chemistry, physics, or math background," and I said, "Good." I signed up for that course, rocks and stars. Jeez, I really loved the geology!

Who taught it?

Richard Foster Flint. He was about seven feet tall, and he had this extraordinary, distinguished air about him. He'd stand up there in front of the class with his great hands, and he'd sort of roll his eyes. He had a lot of slides that he would show and, jeez, this stuff was really fascinating. I got into it and I did well.

You took this as a freshman?

Yes, my first semester. In fact I had some terrific courses. My first semester of freshman year I had Richard Foster Flint for geology, I had Theodore Green for introduction to philosophy, and Archibald Ford for history of Europe. Those three were ultimately three of the most outstanding professors in a faculty of outstanding professors. They were put in freshman courses for obvious reasons. Oh, they wanted to be in freshman courses. Theodore Green had had polio. He walked with crutches lashed onto his forearms. He knew he looked like FDR, and he played on it. He had a cigarette holder and a cigarette in it. He had the same kind of glasses. He talked to people as he came down the aisle. Those were in the McCarthy days, and we were all very impressionable. In 1953 the views of McCarthy were flying high. Theodore Green stood up there and every day he would take McCarthyism and McCarthy apart, a little here, a little there, and the nuances . . . it took real courage, and he did it and he did it beautifully. I don't think anyone would have dared touch Theodore Green. It was good for a freshman to see.

I said, "What am I going to major in?" Well, I still want to be a novelist, but I'd like to take another geology course. So I signed up in the first term of sophomore year for what was called the geology of the United States. That was team taught by the people who had done so much of the great work in the country. There were Chester Longwell, John Rodgers, [Richard Foster] Flint, [Roy] Jensen, and [Carl] Dunbar, who discovered the Permian oil basin. We were hearing all this stuff firsthand from the great people in the field. Oh, this was better than I ever thought, and furthermore, walking down the halls in Kirkland Hall, the geology building, it was fairly old and maybe not so nice, but nonetheless there were these beautiful maps on the walls with all of these bright colors—maps of France and the United States and so on—geologic maps. You could see there was a pattern, but you didn't know what it was. It was fascinating.

Like the map downstairs?

Exactly.

Wonderful patterns!

Aren't they? You know, they mean something really interesting. To learn what that might be, to find out what that might be, that was exciting.

I was heading off to graduate school at Berkeley in geology, but I got a summer job up in the northeastern corner of Alberta with the Research Council of Alberta. We were doing field mapping north of Lake Athabasca, which is about the size of one of the smaller Great Lakes. At that time, uranium was big. Right across the border in Saskatchewan there was a place called Uranium City with all sorts of mines and lots and lots of uranium. We were going up there to map these ancient rocks, the Canadian Shield.

So, these were Precambrian?

Yes, mainly Proterozoic in age, but some of them may have been even older than that; we didn't know the precise age at that time. I was the senior assistant in a party of four. There were two assistants, myself, and the party leader. We'd go out everyday and we'd map. It had never been mapped before. In fact, any place you'd put your foot, there was almost 100% chance that no human foot had ever stepped there. We camped out, and we were serviced by float planes coming in, Cessna 180s. We had a radio with us. It was a tremendous experience. I do love field geology.

At that point, I did not think I was going to be a geologist, but I got this opportunity to go up to Alberta. Actually, I had rounded up a summer job with a very famous, in those days at least, copper and silver mining company, Cerro de Pasco in Peru, but they said, "You come with us this summer, and you stay on and work for us, and we will send you to Stanford University for a master's degree in economic geology, and then you can work in Cerro de Pasco in the form of a job." I mean, that sounds really good, but I wasn't sure at all that I wanted to do economic geology. Everything seemed so constrained in that opportunity, that early on in my life. I was just going to do what my employer required of me. I wanted to have greater freedom, so I applied for this job in Canada. When I got that, I said, "Ah, I'm going that route!" I did two summers up on the Canadian Shield; that made me a devotee of field geology, not by itself, but field geology in a framework. I found that it combined the intellectual challenges of geology with the physical challenges of the outdoors. I really liked that, and I think that's what a lot of the type of students we get at Colgate like also.

I grew up in two different places. Until the age of ten, I grew up in the West Indies, in the Dominican Republic, way out on a sugar estate, where my dad worked. Then, at the age of ten, because there were no schools there, my parents put me in a boarding school outside of New York City, the Riverdale Country School.

So, being out in the field and camping was kind of new?

Yes, I liked that kind of life. I've always been fascinated by the image I'd had of the southwestern United States. If I had had my choices of things to do, I'd have been a center fielder for the Brooklyn Dodgers or a cowboy, but as it was, I had to go to school. But the outdoors, with the attraction to the Southwest, the cowboy image, remained attractive to me. The idea of going into the Canadian bush had that same sort of romantic and romanticized attraction. That's all well and good, but there were mosquitoes and horseflies and all the discomforts of it. But, you forget those things easily. The overall retention that you have is positive.

At Colgate, my real field experiences started. I looked around and I said, "Where the devil am I going to do research around here?" Then I remembered my old petrology professor at Yale, Matt Walton, had all these maps lying around and all these interesting talks he gave about his work in the Adirondacks. So I said, "Gee, yes, the Adirondacks." So then I went over to the New York Geological Survey, which served as a clearinghouse for people working in these areas. I said, "Is there any place that you know of where work needs to be done in the Adirondacks?" I didn't know him then, but my now good friend Yngvor Isachsen, in charge of that part of the New York Geological Survey, said, "The entire central Adirondacks has never been mapped." I said, "Oh, that's for me."

I then set about figuring out where I could start. It turns out the area I chose is only an hour and a half from Colgate. It is wonderful geology, absolutely untouched before. I went in there my first year with my field assistant, David Howell. We rented a motel cabin, a dreary place with no hot water, down at the end of a very nice lake called Canada Lake. We worked out of there and did an awful lot of mapping with precious few good results. The first two or three years in there I wasn't getting very many results. I was getting frustrated, thinking that well, maybe, fieldwork wasn't for me. Then, one afternoon, I just happened to pick the right series of outcrops, and everything fell together! It was a remarkable experience. I remember coming here at the end of that day, I was clearly emotional about it, and sitting down with Cathy and saying, "I'll never have another experience like this in my life." This is the first time that my research has eventuated in a significant result. That can't happen but once in your life. I've had other experiences but never one like that. It was a first, and you can't have more than one first.

From then on, I just rolled through the Adirondacks. It was a particularly significant day on the emotional level when, after having been in there six or seven years—I worked my way over from Canada Lake, where I started the mapping, and progressed and progressed—and finally I came down over the hills into the Lake George area, because that's where Matt Walton had done his work in the Adirondacks back when I was an undergraduate. He was my professor of petrology, and I tied into Matt Walton's work. That was a wonderful feeling. Subsequently, almost ten years ago now, Matt was director of the Minnesota Geological Survey. I asked him if he would like to come and visit me in the Adirondacks, and I would show him what I had done and so forth. That was a splendid three or four days we spent together, out in the field, me and my professor, showing him my work and then we got to his work. I found that to be one of the most rewarding experiences of my life.

The other side to all of this is that that first year, I took young David Howell out with me. He was brilliant and a good field assistant. He also went out the second year with two or three other students. Those field experiences with those students were extremely important in their maturation. Now Dave Howell is well known in geology. He has published several books and articles; he's with the U.S. Geological Survey, and he's a bigwig there. Graham Closs is a professor at Colorado School of Mines. The other two have gone on to do other things, and that's fine. I think that David would say that those experiences were important to him also.

So much maturation takes place, even outside the intellectual discipline lines.

That's right. You learn how to do things and how to think.

To go at the student involvement from a somewhat different direction and to flip back to 1967, gee, this has been such a good experience working with these kids in the field, and my field camp wasn't such a good experience. We had to make sure that growing numbers of concentrators had field experiences, so we then, at that time, evolved our own field camp. We started the geology off-campus semester. Pictures from the last 15 years line the stairwell down there.

I noticed them as I walked up one stairwell and down the other. What a nice introduction to the department and what a nice tradition it sets for future students.

Yes, I wish we had pictures going back to some of the earlier ones. I think the earliest we have is 1981 and really that leaves out 14 years.

We set this thing up, and "the field experience" has several dimensions. One is that if you try to have someone become a pianist and never let him play the piano, you're obviously not going to succeed, that's obvious. In the case of geology, it's similar. If you try to teach geology without having people go to the field and look at rocks, you are not going to succeed. They are not going to have any real appreciation of relationships you are talking about in the abstract.

Aha, so you don't think you can teach how to map in front of a computer?

No. I think that it is an exercise in futility. So many things come into play when you've had an experience firsthand, including getting lost in the woods, being able to find your way around, realizing that relationships just don't jump out and hit you, that it is interpretive, that lots of time you have trouble telling the rocks apart, and that there is a lot of frustration.

The other thing that is also enormously important and has been, I think, key to the success of this project is that when you take 15 or 20 students and take them out into the field with a faculty member or a couple of faculty members, all of those in-house formalities break down, you get to know one another as people, as human beings. When the great man stumbles over some log in the woods and utters some foul imprecation, you are finally all reduced to the same base level, and you can go on from there. You get an enormous sense of camaraderie, so by the time the students do this trip, they are no longer just our students but also our friends, and we can party together and do other things besides geology.

Those relationships that begin in the field experience, I think, are much deeper than you can ever form in a classroom. That's an important noncognitive aspect in the whole learning environment that I think is crucial to our success. One of the reasons that we have succeeded as well as we have, which is pretty good in relationship to other liberal arts colleges, is that . . .

You've not only provided your students with formal training in geology, but you've also given them the confidence and maturity to ask questions and to know that they can seek the answers on their own and through interactions with other people.

Right.

As smart as you can be, when you get out and start wrestling with nature, it's not very simple.

B. Clark Burchfiel (b. 1934)

Schlumberger Professor of Geology
Massachusetts Institute of Technology
Interviewed in his office 4 April 1998

I was amazed to learn that one of the most outstanding graduate programs in geology, including field aspects, is to be found at MIT. Its reputation is based as much on the reality provided by the regional field geologist, Clark Burchfiel (whom MIT had the foresight to hire in 1975), as on the complex, quantitative, theoretical and lab-based geophysical and geochemical capabilities the school had developed in the early 1960s with NSF support. Burchfiel developed as a scientist under the tutelage of two superb mentors, whom he emulates as a teacher. He established a field camp for MIT (made possible through an endowment specifically for field science), and all of his numerous grad students have done field theses. Burchfiel compellingly outlines the history of the decline in teaching and the ascendancy of research in the modern university setting.

I had the option to stay at Stanford and get a Ph.D. or think about going somewhere else. There were no jobs around in the oil industry, but I could continue on as director of the physical science labs, because I had been teaching them as a TA for a year while I was taking a master's degree. I walked into Professor Hubert Schenck's office, and I talked to him about this. He said, "That's a pretty good offer you have here, but you know what? You're going to go to Yale!" I said, "What! Look, my grades aren't all that good," because I had been playing sports. My record was OK, but I wouldn't call it sterling at that time. He said, "You are going to go to Yale and you are going to work with John Rodgers." I said, "Oh you're crazy. (a) I don't know anything about any other university, I don't know anything about John Rodgers, and (b) I've got this very pleasant offer here. I haven't put in an application." He said to do it. I got a reply back saying I was accepted with a full fellowship. I thought, wow, where does this come from? I didn't know about all this, but there was stuff working in the background. It turns out that Schenck and Rodgers had been in the army together during the Second World War and were working on the reconstruction of Japan after the war. It had a very major component of earth science in it: reconstruct their economy in terms of their mineral resource base. I found this all out later, that they knew each other very, very well. My suspicion was that what had happened was that Schenck said, "Look, this is a guy that you should take." Well, I was flabbergasted.

John Rodgers was a marvelous teacher. I would like to say I try to emulate him in some way. His lectures were fine, but he really taught by example. This is the way I do science. That's how you learn it. I was very influenced by John, tremendously so, and changed my field from things that were more oil company–oriented to something that was more academically oriented. It also changed my perspective because John had this global perspective of the Earth, and here I was looking at oil fields rather than more broadly related questions.

When I graduated from Yale, the oil business hadn't come up yet. I had an offer both from Texaco and from Rice. Now the question was: where do I go? I had a real tough time making a decision. I made it based partly on the idea that if I ever wanted to try the academic business I should do it now, because it is easier to move from an academic job to an oil company job than it is the other way around. I took the Rice job, at half the salary they were offering at Texaco. I was never happier.

A supportive environment?

Yes, but also, it is really a very prestigious university if you stop to think about it, a very small population of very intelligent undergraduate students and a small graduate department, where you could really get into things and do your own thing without a lot of external pressures.

I worked there for 15 years. It allowed me to work in an environment where I could do my own thing without a lot of the pressures we have now, which started to develop as big science in universities started to take over from the kind of department we had at that time. It was very academically oriented. NSF was just starting to come in. People were starting to think, maybe we could get some federal money to do a little bit of work, that sort of thing. Whereas, now, we really run on that, and we have for a long, long time. It was a much less stressful environment. It was an easy place really just to do work and enjoy yourself.

I've seen that, since I've started, the pressure has increased over the years in terms of bringing in more money and being more research oriented, which cuts back somewhere else. That cutback is partly in your ability to do the kind of broad teaching that you'd like to do.

Part of my concern and part of my motivation for this book is: when are we going to have to pay the debt for that?

An interesting question, because it's always a balance between the two. When I started, the balance was very clear on the teaching and the individual research without a lot of pressure. It's changed now to almost the other direction, so that the real pressure is on getting the research done and bringing in the funds to do that research. The time and effort to do that has to come from somewhere else, and clearly, over the past 35 years or so, it's been a diminution of the academic end of things.

Do you feel that your students have the opportunity to learn from watching you do science?

That's one thing over the years that has never changed, as far as I was concerned. All of my students have had field theses. I've had almost 50 Ph.D.s now, and every one of them I've been in the field with every summer, anywhere from a week to two weeks and sometimes longer. That's where you really get to know them. They can see how you operate and how you think. I think that's where a lot of the transfer really takes place, probably more so than in the classroom.

You've made the choice to do that?

Oh, yes! In fact, when I came here to MIT, for example, there was no field course taught for undergraduates. This department had changed dramatically. In the late 1950s, it had a very strong field component to it, like most departments did at that time. NSF and the funding agencies were just starting to crank up. It was obvious that this was going to be a major source of revenue in the future. We were developing a lot more sophisticated tools in the sciences at that same time, things in geochemistry, things in geophysics. It looked like the direction that science was heading was more toward a laboratory base. It was not necessarily a deemphasis of fieldwork, but again, time has to go somewhere, and the more you emphasize the laboratory end of things, the less you have for other things.

A shift in the kinds of questions you might ask?

The questions and the techniques and tools you use to answer them. In the 1960s this department became very strong in geophysics and geochemistry. At that time, that was probably the right direction to go, because that's where a lot of the money was being put.

When NSF was established and started to operate, the organization wanted to know what the division of funds should be within NSF to support the various parts of earth science. My understanding of this, although it comes second-, third-, and fourth-hand, is that they sat down and said, "OK, at this point in time (in the late 1950s, early 1960s), what areas need the larger amounts of money?" Well, obviously, if you're buying mass spectrometers and you're buying ship time in oceanography and you're buying a lot of geophysical equipment because you're running geophysical experiments, that takes more money than it does if you are out walking around in the field. When they started to divide up the money, that's kind of how they did it, and it's never changed since then. That division is still with us.

What happened is that this department changed very strongly to be laboratory based. MIT became known and recognized as a major center for quantitative and laboratory research in the earth sciences to the point that, by the time I got here in 1975, they had decided that the field aspects of geology had become so deemphasized that they needed to turn it around a little bit. Plate tectonics got started basically through geophysical observations in the ocean. So it was on a real roll through the late 1960s and early 1970s, but then the question came up: well, what does this have to do with the continents? That became geology again, because we had to get out and start looking at the geology in its context of the plate tectonic environment and how these things were formed. That meant you had to get back and do some fundamental fieldwork, asking different questions than you did before.

You might look at and measure the same things, but you would interpret them differently now.

Exactly. It needed a very strong field base to do that.

Frank Press was chairman of the department at MIT at that time. He called me on the phone. I still remember the conversation very well. He said, "Would you like to come up and interview for a job at MIT?" I said, "Wow, why?"

This was out of the blue?

Completely. I said, "Well, I don't know, tell me a little more." He said, "We're interested in making this change, to do more field-oriented things to interact with the more quantitative science that we do here at MIT." I didn't know a thing about Boston. My introduction to the East Coast, being a Californian, was straight from Stanford to Yale. At that time, New Haven was run-down. I thought, if Boston is anything like New Haven, I'm not particularly interested. I told Press, "There is about a 2% chance that I would come to MIT. If you want me to come up and interview, I'll come up but you've gotta know that's the way I feel about it." It had nothing to do with MIT; it was just the East Coast environment.

When I came here, I was overwhelmed in two ways. One, I thought Boston was a great place to live. Second, I had a very sincere feeling from the faculty here that they really wanted to change direction. That was terribly important.

And these were faculty members who were successful in doing what they were doing?

Oh, yes, very successful. I can still remember people like Stan Hart, PETER MOLNAR, Tanya Atwater, and Pat Hurley. I got this really strong sense that there was the potential here to work with people who were doing things from a very different viewpoint than what I was doing, but it would allow real serious cross-fertilization to take field geology where it was meant to go: to provide very basic data on which to build more and more quantitative understandings of how the Earth evolved.

As soon as I walked in this door, all of the students I had doing fieldwork were working with the best people doing the most sophisticated geophysics and geochemistry that could be done. They opened their laboratories to us. They were up there doing chemistry in Stan Hart's lab, for example. This was a very sophisticated lab. A lot of labs, at a lot of institutions, are basically closed to people who are all thumbs. You just can't do that. That just wasn't the case here. It was all heading in the right direction. It has continued to be that way ever since I've been here. People still come here (I've been here more than 20 years) and say, "Wow, you do a lot of fieldwork!"

I have to say, I was amazed when STEVE PORTER *recommended that I interview someone at MIT. I said, "At MIT?"*

You see, this is exactly the rub we had in the 1960s and 1970s, and it hasn't gone away.

It's always been very clear to me that no matter how sophisticated your science gets, most of the sophistication is in the tools, not in the science. Therefore, if you don't understand the very basic relationships among things as they exist in the field, you can apply all of the most sophisticated tools in the world to those problems, and you won't come up with the right answers. That's why field geology is always going to be there. Because the evidence you need for many of the problems (but not all of them)—in the shallow crust and on the surface, and in the interaction of the surface with the atmosphere—all of that basically comes from observational science in which you produce data which then can be worked on in various other ways with more sophisticated tools. It all comes back to looking at things in the field and asking the right question.

How is that translating for the MIT undergraduates in geology?

OK, I'll tell you how it is. When I came, we started the field course, and we've had one every year since then. It's become more diverse and drawn a greater spectrum of both undergraduate and graduate students as the years have gone on. This is exactly what we wanted to do. It started off as being just hard field mapping, and it still is. And only geologists. But within a year or two, we had a geophysics add-on to this. We would work in the same field areas in California and southern Nevada. We'd camp together; we'd eat together; we were working on adjacent problems. They were working in the flatland where you couldn't see anything, and we were working in the mountain projecting these things down saying, "OK, you guys are down there, we're running this stuff at you, what does it look like down there?" This is when Peter Molnar was here on the faculty. This worked very nicely because geophysics people could see firsthand what we were doing, and we could see firsthand what they were doing and see how we complemented one another.

As soon as I came here, all of my graduate students went into the field to teach the undergraduate field camp with me. I always run into things in the field I have never seen before. What you have to teach is how to solve those problems. I tell my students this: you never really learn field geology. Hopefully, you get better at it, but you never really learn it. That's why I would bring all of my own graduate students out there two or three times with me, to help teach but also to learn more about field geology. We would move the field camp from area to area every year, so we would look at different things and that means they would get experience in different kinds of geology.

At the undergraduate level, it's very important, I've found, to have graduate students around because for many of the undergraduates, it's the first time they get out to the field. There are two things that happen. One is that all the MIT students are very bright; everything is like falling off a log. You go out into the field, and that ain't the case! All of a sudden they start getting confused, and they start saying, "I don't understand this. I don't know how to do this." That's a real eye opener for a lot of these kids. That's when you have to show them, and the graduate students tell them, "I was the same way when I walked into the field. I didn't understand." They'd see me out there saying, "Jeez, I don't know what this is." I'd be just as confused as they were. That was the most important lesson I think they began to learn about fieldwork. It's that anybody that's had 40 years of field experience could be just as confused as they were, but the advantage the experienced workers have is that they know how to go about solving their problems. That's what you try to teach them: how to get to the answers.

So you can humble them but also give them some hope at the same time.

Exactly. That's why I've been dead set against sending students to camps that go to the same place every time, because the faculty know what the answer is already, and that's not what you want to teach them. One experience that the students learn is that as smart as you can be, when you get out and start wrestling with nature, it's not very simple. That's a great lesson to learn.

The other thing is that there are a lot of conditions that you work in. We go out in January; the conditions can be nice at times, and other times they can be pretty awful. It can be raining; the wind can be howling; it can be cold; it can snow. You only have so many days, and we're in the field every day, unless it's a total wipeout, and that is very rare. The advantage of having the graduate students along is that these people are motivated, they've made a commitment, they're interested in fieldwork. It's pouring rain, and these people are beating themselves on the chest saying, "This is great! I'm having a great time!" and the undergraduates are looking at them and saying, "Wait a minute, is this really what I want to commit my life to?" They find out that (a) you can operate under some rather unpleasant conditions, and (b) you find out at this point whether you really want to do this or not. You end up with a small group of very enthusiastic people, no matter what the conditions are. That's why you need the graduates; otherwise you could end up with a group that would get very depressed. They don't want to be here, and they can't see why anyone would want to be here; the only person who's happy is the faculty member who's out there because he or she likes it. When they see a third of the people out there are happy, really enjoying themselves, whatever it is, then a little bit of that rubs off. It helps a lot for the social environment, the interactions. I think those are the important aspects, besides learning how to get things on maps properly. It's part of the sociological part of the science.

Now, I think, in many respects, at least in this department, we've reached a really good balance, but there are a lot of other places that haven't and, in fact, it's across the board in science. It's not clear that we've reached a good balance. One of the things I suspect that you've found out is that a number of schools are really cutting back on field-oriented programs because money is tight. There are departments that have deemed that it is not particularly important any more, now that we've gotten off on this quantitative and sophisticated science. You see, we were very lucky here. I go back to the late 1950s, when Bob Schrock was chairman of the department here. He had gotten some money into this department from the Lyon's Fund. The Lyon family had given some money, a fair bit, as an endowment for fieldwork. The letter they wrote was very specific. This money is to be used for teaching field geology and, it says, very specifically, not for geophysics. I don't know how Bob got them to write that in there, but he had it in there. That one statement has been so important. This is a fairly large fund because it creates about $45,000 a year, so it's got to have a fair-sized principal involved. There was a period there, 20 years, when MIT did no fieldwork.

And so they couldn't tap the fund.

It just grew larger. The day that I came here, Bob walked into the office and said, "God, I'm glad you're reinstituting the field camp. You know we have this Lyon's Fund." Then he said, "Protect it. Every year, they are going to try to take it away from you," and they have. They've tried to use it for other things. And all you do is, you just bring this letter in, and there it is and they can't do it. So that has been an enormous boon to us, because we have endowed funds to run a field course. If it hadn't been, when we started cutting back on expenses, as every department has in the last ten years, I'm sure that's one of the things that would have gone. I'm sure it would have gone. I mean, I would have fought like hell.

You might have lost that one.

I would have probably lost that one.

The bean counters don't understand how someone develops into a scientist.

That's right. There is no better way to get ideas, philosophy, and so forth across as when you are with students 24 hours a day. They basically see how you live in the field, how you think in the field, and that's part of the game. I'm sure the other faculty members have different philosophies about the way one does things than I do. The one thing we have in common, though, is that this is damned important! It's how you do your science.

A theory of climate change is much more difficult to formulate because it involves the entire planetary system: the solid Earth, the atmosphere, the oceans, and life.

Stephen C. Porter (b. 1934)

Professor of Geology
University of Washington
Interviewed in his office 12 May 1998

Steve Porter demonstrates that persistence in a field of intellectual interest (following your own curiosity) is sometimes finally rewarded with acknowledgment by peers and by funding sources when your expertise proves to be a crucial piece of a larger puzzle whose features are beginning to emerge. Porter focused his interests in Quaternary geology during a period when he had time to read and think, when he served aboard ship in the navy. His work in interpreting surface deposits of sediment (moved by the forces of ice, water, or wind) has taken him to sites around the world and led to questions about global climate change. I am also indebted to him for taking me on my first geology field trips in junior high, from which I can still recognize glacial features in the New England landscapes.

I was born in Santa Barbara, California, and was strongly influenced by the natural environment. When I was young, I used to hike and collect rocks and minerals in the mountains with my grandfather. In my junior year of high school, I received a scholarship to attend Explorer's Camp, a summer camp for boys run out of Mancos, Colorado. It was organized by the former head of the National Park Service, Ansel Hall, but the day-to-day operation of the camp was handled by Ken Ross, an archaeologist. During the six weeks of the camp, we excavated part of an Indian ruin, climbed mountains, and explored canyons tributary to the Colorado River.

I couldn't take an archaeology course as a freshman at Yale, so instead I took geology. I did well in that class, and it increased my interested in the subject. As a sophomore, I took a course in the classical archaeology of Greece, Rome, and Egypt. However, during the summer between my freshman and sophomore years, the two most prominent archaeologists in the department died, one of natural causes and one accidentally. This severely impacted the archaeology program. So I decided, after my sophomore year, to major in geology, but that left me with a problem, because I didn't have much time to take the necessary basic sciences. I needed many geology courses to satisfy the requirements for the major. In those days, during the Korean War, a student had to be enrolled in ROTC or face the strong likelihood of being drafted out of college. I was in NROTC, and this meant that a fifth of my total course load, during all four undergraduate years, was in naval science.

Steve Porter pauses by the Sollas Glacier in Taylor Valley, Antarctica, during the Dry Valleys Drilling Project, United States–Japan–New Zealand joint research program in 1974. Photo by S. C. Porter.

One-fifth!

Yes. This significantly reduced the course options I had.

After graduating from Yale, I was in military service aboard a destroyer. We traveled all over the Pacific region. That further whetted my appetite for geology because we visited many tropical islands and atolls, areas of volcanism, tectonic movements, and sea-level change. During those two years in the navy, I debated whether to return to graduate school in archaeology or continue in geology. After reading a number of geology textbooks and giving it serious thought, I decided the best solution was to focus on Quaternary geology. In the end, I elected to return to Yale, which had the best program in Quaternary geology. I wanted to work under Professor Richard Foster Flint, who was a leading figure in glacial-geologic research.

I was employed for a summer by the state of Connecticut to map the Wallingford Quadrangle as part of a cooperative program between the state and the U.S. Geological Survey. Professor Flint mapped an adjacent quadrangle, and sometimes we'd go in the field together. I learned a good deal about field mapping from him. Flint was meticulous in field mapping. He was very precise when he placed a line on the map that separated two geologic units. If I'm a careful field mapper now, it is partly due to his example. I think he liked having a student who was interested in the same glacial-geologic problems that interested him.

As an undergraduate, I had joined the Yale Mountaineering Club and spent part of every summer climbing in the West. During the school year, we climbed on weekends in New England or New York. Thanksgiving and spring holidays were spent at Mt. Washington or some other high snowy area in New England. In the summers we climbed in Yosemite Valley, the High Sierra, and the Canadian Rockies. Three of us made the 35th ascent of Devil's Tower in Wyoming. I enjoyed being among mountains and glaciers, and so I think that was a major factor that led me to my focus on Quaternary geology, particularly alpine glacial geology.

At the time I went into glacial geology there were many such studies being done in the western mountains, mainly by USGS geologists, both in the western states and in the territory of Alaska. One of Flint's graduate students, Don MacVicker, had worked in the Brooks Range. The summer before I entered graduate school, he had died in a boating accident while in the field. Flint had a continuing interest in the area and encouraged me to consider a research project there. I also considered an area in southern Alaska near Cook Inlet on a large glaciated volcano. That area would have been more difficult logistically, and furthermore, it was inhabited by large ferocious bears, which didn't appeal to me.

There are bears in the Brooks Range also?

Yes, there are tundra grizzlies, but all the time I worked there, I never encountered one.

Flint was on the board of directors of the Arctic Institute of North America [AINA], which had funded MacVicker's work, and so I wrote a proposal for fieldwork in the central Brooks Range. I received support for two summers. During the first field season, in 1959, I was supplied from Point Barrow, which was the staging base for AINA research. We flew by float plane to a lake at Anaktuvuk Pass. With me were an undergraduate and a graduate field assistant. We worked out of the Eskimo village, mapping an area of several hundred square miles that had a good deal of relief. On a typical day we hiked 10–15 miles and climbed several thousand vertical feet. I mapped both the bedrock and the glacial geology. I thought it would work out about 50:50, but because the bedrock was extremely complex, I ended up spending 85–90% of my time those two summers just working out the bedrock geology. It had been studied in 1915 during a rapid reconnaissance by a USGS geologist who passed through the area in several days, but basically, very little had been done.

I was coping with the problem of how to explain the deformation of the Paleozoic formations. How could I adequately explain the source of sediments in the sedimentary rocks when the evidence implied a northern origin, yet to the north lies the deep Arctic Ocean basin? The answer later became clear, but only several years after I had completed my study, when the plate tectonics theory emerged. The Brooks Range was an area of large-scale alpine glaciation, much like the Alps. Although some of the basic story had been worked out, there was virtually no dating control. One of the contributions that I made was to find several sites from which we could obtain wood and peat samples for radiocarbon dating. I published a paper that had the first Brooks Range glacial chronology based on radiocarbon dating. Although by modern standards the dating was minimal, it was much more than had been known before.

Before my first summer in the Brooks Range, I had married Anne Higgins, your charming older sister, who continues to do fieldwork with me in exotic places. At that time, it was not possible for women to work in the field on such projects. This was largely because the Arctic Research Laboratory, which was supplying us, was a navy-sponsored laboratory, and the navy was run by admirals, who did not approve of women on scientific field projects. Both summers I was based at Anaktuvuk Pass with an archaeologist, Jack Campbell. While writing up my Ph.D. research, I learned that Jack was going north for a third summer with NSF support, and he invited Anne and me to go along. This was an archaeological expedition, and I worked on the Quaternary geology around the archaeological sites. I have not been back to the Brooks Range since 1965. Many of the Eskimos we knew are still alive, but the children are likely parents or grandparents now. The Eskimo village, which was rather a temporary-looking settlement when we were there, is now a permanent town.

When I received a Ph.D. from Yale in 1962, there were few job opportunities available. One was with the USGS in Menlo Park, California. Yale had a long tradition of supplying the geological survey with graduates, many of whom became leaders in the survey. I think if I had been offered the opportunity to continue work in glacial geology in the western mountains, I likely would have accepted the position. Then I received a phone call from a Yale classmate, Bates McKee, who was on the geology faculty at the University of Washington. Bates told me that Hoover Mackin, a distinguished geomorphologist in the department, had just accepted a job at the University of Texas. Therefore, Washington was looking for a replacement, and would I be interested? I said, "Of course!" Several persons had applied for the position, but I was hired after a phone call from the chairman. We arrived in Seattle with the World's Fair in 1962 and have been in Seattle ever since.

At that time, the department was small. I think I was the sixth or seventh member of the department. We have about 20 now. I started out teaching the beginning geology class. I had taught both physical and historical geology classes at Yale before I left, and so it was fairly easy for me. It also was useful, because I've written several textbooks on physical geology during the last decade, and I'd had that valuable teaching experience. I also organized several other classes, primarily in geomorphology and glacial and Quaternary geology. I joined the University of Washington faculty together with two other young faculty members, whose fields were geochemistry and economic geology. We drafted an NSF proposal with the hope of getting some badly needed equipment and funding for research. It was the first NSF proposal ever submitted by the department, and it was funded! It came to more than $100,000, which was big money in those days. In the proposal, I had requested support for several years of research in the Cascades because little was known about the glacial history of the range. Here I was in Seattle with some great research opportunities in the Olympics and the Cascades, and few others were working in these areas. I began work in the southern North Cascades, where I mapped the glacial geology. Over the last 30 years, my graduate students and I have mapped a substantial part of the Washington Cascades.

These are the places I've worked during my career [Steve hands me a map showing his field sites all over the world].

Thus far, you've only talked about North America, but I think of you as having worked on all of the continents, except maybe Africa.

No, I haven't worked in Africa, but I may do some field research there next summer. The first non–North American research I did was in 1964 when I worked in the Marshall Islands. I was a member of an Atomic Energy Commission expedition. Some of the H-bomb test fallout landed on Rongelap Atoll, and so the AEC team was going back to see if it was safe enough for the native islanders to live there again. We spent several weeks at Rongelap, during which time I mapped the geology of the islets that surround the lagoon. I was looking for evidence of Holocene sea-level fluctuations but found scant evidence there.

At some point you went to Pakistan.

This was due in part to my association with an archaeologist, Walter Fairservice, director of the Burke Museum at the University of Washington. For many years he had worked in Pakistan on early cultural history. He was organizing a party to be involved with archaeological excavations at Timagura, northwest of Karachi, in the summer of 1966. Anne and I were invited to join him. I was interested in doing some paleoclimatic research in the adjacent mountains, a topic also of interest to Fairservice. We drove north in a Land Rover from Karachi to Quetta. From there we crossed into Afghanistan, passed through Kandahar, and traveled north past Ghazni to Kabul. From Kabul we drove over the Khyber Pass to Peshawar. My intent was to study some of the glaciated valleys in northern Pakistan. That summer we made a reconnaissance of Swat, Indus Kohistan, and Chitral. I considered Swat to have the best glacial record, and it was easy to get to by car. However, there were a lot of armed brigands in the upper valley who posed a danger. I went to see the secretary of the Wali, the local ruler, and explained why we wanted to work in the upper valley. He said there was no problem and gave us permission. Later that summer, we flew to India and spent some days in the Vale of Kashmir, walking into the Pir Pinjal and the Himalaya. We reached some glaciers in the higher valleys that I thought would be worth studying. However, such a study would amount to a different project because Kashmir lay in a different country, where as yet I had no close contacts.

Did the glaciers there experience recent retreat like those in the Alps?

Yes, the glaciers have been retreating. The older glacial record is not yet well dated. I would like to return there, but in recent years it has not been safe because of armed conflict along the India-Pakistan border, and it is an area where foreigners have been kidnapped and murdered.

I know you've worked for long periods of time in the Italian Alps, Hawaii, New Zealand, Chile, China, and in the Himalaya. Why, from your perspective, was it important for you to work in all of these places?

Until rather recently, it was not known whether the Northern and Southern hemispheres responded to major climate changes during the glacial ages in the same way.

Because of the unequal hemispheric distribution of solar energy, which results from the Earth's axial tilt and irregular pattern of its orbit over tens of millennia, Northern and Southern hemisphere glaciations should be out of phase. Yet, data now suggest that the hemisphere climates were in phase. When I began my research, the evidence was insufficient to show whether they were in phase or not.

In 1973, I took a sabbatical leave in New Zealand. I did fieldwork in the southern Alps near Mt. Cook, where I studied the glacial chronology and made snowline reconstructions for several major glacier advances. I also spent several weeks in the Antarctic dry valleys working by helicopter out of Scott Base, the New Zealand facility. My work in New Zealand, Argentina, and Chile was concerned with working on Southern Hemisphere glacial records and dating them as well as possible. After working in the Chilean and Argentine Andes for several field seasons, I shifted my focus to several other areas. One that really intrigued me was the Hawaiian Islands. Hawaii has the only high land area in the vast North Pacific Ocean basin, an area larger than all of North America. Furthermore, it had been noted as early as 1912 that the top of Mauna Kea volcano was glaciated. In the 1930s, several geologists claimed there was evidence of several glaciations, but in those years there was no good way of dating glacial deposits. By the 1970s, there were two interpretations: one, that the mountain had been glaciated only once, and the other, that the mountain had been glaciated four times. These conflicting interpretations needed to be resolved.

In the late 1970s, I spent three years in the Italian Alps working on Holocene glacier variations with Giuseppe Orombelli, a friend and colleague from Milano. All of the Porters were in the Alps each of three summers. We lived in Planpincieux, a small village up valley from Courmayeur. Although the focus of our study was glacial deposits, we quickly learned that many of the deposits that had previously been interpreted as moraines actually were deposits of large rockfalls. Here was a case where the rockfall deposits closely resembled glacial deposits. Our study showed that many of the small villages and larger towns were in potential rockfall zones, and so we prepared a rockfall hazard map outlining the most dangerous areas.

One of the challenges in my field is recognizing glacial deposits for what they are. Some people have reported glacial deposits that have turned out to be nonglacial in origin. They may be mudflow or landslide deposits, for example. Although in some cases the misinterpretations are related to a lack of experience, in fact, the interpretation is not always easy. When I worked in a glaciated valley in northern Pakistan, for instance, I identified till [a glacial sediment] as well as four or five nonglacial till-like deposits that might be mistaken for till. If misinterpreted, the wrong story results. However, with experience, one might conclude, this looks like till, but it can't be till because it makes no sense with regard to the overall distribution of glacial deposits in the area.

Fieldwork is not just a matter of putting on paper what you see in the field. Accurate recording of what is seen is certainly important. However, it is also necessary to make sense of what is seen in the field. One of the most satisfying experiences I get from fieldwork comes from integrating data obtained through mapping and stratigraphic work and then anticipating what I will see over the next hill or around the other side of the mountain. If I don't find what I anticipated, it means my reasoning and deduction are faulty or some of the crucial evidence is missing. Basically, I am testing hypotheses. One generates hypotheses while doing fieldwork and also tests them. Without some experience, insight, and knowledge, one cannot ask the right questions and anticipate the most likely answers.

It has been a gratifying experience working in China for the past dozen years. One of my first projects in China was to map the terraces in a drainage basin near Xi'an. In this study, I showed that the loess-paleosol stratigraphy, which the Chinese have worked out in great detail, can be used to date the history of stream terracing. It became apparent that the main intervals of alluvial deposition occurred during glacial times, and intervals of stream dissection occurred in interglacial times. There is a complete record extending back nearly 700,000 years. The Earth-orbital variations that control global climate also influenced the cycles of stream activity in this nonglaciated drainage basin, just as they controlled the timing of the primary glacial cycles in glaciated drainage basins.

The study of climate change is very complex. I've witnessed many of the major advances, interacted with a number of the major players, and made a few modest contributions myself. Yet there is still a great deal we don't know about climate change. There is no way to predict with confidence what the climate will be a hundred or a thousand years from now. We don't understand all of the mechanisms involved in climate change, and that is where a good deal of research is moving now. We know that the timing of the ice ages is related to the orbital variations of the Earth. We can demonstrate this for the past and can project the changes into the future. We can say with some certainty that about 20,000 years from now the Earth will experience another glaciation. However, we can't say with any confidence what the climate will be during the next few hundred or few thousand years. This is because we do not yet completely understand how the climate system works on shorter time scales.

Here is a logarithmic time scale from one year to a million years in the past, and these are the frequencies at which cycles have been identified. Some are related to variations in the Earth's orbital pattern. The primary ice-age signals occur at frequencies of about 100,000, 41,000, and 20,000 years. We know far less about cycles of shorter duration. Here is a 1,500-year cycle that we know little about. Some have suggested that it is related to solar variations, but at this point that's just a guess. Here is a 2,500-year cycle that appears in many records. Its origin also is not clearly understood. Volcanic events are also important at the scale of a year to five years. However, volcanic eruptions are random. When several volcanic eruptions occur almost simultaneously, then a large climatic perturbation could result. Because eruptions remain unpredictable, we cannot anticipate future climatic changes related to volcanic events.

I have greater interaction with nongeologists in some of my work than I do with geologists. For example, I have worked with zoologists, paleobotanists, anthropologists, oceanographers, and atmospheric scientists. For instance, I try to keep up with coral reef research as it relates to paleoclimatology; I try to keep abreast of archaeological studies in Asia and South America concerning early human populations; I keep up with climate-modeling studies concerning atmospheric circulation during the ice ages; and I especially keep abreast of what oceanographers are learning about the role of the oceans in global climate change. In my area of research, one must be alert to scientific advances in many different fields. Otherwise, it is easy to miss important evidence and fresh ideas that can help me understand and correctly interpret my own work.

You couldn't have predicted back in the early 1960s that the field you were interested in, which you have described as being multidisciplinary, would be such an important field today?

No, I wouldn't have predicted it.

But you persisted in it, because you liked it.

Oh, I find it fascinating, and the evidence is global. I have been able to do research in many different areas. To understand climate change, we must have a global and interdisciplinary perspective. Unlike the theory of evolution, which is logical but took a long time to be accepted, the theory of plate tectonics took a leap of imagination, but the evidence is convincing, and it makes so many otherwise confusing things fall into place. A theory of climate change is much more difficult to formulate because it involves the entire planetary system: the solid Earth, the atmosphere, the oceans, and life. Although we are confident about our understanding of some aspects of the climate system and the history of the Earth's climate, there remain important gaps in our understanding. I think this exciting research will continue well into the future. There will be many opportunities for bright people to make important contributions.

If you are going to do science somewhere on this planet, I think the United States is by far the best place for a young person, especially for research.

Daniel Jean Stanley (b. 1934)

Smithsonian Senior Oceanographer
National Museum of Natural History
Interviewed in his office 3 November 1997

One could characterize Daniel Stanley's career pathway as full of unexpected twists and turns. It is clear, however, that it was a conscious decision on his part to conduct his science in the United States. Stanley works on sediments deposited in the marine environment, combining studies in ancient and modern settings, particularly in the Mediterranean. His interest in deltas throughout the world has enabled him to address questions in climate change. Also, paralleling PORTER's experience, these studies are presently leading him to questions that combine knowledge of human evolution, human archaeology, and geology.

My earliest memories go back to 1938. I am going to the Vosges and walking up and down those lush, gentle hills. But the hiking didn't get really serious until 1940 and 1941, when I was seven years of age. I have an aunt, a great, lively outdoor person. She's still alive and lives in the Alps. She took me for walks outside of a town called Le Puy in the volcanic central part of France. Although those times were bad times in Europe, I didn't see it quite that way. Sometimes we couldn't go to school, so we were taught in the fields.

About nature?

No, math and literature, history, and things like that, readings about our ancestors, the Gauls. Then, coming to the States, I became a Boy Scout. I think that played a real important role, because my mother was raising me alone at that time. I think that was the male extrovert part and being surrounded by other kids. I was a lonely type and not particularly socially adept. I wasn't into bad behavior or anything but was sort of a loner, shy, and timid

When I was in high school, I knew that I was attracted by science, but I assure you, Betsy, that I wasn't that good in science. I certainly had more difficulty with physics and math than I did with biology. Fortunately, we weren't raised at a time when adults would frequently ask, "Where are you going to go to college?" It wasn't anything discussed at home.

There were no expectations that you had to do this or that?

No, at least, not in the direction of science, none. I started thinking that I would like to go to college, and people at school were pushing me. I won't say pushing. They were encouraging me. I very much liked a history teacher, Mr. Fagan. He made me think about the Mediterranean. He made it alive. We had an excellent Latin teacher, oh, she was great. We could dress in togas.

We could too!

Three years of that! I must have been intrigued by the Med, because, of course, that's where I ended up.

Then I thought, well, if one does go to school, some fields have financial limitations, and one does have to eat. During the war, my family had lost their holdings. My parents were divorced by then. I suppose I would have been the natural inheritor of the company in France, but it just wasn't something I was being pushed into doing, and I don't think it would have really appealed to me anyway. I didn't visit colleges. I don't think it was as common to do that as it is now, or as I've done with my children. While in high school, I didn't know exactly what people in natural sciences did, but I registered in the Cornell ag[riculture] school. That was in 1951, when I had just turned 17. I liked the courses. I took an ornithology course taught by a chap named Arthur A. Allen, and his sidekick was a man named [Peter Paul] Kellogg, and they used to listen to bird songs.

Very famous.

Very famous guys. So I decided I wanted to be an ornithologist.

Were they good teachers?

They were. They were good teachers, but I don't think they were especially interested in a freshman like me.

Then an exciting thing happened. I contacted my father, who was then working for GE in Venezuela, and he said to come on down. In 1952, I took a banana boat, the cheapest way possible, and I ended up spending the summer down there. There was a very famous park in northern Venezuela, written about by a chap named [William] Beebe, the famous North American Beebe of natural history fame. He went down in a bathyscape and he went up in a balloon; that was the sort of thing he did in the 1930s and 1940s, I guess. I liked his book a lot. When I got down there, the director said I could work there. It was called Rancho Grande.

We've gone birdwatching there four or five times.

Portuello Pass. It was the escape holding of a previous dictator. It was nice, in the middle of the jungle, a good hideaway. The guy who was running it was an ex-Nazi named Schaeffer who was on the run. I know it sounds very strange, but during the war he had a tie with the ex-king of Belgium, Leopold. Of course, after the war, these royals were on the outs. The ex-king had, as a hobby, natural history, so Schaeffer had invited Leopold to Rancho Grande. My job was to be a gun bearer and bird collector. They had me stuff birds, cut them up, and put cotton in them. I had taken this ornithology course at Cornell the previous year, so I knew how to do that sort of stuff. I shot down, by pure accident, a new species of parrot, in fact, not far from where you were, Portuello Pass. This was in 1952; for an 18-year-old, all of this was pretty exciting.

So, by then, I really liked the field of natural science. Things were getting sort of serious at Cornell. But I wasn't really convinced that you could make a living doing that: looking at birds and butterflies. I mean it's sort of nice, but in my view, it was sort of like what children of a well-to-do family might do. I thought I should go into a more professional field, something serious, so I thought about medical school. Since you know Cornell well, you can imagine me sort of naively running around there. I had started taking courses at the ag school, which didn't cost me anything because I was a New York state resident. Now I added courses in arts and sciences, running from the ag campus down to the arts quad, in under ten minutes, huff, huff, huff.

Probably down to Stimpson Hall.

Right, to the zoology department, and I took the courses in comparative anatomy, histology, embryology, and all this neat stuff.

Probably you had Perry Gilbert for Comp. Anat.

Who used to lecture and at the same time draw on the board using both hands? Oh, it was dreadful! Who could keep up? My grades were OK, in fact, much better than in high school. I started doing very, very well. Then during my third year, I felt like I was being entrapped into something I still didn't want to do. Because then I felt, oh my God, here you have sick people, and if you were a doctor you'd probably have to look at sick people—it's the sort of thing they do—and worse, be in a hospital.

I wrote to my father and said, "I think I'll drop out of school." Well, that was not done in those years. I'll tell you, you didn't drop out of school in the early 1950s! I said to my father, "Something's wrong, and I don't know what to do."

You really felt that psychologically you were getting trapped in it?

I felt like I was getting trapped, and I didn't know how to get out of it. I was doing well, but I didn't like it. He very wisely said, "Take an aptitude test." I did that, but I didn't really like the results. I felt a little bit insulted because the results said I should be a landscape architect. Now, it's turned out—we jump many, many years later—I have a garden, I love my garden, and probably that test was on the mark! I had told the dean that I was going to leave. He said, "Now, young man, you don't just leave school, what is going on? I'll give you one year, young man, or your career is over here." You didn't go around "finding yourself" like you do in the 1990s.

But it was very good that I left because here's what happened. In 1953, I went to France with nothing, virtually. My father and mother said, "We'll give you $50 a month." So I arrived there, and I taught English to a family. I went to the Museum d'Histoire Naturelle. I was still interested in natural history. I improved my French writing, and I read Sartre in cafes. Then the thing clicked! I went and joined my aunt, the one who had taken me during the war years on those hikes in the Alps. She lives in the Savoie. I just love it, just that air, and it's so beautiful. There was a family reunion high in the Alps. My uncle's brother showed up. At that time, I was 19 or 20; he seemed old to me. He must have been all of 40, or maybe 50! He was dressed with golf-type mountain pants with high knees, and he was a geologist. I asked if I could go with him for the day. There he was. He took these loping walks over the crests of the mountains, smoking his pipe and talking about the Urgonian. I thought, he's paid to do this. Walk around and make these colorful maps and drawings. I said, "That's what I want to be."

I came back to Cornell and in my senior year, in 1956, I continued to take pre-med courses like cytology, but I also took geology I, mineralogy, and stratigraphy, a complete mix. I applied to Cornell Medical School, and I was accepted. At the same time, I met a professor, Bill Oliver, on a New York state geological field trip in early 1956, and I didn't know beans about what I was looking at, because it was my first year in geology. Oliver asked me what I was going to do, and I said, "I want to go into geology, that's what I think I'd like to do." He said, "Why don't you come to Brown? That's where I teach." But in the meantime, I had to finish at Cornell, and that meant that I had to spend a summer on a farm to answer the Cornell agriculture requirement. I have a degree in agriculture from Cornell, but really mostly natural sciences with pre-med as the main thing.

So you did spend a summer on a farm?

I had to do that! I knew before I even went to the farm that I didn't want to be a farmer, and of course, this confirmed it. There were no tractors. I had to get up at five every morning and brush down horses. By 6 o'clock at night, I was so tired, I was ready to go to sleep. Thank goodness, the summer went by very fast.

In 1956, I went to Brown, and I was making up in the master's program the elementary undergraduate courses. I asked permission from Professor A. Ronzo Quinn, who was then the only full-time geologist on the Brown faculty, to do my practical geology experience in Venezuela. My father put me in touch with the Esso Company, which was called *Creole* down there. They took me on as a volunteer. They'd pay my food and provide a tent, and I'd go into the field with two company geologists. They sent us to the jungle, near Aroa and Barquisimeto, in southwest Venezuela, just at the foothills of the Andes. So there we were, Brunton compass and pace mapping. In other words, you have to know how many steps you were taking, in what direction you were taking them, and record all of this. At the end of the day, after returning to base camp, you would plot where you started and where you had gone all day, using aerial photographs of the canopy. Then you could just sort of get a feeling from the photograph of where a small stream bed or rock exposure might have been. That's all you'd see, a break in the canopy. There were no maps, per se. So that gave me a real feeling for what geology is and what geologists do. They don't just sit on the Alps. That was pretty exciting.

I graduated from Brown in 1958, and I got accepted by Columbia to work with Marshall Kay, who was a very well known stratigrapher, and by the paleontological group at Harvard. Once again, I ended up accepting neither fellowship and going to

France with nothing. It just was calling for me. I guess I kept thinking of the Alps and my uncle's brother smoking his pipe on a mountain crest. An uncle put me in touch with the French Petroleum Institute, the IFP. They felt that, with a master's degree, I could qualify to enter this high-level research center and work on a thesis. France had recently discovered a major petroleum find in Algeria, which was still part of France. Until then, French petroleum companies emphasized carbonates, and no one at IFP was doing any major work on sandstones in France itself. So a director of the institute, a Madame Yvonne Gubler, said, "OK, Stanley, we'll put you on the problem of the Annot sandstones located in southeastern France. Are you willing to study flysch?" Well, yes, I would like that, but remember, I had no money. I looked up what flysch was. It's a catchall term, a German-Swiss name for formations of alternating sandstones and shales.

Indicating differing depositional environments? Pulses?

Yes, that's it, sedimentation-depositional pulses. I started taking courses at IFP. Madame Gubler, a very dynamic and strong person, another life marker, like my aunt when I was a child, became my thesis advisor. She would ably lead 20 or more male graduate students on a 2,500-meter mountain crest, keeping us well in line, as well as teaching us the basics of Alpine field geology. She was just terrific.

So, in the following May, I was in the field, on the Annot sandstone. Wonderful thesis area, in Provence, north of Nice. I worked on problems of turbidites. What I came up with, after three years, is that it was a deep-sea environment in the late Eocene. So I changed the whole concept of the development of eastern France because before, they always thought that the deposit was a shallow-sea marker. The reasoning was that it was coarse grained, coarse sediment, therefore close to shore. Well, I had trouble with my thesis defense. It took three hours to defend, but it was a tremendous learning experience. I graduated in 1961 with a docteur-es-science.

With a doctorate, I thought of going back to the petroleum industry because I had liked my experience in Venezuela. I went with Pan American Petroleum in Fort Worth, Texas. I had just started when I got called by the army. At Cornell, while an undergraduate, I had taken ROTC. I thought that I had gotten out of the woods and out of active service. I was in the reserves. But remember in 1961, they put up a wall in Berlin. Somebody found me, I don't know how. I was called up and sent to the Corps of Engineers at Fort Belvoire, Virginia. I had to go to basic officer training again: running, doing push-ups and other exercises, and camping. It seemed a little strange. I was tired of doing this stuff. It didn't seem like it was adding up to very much. So I went to near what is now Washington's National Airport, where the Army Corps of Engineers had its major headquarters. I asked, "I'm sort of looking for career possibilities in the corps when I get out of basic school." The officer said, "Well, there is digging out mines, putting pontoons across rivers, building bridges." I said, "Those are nice and so is blowing up bridges, but I wonder if they might have something more suitable. I've been reading this stuff by Fisk, these corps reports, and it looks like the Corps of Engineers studies the Mississippi, the delta, and that looks like something interesting." At the end of basic, they post your assignments. I was named special assistant to the director of the Waterways Experiment Station, a research facility in Vicksburg, Mississippi. What a plum! So I learned about the Mississippi Delta, did work in the field there.

So that was your introduction to deltas?

Yes.

You had dealt with sediments before, but the ones that had turned into rock?

Ancient, that's right, into rock, and well into the sea, no deltas. Now, I'm finishing up my military service, a postdoc had fallen through, and I had nothing in hand. At the last minute, a position opened up at Dalhousie University in Nova Scotia. So I looked on the map. OK, where the hell is this? I had no choice. I needed a job, and it was geological oceanography. So I went there, never having taken a course in oceanography. I team taught marine sediments with someone who has become a lifelong colleague, Donald Swift. We taught all aspects of marine sediments. Swift was nearshore and shelf, and I was deeper marine environments. I was applying what I had learned from the rock record to the modern record. To get oceanographic shiptime, they said, "OK, Stanley, we'll give you and your students a ship. We'll give it to you January, February, and March every year." Well, I don't know if you've ever gone out off the coast of Nova Scotia in January, February, March.

Cold?

Worse than that. You can put up with the cold, but there are waves out there! It was just awful. The ship swung madly back and forth while you worked, cored, surveyed, and sampled. We were mapping the continental shelf and slope and worked on some deep-sea canyons off the Nova Scotia coast. So it was a great hands-on field experience.

I realized somewhere along the way that I had to come to the United States to work. What am I doing in Canada? If you are going to come to the New World, my view of it—and I think really it has been reinforced over the last 30–35 years—if you are going to do science somewhere on this planet, I think the United States is by far the best place for a young person, especially for research. It's harsh; it's not necessarily a fun place to be a scientist. It can be difficult, but it's competitive, and it's a good place to keep improving, honing your talent, growing. There are resources, and intellectually, it builds a very good pyramid of people, so that you can find your niche on this dynamic pyramid. Scientists do not usually talk about it, but it exists. In spite of the so-called egalitarian phase we are experiencing, the hard-nose science pyramid is there, and it leads to powerful science. It may not always be equitable, but it is a good thing for the United States. I've never seen it overly written up. So, I wanted to get back to the States. Dalhousie was a very good school, and the graduate students were good also. And the ships. Even though it was January, February, and March. I mean, I was lucky to have them, I now realize. It was a good place for a starting experience.

In the meantime, I would like to show how important human contact is. It is most critical in our lives. Ken Emery, whose book *Sea of California* had been an inspiration to me, had moved from the University of Southern California to Woods Hole Oceanographic Institution. He was a science hero of mine. I knew him through his prolific publications, and I met him once in 1962 at GSA [Geological Society of America], one of the first big geology meetings that I attended in America. I had asked him if I could work with him. He looked at me and asked, "How old are you?" I said, "I'm 28," and he said, "You're too old." I said, "I'll do whatever you want, so that I can learn marine sedimentology." He sort of liked me, I guess, because he said, "OK, you can be my postdoc." I spent the summers of 1963 and 1964 at Woods Hole. I worked with his USGS group, then mapping the continental margin off the East Coast of the United States. I published a number of things with him. I was very impressed by the way he worked. He had several statements he repeated, although I'm not sure that he would agree that he had said them. For example, I might say, "Dr. Emery, I had an idea to map those rocks on Highway 128 around Boston on the weekend." "Well, Stanley, if you don't do it, somebody else will!" He would say things like that, and I'd think, I'd better be moving my tail. He'd say, "Give that job to the busiest of the scientists" and "Don't forget, time is moving on, you know!" I'd think, jeez, I've really been wasting my life. Here three years have passed since I've arrived from France, and I've really done nothing.

So, with Emery's keen drive and enthusiasm as a guide, I knew that I really wanted to get back to the States. I could have worked with Ken at Woods Hole. I was still sort of impressed by an academic life, even though I had begun to realize the limitations of universities for me. I mean they'd be OK if you didn't have (1) the students, (2) some obnoxious staff, and (3) the teaching. I was becoming less naive. I called around. I wrote a letter here [to the Smithsonian]. That original professor from Brown, Bill Oliver, had ended up here in Washington, D.C., with the U.S. Geological Survey, but he was based at the Smithsonian Natural History Museum [NHM]. Another professor from Brown, Dick Cifetli was here too, and they responded, "We don't know of any jobs in the States, except for one here at the NHM. Do you want to come here?"

I had the feeling that guys who come to museums are basically weird, introverts who putz around in lab coats in dank corridors filled with fossils. I didn't really know what they did in a museum, but it didn't sound very dramatic. But they encouraged me to come down and take a look. So I went down and gave a lecture on turbidites and marine sediments. Then a sedimentologist who was already on the NHM staff, Jack Pierce, showed me around. There was this whole lab downstairs that was empty, and I was told that I could do whatever I wanted, it was my show! It sounded too good to be true, but I took it. With my young wife and three babies, I moved to D.C. and I started here at the Smithsonian the first of October 1966. I completed some Nova Scotian continental margin studies, because I like to finish projects I start, but I was anxious to go back to the Mediterranean. I started working on the rotation of Corsica, an outgrowth of my flysch–Annot sandstone thesis work.

You must keep in mind, this was the 1960s, a different America, where oceanography meant something, and you could easily obtain funding for research and space on cruises. As I'm hearing myself, it sounds a bit negative, but as you know now, it is much more difficult to get an expedition underway. Then, it was ideal for many of us because the U.S. Navy was really neurotic about Russian submarines, and I was interested in the deep sea and the slopes and canyons, ideal places for submarines to hide. So we had a series of cruises off the eastern United States and in the Mediterranean. Even the submersible *Alvin* was used.

The more I learned, the more I realized that there was a lot to do in my field. Remember, I'm not a geophysicist, a deep crustal person, but into sediments. So I got together with a bunch of people in 1970, and we got our first book together on the sedimentology of the Med, published in 1972, *The Mediterranean Sea: A Natural Sediment Laboratory*. Another eight books and several hundred articles have followed. Cruises continued during the 1970s, and at that same period, I continued to go to the Alps, connecting the two: the modern deep sea and the ancient rock record. I also started working on the Paleozoic sedimentology of St. Croix. That's where we met. Somewhere in the late 1970's, early 1980s, I was working in the eastern Med, the Hellenic area, Greece, where it is very, very tectonically active. One thing led to another and I got some money, some Egyptian pounds from the U.S. State Department. You had to use it in Egypt. It was a very exciting time. Sadat had just gone to Jerusalum. I went to see the Aswan high dam. It was pleasant, but nothing ever came of it at least on land. Instead, I started working offshore, off Egypt and the Nile, which is easier in some respects than working on land.

I kept going to the field all the time in the maritime Alps, and it had been two decades since my thesis work. I was getting a little tired because it was beginning to be repetitious. Periodically, I need a change of pace, a shot in the arm. Some of my Smithsonian colleagues, who are very good in their fields, have a stick-to-it-iveness. They wanted to be a doctor of fossils since they were six, and they have stuck with it. But, after a while, I like the injection of a new problem. Usually when I get tired and bored, I start down a new road. I usually do not know where I am going, but I strike out, trying one thing or another, and soon things begin to evolve and move ahead. Sometimes, things just happen.

For example, a man came to this office door one day in 1981. I didn't know who he was. A man in a green loden coat and an Austrian Tyrolean hat, a veritable munchkin, who asked, "Are you Dr. Stanley?" and I said, "I am." Then he talked to me about the biblical Exodus, and frankly, I was getting a little nervous. I said, "Oh I think you want the archaeologists, they are up in . . ." and, you know, there was a Stanley in archaeology. But my visitor, who turned out to be Hans Goedicke, chairman of Johns Hopkins University's Near East department, said, "No. You are the one I want." For two years he said, "We've got to go to Egypt." So finally, I accept, and we go to Egypt. He's an Egyptologist and he's an epigrapher, in other words, reading and interpreting hieroglyphics. But, he had a bee in his bonnet about the biblical Exodus, that there was a geological explanation for the plagues in Exodus. I had been working off the Hellenic Arc, and I had looked at the volcano Santorini, and it was that way he got my name.

So his idea that the plagues had a geological explanation—a blast of Santorini, the biggest blast between the second and first millennium B.C., something like 13–18 cubic kilometers spewed up. It makes St. Helens look like a little popgun, and don't forget, there are only 800 kilometers between Santorini and Egypt's Nile. So, for two years, I had put off going to Egypt since I had gotten involved in a new book on the Mediterranean that tackled problems of the crust, geological history, and evolution of this sea. I felt it was my duty to finish this volume before striking toward the Egyptian coast. So 1983 comes around, and Hans Goedicke and I finally begin looking into the Nile Delta, where he thinks there may be some evidence of a tsunami. While there, I realize this is a magnificent place, and nobody has ever done a systematic study of this famous delta. My thoughts are, what the hell am I doing that is any more interesting than looking for the biblical Exodus, on the one hand, and beginning a long-term survey of the Nile Delta? So, in 1984 and 1985, I got funding and people together. It was still decent timing: the oil companies were moderately generous as it was just before the petroleum crash. Sure, here is some funding, get off our backs, but here is $30,000 for the work in Egypt. It was enough to get started on what turned out to be a very large project. "Fools rush in where angels fear . . . !"

We start doing drilling in the delta, and I realize we have never done any of this before. I'm doing something new, totally new, but that was fun! I had never seen anybody drill. You need to understand that the delta surface is flat, and you aren't going to see anything on the surface. So we did drilling from one end to the other of the delta plain, from 1985 to 1990, leasing gear from a foundation engineering firm, civil engineers. I explained to them what I needed, and they modified it. We worked really well together. Here is the picture of our drilling crew—talk about fun and productive fieldwork!

We had five expeditions; we took 100 cores. That's a lot of work, down 50–60 meters' recovery, then working this up, analyzing all the cores from east to west. Then I got interested in the even more modern lagoon-marsh wetland systems. Obviously you've got something that's completely changing at the present. You've got a delta; it's been modified by the Aswan high dam. You've got 60 million people, one million new ones every year, creating a human density 500–1,000 per square kilometer. This is something important, the anthropogenic part as well as the natural factors that formed the delta in the first place. By the early 1980s, I became involved in sea-level change, tectonics, structural integrity of the delta, climate change, processes on land, the coastal processes offshore, and how what's happening in Egypt is affecting Gaza, Sinai, and Israel. So, since that time, the middle to late 1980s, I've continued eastward along Sinai and off to Israel, along the entire Israeli coast, and I've also gotten a better handle on the Nile itself. We've also had excursions down the Nile to northern Sudan, sampling every wadi that enters the Nile Valley from Egypt's southern border to the mouth of the delta.

By the way, we *did* find evidence of Santorini ash in Egypt, which was a most exciting discovery, I think. It doesn't prove or disprove the Exodus or the plagues, but at least it demonstrated that ash got there.

And the timing was right?

Well, I don't know who was the pharaoh then. It was perhaps a woman, maybe the times of Hatshepsut. We've dated the event, and the ash reached Egypt around 1500

B.C. This is earlier than the reign of the pharaoh usually mentioned, Ramses II. So as you can see, with this interest in the Middle East, I've become interested in the archaeological side, in geoarchaeology. You can't help but see all of these things and putting together two plus two. We explored the chronology of the Nile Delta. It began 8–7,000 years ago.

Are we talking sea-level change?

Yes, in large part. I made a survey with one of my postdocs, Drew Warne, of every dated delta in the world. We discovered that all had begun at the same time all over the world.

I'm thinking these are the same dates as Holocene reefs all over the world.

Yes, at about that time. Deltas start at the time of deceleration, or slow down, of sea-level rise. It doesn't matter if the deltas are up in the Arctic or in the tropics, in tectonically active or calm and peaceful areas of the world, thus our study extends farther geographically than reef areas. Reefs are in the low latitudes; so are deltas, but they are also in the middle and high latitudes. Having determined the initiation of so many deltas at 8–7,000 years ago, we published the discovery in *Science*.

Although I'm older, 63, I still have some vitality left as far as trips and work in the field. In view of what I've heard myself say, I would be hard put to say where all of this will end up. As one gets older, there is a limited energy, I suppose, but I think the archaeological part is very appealing to me.

It brings you back to what you said was one of your big interests in high school, or big inspirations.

Yes, the history teacher. I never took history or archaeology at Cornell.

But archaeology certainly is a story of history, humanity's history.

And I like the juncture with geology. It's helpful both ways, to provide something to the archaeologist and vice versa. Theirs is a very precise past time record, which we geologists don't have but can use. To us, their information is so temporally precise. Interestingly, we have on a submerged village off of Israel, a prepottery Neolithic village that's also 8–7,000 years old, at a depth of 8–10 meters. This fits the sea-level curve well, and in one respect, the archaeologist is confused: they find that 95% of the fish are *Balistes*. Well, *Balistes* still lives in the Med, but it certainly doesn't account for 95% of a catch, or anywhere near it. It ties in with a hypothesis that I had some years ago of a major change in physical oceanography, in the circulation of the Mediterranean at about 8,000 years ago. I had water masses flowing completely differently than today. Perhaps that change might explain the altered fauna at the site today. Also, the clays that I've identified show a completely different clay mineral assemblage at the submerged site. Look, most new archaeological projects can benefit from geological input.

And I can see myself, as an aging geezer, I won't always be able to go in the Alps. I went this year. I found the mountain slopes seemed higher and steeper. I will travel less. Today I saw one of my elderly colleagues, a rather charming man, really he's 85 or 90, he can barely walk, he was coming in to work at his office, and I'm in admiration of him. Probably, I will use the Dibner Library, the old book library, and with my knowledge of languages I will find something to translate. So, when I can't walk and I can't go into the field any more, I'll do my best to exercise my brain.

To conclude, I've been really lucky. But [my life is] probably not really accidental, since what we do is, in so many ways, based on the influence of others early in our lives. I acknowledge a nurturing family start, with excellent mentors along the way, and the opportunity to work in one of the world's best organizations, the Smithsonian. Each encouraged a lifelong disciplined career involving field observations in sedimentary geology, oceanography, and now, archaeology. The Smithsonian has been unique in giving me freedom to follow my whims. I've been encouraged to integrate field observations with laboratory analyses and to write these up promptly. In an otherwise hectic world, I've been able to live the life of an eighteenth-century naturalist. What luxury! Looking back, the best part has been to have had ample time to integrate findings in several disciplines, to learn to be reflective, and especially, to allow room for serendipity—and to follow hunches.

It is very important to have people who turn people on.

Ian Gunn Macintyre (b. 1935)

Research Scientist, Department of Paleobiology
Smithsonian Institution, National Museum of Natural History
Interviewed in his office 3 November 1997

Ian Macintyre became a natural historian at an early age. Although he never had a formal course in biology, his observations on coral reef fauna and flora are far more perceptive than those of many reef ecologists I have encountered. Macintyre might today still be working for an oil company had he not been introduced to the modern carbonate environment by Bob Ginsburg. He has shown both flexibility and versatility in picking up the conceptual and field skills necessary to answer the questions that arise from his experiences in the field. His work through the years has spanned an impressive range of spatial and temporal scales.

First, I should maybe explain the complication of my background. I'm basically Scottish. I was born in Maricaibo, Venezuela. My father was in the oil business. We were in Britain briefly for a while and then, before the war broke out, he was offered a job running a small oil company, British Union Oil Company, in Barbados. This was an opportunity to get the family away. He had been in the First World War, and he knew there was a war coming, so he took that job. We left England in April 1939. We ended up being stranded in Barbados for the Second World War, surrounded by German submarines. It was a wonderful place to be stranded. I grew up in Barbados from about four 'til eleven years old. It was a magical time. I spent half my time in the trees and the other half in the water. I lived opposite the golf course, which gave me a big area to expand my interests. I collected, well, everything. Birds eggs is something that people don't talk about collecting any more. I never, of course, completely emptied the nest, and I kept a record of all these things. I collected butterflies. I collected seashells. When you're in a place like that, on an island as a child, you are just exposed to the natural environment and interested in gathering things like that.

My interests in geology were very strongly influenced by my father. He was a chemist, but he had working for him Dr. Alfred Senn, the geologist for the British Union Oil Company. I used to spend a lot of time in his lab watching him collecting foraminifera and putting them in vials. So I really never thought of being anything other than a geologist; I really wanted to be a geologist.

When you were just a little boy?

Yes. When I was eight or nine. I mean, this was the ultimate. Here I was collecting things in tins and boxes, and here he was with nice, neat little vials. He was the ultimate collector, as far as I was concerned.

After school in Scotland, I came to Queen's University in Kingston, Ontario, which actually has a close tie with Edinburgh University. I came to Canada because I thought the future of geology would be in North America. I didn't see the point of getting a degree in a country where you are not going to work. I almost went to MIT at one point, but I ended up at Queen's because of the Scottish connection. I did a B.Sc. in geological engineering and after that I worked for six and a half years with Shell Oil Company. But before I finished my bachelor's degree, I worked one summer mapping in Sudbury for the Ontario Department of Mines and then another summer working in western Canada for Shell. I think it was great. At that time, in engineering especially, I don't know about arts, but in engineering you were required to work in your summer in the field that you wanted to be in as an engineer.

With Shell Oil Company, I ended up thee years in Regina, Saskatchewan, in oil exploration and three and a half years in Calgary, Alberta. Part field, part office, because at Shell, you became an expert in a section, like the Mississipian or the Devonian section. I would be working in the office, looking at samples from different wells.

In the lab?

Yes, putting maps together. But I would go out when wells would be hitting my particular section. I would look at the samples as they were coming up and particularly, of course, look at the characteristics that might indicate that we might have an oil reservoir. I was the Mississippian expert in Regina, and then in Calgary I was the De-

Ian Macintyre (*center, with plaid shorts*) and a student assistant in the late 1970s, coring an ancient relict reef at a depth of 50 feet off St. Croix, Virgin Islands. Photo from the files of I. Macintyre.

vonian expert. I would produce facies maps. For instance, if land sections were available in a certain area, I could tell them if any reservoir rock could be found in that area. The seismic data would be a very important factor too—whether you saw a domal structure there that rose up—but also I could tell them if it was the type of rock and porosity. Porosity is all important, and permeability; if you don't have porosity and permeability, you don't have a reservoir.

This then leads to the next stage of my life. Bob Ginsburg was running Shell Development Company down in Miami at that time. He had a course called the Stage C course, where oil exploration geologists and the production people were sent down to look at modern reefs and oolitic deposits, because oolitic deposits and reefs are oil reservoirs in the fossil record. What Bob Ginsburg would do was show us the overall geology of modern reefs, facies, patterns, and oolitic banks, looking again at the sort of patterns that you would see on the surface, so that when we were drilling holes we would look for similar patterns in the fossil record. I just went crazy because I came from the wintertime in Calgary, and I came down to this nice, sunny Florida, and we were looking at reefs. It was at that point I thought, I'd rather be looking at modern reefs than looking at fossil reefs, and so things developed. Bob actually came up at one point, and I gave him a field trip to Devonian reefs in the Rocky Mountains. It was on that field trip in the very early 1960s that I mentioned I probably would much rather work on modern reefs. He was the one who said, "If you are going to do it, you better do it now." He was the catalyst that got me going.

I ended up leaving Shell in 1963 and going back to university at McGill. This is where Barbados comes into my life again, because at that time they had the Bellairs Lab run by John Lewis. It was all biology at that time, but they had the equipment to do geology: bottom cameras, grab samplers, and dredges. So I wrote a proposal, working with Eric Mountjoy of McGill, to do a project at Bellairs. That was an exciting project because that's when I discovered submerged reefs for the first time. We discovered relict reefs related to changes in sea level and a very interesting pattern. I was looking both at deep sediments and the transition up to shallow-water reef sediments and looking at the whole West Coast.

When that was finished, I was trying to think again, where are you going to have to make the future? In Canada it was going to be either British Columbia or the East Coast. Then I heard about Duke University Marine Lab. I'm not sure quite how I got hold of Orrin Pilkey, but somehow I did, and he gave me information on the shelf of the East Coast of the United States. There were some interesting features at the shelf break all along the East Coast, topographic features. I wrote a proposal to study the shelf edge of the East Coast all the way from North Carolina to Miami, and it was accepted. At that time NSF had an oceanographic program that had scholarships. NSF supported it, but it was a Duke Marine Lab program, and it was to use the research vessel *Eastward*.

The opportunities down there were just tremendous in those days. Lots of NSF money. I had as much boat time almost as I could use; I had travel money; I had lots of support. It was really fantastic! I studied the topographic features. Some of them turned out to be actually erosional features rather than constructional features, and some of them were reefs. There were also sediment studies on the shelf, and then, suddenly, they started to have a Caribbean program where you could take the *Eastward* down to the Caribbean. So I wrote another, follow-up proposal to study the submerged reefs of the eastern Caribbean, to see if there are similar features to what I found off Barbados, off other islands. I had several trips in the *Eastward* down to the Caribbean and found all sorts of evidence of relic reefs, including the ones on St. Croix.

On these trips, were you sometimes chief scientist?

Yes, but also I took the opportunity to go on one or two other trips. There were training trips, as well as research trips. So I went on a chemical oceanographic trip just to see what chemists do at sea. Dr. [Unnsteinn] Stefansson from Iceland used to come and take training trips, collecting data with the students. You always had a group of students. Even on our research cruises we had a group of students helping us. I think the scientific crew on *Eastward* was about 10 or 12.

And about half would be predoctoral students?

Or even undergraduates. Some of them learn very quickly; there were some of them who would plead with me to get off the boat, you know, seasick. I think some of them sorted it out very early that they never wanted to see a boat again.

It's quite interesting in that when I came here [the Smithsonian], I was actually returning from an interview at the University of Delaware, where my former boss at Shell, Chris Kraft, was offering me a job. It probably would have been more looking at oyster beds than reefs, and it had a very heavy teaching load. I came by here, and Porter Kier was chairman of paleobiology. He said, "Why don't you join us?" This was in 1970. They had a position in carbonate sedimentology, which had been closed, and they wanted to open it up again. In those days they looked for people they wanted and brought them in and then found a position to open. It wasn't like there was broad advertising and 200 people would apply. At that point, I was still a Canadian citizen. That's another thing, too, in the bad old days you could hire all sorts of people from around the world like Klaus Rutzler from Austria, and Marie-Helene Sachet from France, and Dave Pawson from New Zealand, but it was very important to do so.

And presumably you could mold a broad department with the specialties that you wanted?

Right. It is fairer today, opening it to more people, but you don't have control of staffing your department the way you'd want to do it.

One thing that happened, it still happens, is that when you come on at the Smithsonian, they usually give you some start-up funds for equipment, like a computer. In those days, I was very interested in looking inside reefs, to look at the history as to how reefs form. At that time, very little work had ever been done on that. Everyone was describing the surface morphology and surface communities knowing nothing about the previous record. So people like [Norman] Newell back in the 1970s were saying, reefs are just thin veneers, almost no accumulation. The relief was all inherited from the substrate they were growing on and that they were insignificant.

But there was no knowledge about the substrate they were growing on?

They would see a little Pleistocene exposed in places like the Bahamas, and it was understandable that they would think it was just veneers. My request for equipment was to put together a hydraulic drill. I got support for that. It was a diver-operated drill, as you know, and one of the people I met at that time was PETER GLYNN down at STRI [Smithsonian Tropical Research Institute]. Glynn is—I always kid him—a frustrated geologist, and I am a frustrated biologist, so we work well together. He encouraged me to come down to Panama to start doing some drilling down there. We went down and we worked both in the eastern Pacific and at Galeta Reef in the Caribbean. Galeta Reef was a big surprise because it looked like a very unimpressive reef, an algal-dominated surface with very few live corals. Then we drilled it and found that in the past it was a beautiful *Acropora palmata* reef! That's how I got into looking at the history of reefs and very much into scuba diving, of course, because this was all done with scuba gear. So we worked both on the Caribbean coast of Panama and the Pacific Coast of Panama.

In the late 1980s, after the El Niño, I was looking at reefs off Cocos Island and also off the Costa Rican coast, and that was presented as a paper at the Guam [International Coral Reef] meeting. What it really amounts to is that a few reefs are protected from the El Niño, behind islands. Some of them, like those in the Golfo Dulce in Costa Rica, are in a gulf and well protected from El Niño, but with changing river patterns and all the deforestation, they are now almost all dead, smothered with mud.

So they have a long history?

Yes, the one in Golfo Dulce goes back 5,000 years but now it is only 2% live, so they are caught between a rock and a hard place. They can be away from the oceanographic influence and then the terrestrial influence does them in, and if they are out in the open ocean, the water conditions there will pretty much wipe them out. So it's very difficult.

Then I got into a phase of looking at the mangroves, looking at peat. We found spectacular accumulations of over 30 feet of peat under some of these islands in Belize.

Can you date those?

Yes, very well with carbon. We can show that for some of these islands, peat has been accumulating for 7,000 years; in some of these islands, it is 10 meters or more thick.

After that I really got involved in more detailed work. Actually, it's the diagenic alteration of carbonate grains. I got involved in this with a postdoc from the University of Miami, PAM REID. At first, we looked at the mud that is in Chetumal Bay in northern Belize. It's all high magnesium calcite. Most of it is formed from the breakdown and alteration of sand grains. Everything is altering to high magnesium calcite in that bay. Even the oysters that are calcite are altering. Then it breaks down and gives you this mud.

What is turning it into mud grains?

It is bioerosion, endolithic microborers, and abrasion. We were very interested in this alteration. However, there is always a problem in our field. If you get a big name making a strong statement, it tends to be followed. Now Robin Bathhurst back in the 1970s said that the microcrystalline alteration of the sediment grains was related to microborers boring the surface of grains and then cement precipitating in the borings, and that it wasn't recrystallization. He called them micritic rims. The result was that a lot of the previous workers, people like [Edward] Purdy, were all sort of ignored.

What we found was that, in fact, Purdy was correct. A lot of carbonate sedimentologists always thought that grains are stable in the sea environment. It was not until you brought them out of the water and exposed them to freshwater conditions that you'd get the alterations. Well, now with our new techniques, with the microprobe and the SEM [scanning electron microscope], we can show clearly that grains we have now found with this detailed work are altering extensively.

You could probably show a nice time series.

Yes, but the thing that is really surprising is that the alteration starts on the organisms before they die! We have one paper out on *Halimeda,* for example. *Halimeda* originally forms skeletal needles. Those needles alter to micrite, microcrystalline aragonite. And then, they alter as you go down the stem of a *Halimeda*. It is a process that we think is related to the changing ratio of photosynthesis and respiration. In other words, at the tip of the plant, the ratio of photosynthesis to respiration is higher than when you get farther down to the older portion of the plant.

So, I would say, you are really not a frustrated biologist, you really are a biologist!

I have not had a course in biology. You know [Robert] Barnes's *Invertebrate Zoology* text?

Of course.

If you look in that, you'll see that I'm acknowledged for helping him. For a geologist who has never had any biology, I thought it was quite an honor.

I would never had guessed that you had never had any biology, because among the geologists I know, you are one of the most interested in biological problems.

It is funny that you say that, because I am just saying that about Peter Glynn, writing a letter on his behalf. It has always been exciting to go on a reef with Peter because the two of us are interacting all the time. We are always describing some little thing, whether it's a little mollusk that's feeding on the coral or whatever. Peter has a tremendous insight on the ecology of reefs, so when I go to symposium field trips I nearly always try to get him as my dive partner. I always kid him that he is a frustrated geologist because he is always interested in geological things, too. Peter is a wonderful individual as well as a great scientist, so it really is a pleasure just to work with him. If you work in reefs though, Betsy, you just have to be interested in all the different fields.

When I think back to the people who came to West Indies Lab, you stand out as one who really tried to coordinate people in a number of different disciplines looking at a problem, which to me is a most exciting aspect of doing science.

You know, it's a funny thing, Betsy, but here at the Smithsonian, of all places, in my last NHM peer review, I was criticized for this. To me, working on reefs is interacting with different people who understand what is going on. If you are interacting with those people, you recognize them, and you have coauthors. But one thing the peer reviewers were saying is that I'm not doing enough single-authored articles. The nature of my work is such that—sure, occasionally, I might do a review or something like that and that might be single-authored—for the most part I'm working with people: Pam Reid, Peter Glynn, Bob Steneck. We're interacting to try to understand a reef process. For me, it's idiotic to be criticized by my Smithsonian peers; it is like saying, ignore all the people who have helped you and just put your name on all the papers. I don't know what they're up to. I really lashed out at them on that. I won out.

It is a strange business all around. In some cases, good teachers have a problem in that they don't have enough research and they won't get tenure (and some very good researchers are very poor teachers). We have the same problem here in terms of people who are very good with helping at exhibits but who are not high-powered researchers. They tend to get slapped on the wrist because their research isn't up to par, but they are not fully recognized for their tremendous contributions to what the public sees, the exhibits. So it occurs all around in different areas. They are talking here about giving more recognition, but they don't. Fundamentally, it's your publication record, it's your research that is going to count in your review here. Similarly in universities, teachers who have established a good record and don't have to worry about building on it can do more in terms of spending time with students.

It is very important to have people who turn people on. I'm sure that all of us look at individuals who have made a major impact in terms of our career decisions. Bob Ginsburg is, for sure, one that I always recognize. His whole job was to show us how modern reefs might relate to ancient reefs, but if I hadn't gone on that field trip I probably would not have gotten into modern reef work. As we've gone through this, there are so many different interests, the stromatolites, the mangroves, let's face it, there are all sorts of aspects of reef geology and biology, and I've gotten into a lot of them.

I think the reason why is that you are the same little boy you were in Barbados.

I think that's part of it. It is kind of fascinating.

One of the things which fascinated me about reefs in the eastern Pacific was that people denied they existed.

PETER W. GLYNN (b. 1933)

Professor of Marine Biology and Fisheries
Rosenstiel School of Marine and Atmospheric Sciences
University of Miami
Interviewed in his office 2 May 1998

A firm footing in the biology of many groups of invertebrates, started in childhood but cemented through taking Abbott's invertebrate zoology course, has enabled Peter Glynn to study a variety of ecological questions in coral reef environments. He began investigating trophic relationships in the central Californian intertidal. Since then, all of his work has focused on tropical systems, particularly coral reefs, concentrating on determining the mechanisms underlying community organization. Collaborations with IAN MACINTYRE have addressed the question of how reefs change in time and space. As important as the advancement of knowledge through his research, however, is Glynn's contribution to producing the next generation of scientists, not only in the United States but also in the tropical countries for which coral reefs play an integral role in the environment and in the economy.

I had a great-aunt who was a schoolteacher. When I was 5, 6, 7, 8 years old, she used to take me on nature walks. Since I'm from Coronado, California, which is an island, we used to go walking on the sandy beaches, collecting seashells and looking at sea life which had drifted up. She got me started making a shell collection. From that point on, my interests began to grow. I began to explore that on my own by getting library books to identify different organisms. Then I started a little museum in a house that my parents had in the backyard. I just continued to pursue my interests and that's how I got interested in university.

At that time, did you realize people did that kind of thing for a living?

I think so, because my great-aunt would take me to museums and zoological gardens in San Diego and Los Angeles. I saw people working on collections there, interested in sea creatures, so I knew there was a field there. She also took me to Scripps [Institution of Oceanography], which had a small aquarium. I met some of the scientists there. Serendipitously, when I was in high school, I went out for the wrestling team. One of our wrestling coaches was Jim Kittridge from Scripps, who was working with Dennis Fox. Jim right away realized I was interested in marine science. He invited me over to Scripps a couple of times. He was a graduate student at that point. I actually got into their laboratories, saw them working, and saw how happy they were at their professions, and I decided, that's where I want to go.

I find most people who do field science love it. They wouldn't put up with the hassles, I don't think, unless they did.

I was talking with my wife not long ago, and she reminded me at one point, "Remember one time you said when you get to the point where you retire, if you really do have to retire and leave the field and marine science, that you'd rather put a bullet in your head?" And I said, "Yes, thank you for reminding me of that!" So it's something you want to do your whole conscious life.

My grandparents were interested in my pursuing a medical education. Of course, much of what you take as an undergrad can be applied to medical school or to basic science. When it came time to apply to medical school or to go into straight science, I opted for straight science. Already I had been thinking of ichthyology, working with Carl Hubbs at Scripps. Also, I took the invertebrate class with Don Abbott at Hopkins Marine Station one summer. I decided to go to Scripps. Carl Hubbs invited me over to his home in La Jolla for an interview. He said, "Really you should go to UCLA first for one year before you come down here, because I want you to take the basic course in ichthyology and some other courses." So I did that. In doing that, Boyd Walker turned me loose on a large collection of pickled carangids to try to work on the systematics of those. I got highly bored with it. So I decided to leave UCLA for Stanford, remembering Don Abbott's course and how exciting that was.

Don Abbott was a very dedicated, excellent field scientist. He knew his invertebrates like no one I had ever met before, and he had really good insights into the phylogeny of the invertebrate taxa, into the ecology of invertebrates, and so that's where

In 1989, Peter Glynn (*center*) posed with students at the Uva Island, Panama, field camp of a graduate student. Photo from the files of the P. W. Glynn coral reef research group.

I stayed. I talked with Don extensively about what kind of community I might work with at the Hopkins Marine Station. I decided on describing a high rocky intertidal community dominated by algae and barnacles in terms of its trophic structure, feeding interactions. I began my fieldwork, I think, about 1959.

Your project has a history. It was redone more than 30 years later.

Oh, yes it has. The results demonstrated a shift in the composition of the community from more northern species to more southern species, reflecting warming of the Californian coast.

I was lucky in having a very diverse committee. Cornelius van Niel was a member of my committee.

Van Niel really inspired HOLGER JANNASCH, *too.*

Oh, he was a fantastic man. Also, Arthur Giese at Stanford had a tremendous influence on me. And then Lawrence Blinks. Blinks really did a magnanimous thing for me one time, I remember. I was working in order to pay my way through Stanford. I was either working as a TA or as an oceanographic technician. The oceanographic technician job meant going out on the *Tage* [the small Stanford coastal research vessel] and working up plankton samples and all of that. It was a whole lot of work. It meant doing endless salinity analyses. I remember one time we had registration, and I was over in the Loeb Building, and I didn't have enough money. Blinks had his office on an upper floor. I was sitting on the steps pondering how I was going to get the money to pay for my registration. Miss Stovall, the secretary at the marine station, must have told Blinks that I was having financial problems. So he came over to me and put his hand on my shoulder and said, "Peter, do you need money? Do you need help?" I said, "Well, I'm $65 short on my registration fee," and he said, "Oh, here, take this." He counted out $65 and said, "Don't worry. Pay it back when you can," and he went bounding up the stairs. I thought, my gosh. Not only interesting intellectual conversations with Lawrence but also learning the humanity in the man.

I left Stanford in 1960 because I got a job in Puerto Rico. I came back to defend in 1963. I've always been interested in the tropics. I guess part of that is because I'm from Coronado. It's not subtropics, but it is lower temperate.

Did tropical stuff come up sometime?

Occasionally, not very often. The closest tropical biota to us in Coronado was in the Gulf of California. Occasionally we would make trips down there in our cars. San Felipe was one of our first places. And then we made trips down to Guaymas. A lot of experiences of getting stuck in the mud with the 25-foot tide and worrying about how to get out before the tide came in! But we began to see some corals in the northern part of the Gulf of California and that kept inspiring me to get closer to the tropics.

With that experience I planned a trip with a friend of mine. We took buses mostly through Mexico, and we ended up in Belize. My friend wanted to go to Jamaica, and I was interested in looking at the barrier reef in Belize. So he got on a schooner and left, and I went down to the wharf area in Belize, and I started talking with some fishermen.

You spoke Spanish at the time?

I took a course or two in it. I didn't do very well, but when I went into Mexico, that's really how I picked up Spanish. I got on a Mexican fishing boat. They were fishing for lobsters and spearfishing for large groupers. So I made a deal with them that everything I caught I would contribute to their catch as long as they would provide my upkeep for the trip. So I went out with them for 30 days. We went up to Ambergris Cay and fished all along the barrier reef. I was spearfishing grouper and catching lobster like you wouldn't believe. So I gave all of my catch to them. They took care of me. They fed me. So that was really my first view of a well-developed coral reef.

With that in mind, I eventually ended up at Hopkins Marine Station and finished my degree. Just before finishing though, I met my wife. She was teaching Spanish in Pacific Grove High School, and I was a substitute teacher at the high school to make some extra money. She started telling me about the coral reefs in Puerto Rico: "By the way, in my home town there is a college, and they are interested in marine science. Why don't you write a letter and see what might be available?" I did that. I got a letter back from the dean saying, "We have a position open right now. If you come now, it's yours, otherwise I can't promise anything." I went to Don Abbott with this letter. I was about a year away in my research from finishing my dissertation. I said, "Don, look at this opportunity. I want to go to the tropics." He said, "Normally, I would not support any student who had not finished [his dissertation] taking a job, but I think you have enough perseverance that you will finish, so go ahead and accept it." And, so I did.

I ended up in Puerto Rico and started doing research on the southwestern coast in the La Paguera area. Actually, they promised me that if I would teach on main campus for the first year, they would transfer me to marine science for the second year. Then I continued teaching marine courses and really getting into research in Puerto Rico. I applied for National Science Foundation research money and was successful a couple of times, and that helped me a lot in getting my laboratory set up. At that point there wasn't much known about the community structure of coral reefs in terms of all of the invertebrates, fishes, and things associated with reefs, including feeding pathways. So I thought I would try to carry over my Ph.D. approach into a coral reef area.

I don't know if you wonder why I went to Panama.

I do.

I started looking around, but in those days jobs were easy to come by. Ira Rubinoff had just finished his Ph.D. at Harvard. He came through Puerto Rico, and I was taking one of my classes out to the reef. So Ira came out with me, and I went over my research project with him, and we chatted about science, marine science, and where it was going. A couple of months later, I got a letter from Ira saying, "Would you be interested in coming to STRI [Smithsonian Tropical Research Institute] in Panama?" It was an ideal opportunity to move on. My interests never flagged in Pacific coral reefs. I knew that being in Panama would give me an opportunity to study Pacific reefs. We moved to Panama in August 1967.

You still did some work on Caribbean reefs?

Yes, but most of my interest shifted to the eastern Pacific side. One of the things which fascinated me about reefs in the eastern Pacific was that people denied they existed. I decided I had to look because many of the people who said that coral reefs don't exist in the eastern Pacific were people who weren't even there, albeit famous people: Charles Darwin, for example, James Dana, and some later people who didn't dive, who just observed things from the deck of a ship. I went there and I talked with some of the local geologists, who were working with the Panama Canal Company, but they were interested in diving and in reefs. They told me, "We've seen reefs around here. They are small and subtidal, but they are here." I started arranging some trips with these people, and indeed there were well-developed reefs off the coast of Panama, especially in the Golfo Chiriqui, in an upwelling environment.

At some point you seem to have shifted from community focus to species interactions. Am I right to say that?

I've been interested in key species interactions, yes, because very often they have an effect on the trophic structure on reefs. The community properties of reefs in terms of species composition and so forth, this was one of my interests in the crown of thorns starfish problem because the crown of thorns attack scleractinian corals, zooxanthellate corals, and obviously these corals are what form coral reefs. I became interested in what regulates the population abundance of *Acanthaster*, and I found that there were certain key predators, for example, the little shrimp that feed on the starfish and the scavenger worms.

I remember when I first heard you give a paper on that shrimp, in a coral reef symposium.

Oh you do! Was it Tahiti?

Maybe Tahiti or maybe in the Philippines.

I'm glad that you remembered. That's why I got interested in species interactions because it is often just a few species that are important in processes that affect the whole trophic structure on a reef.

Ultimately my main interest in coral reefs is: What are the conditions that promote reef growth? How is it that we have coral reefs? How is it that they grow and survive through time? This is the big question. Naturally, geologists are also interested in the big question, and they have an added perspective of knowing what reefs have done through time in the geologic past, through the record. This is where Ian Macintyre has so much knowledge: how diagenesis occurs on the reef, how carbonate minerals have changed, cementation, reef growth, sedimentation, sedimentary processes, bioerosion. These are so vital to reef growth.

The interaction of biological and geological processes that are operating at the same temporal and spatial scales.

Really. Since coral reefs form geological structures, and they do this through the interaction of species and physiological processes, this is the appropriate place for biologists and geologists to come together. I think they can supplement one another, each in their respective fields. This is why Ian and I have so much fun working together in the field. He teaches me geology and the long-term processes, and I try to teach him about biology and the shorter-term processes. He is a fine man. He is a real gentleman.

There is a synergistic interaction with the two of you.

Yes, and I certainly appreciate it. One of the other things that I really liked about working with Ian in the field is that he is the sort of person that when things go wrong—the equipment breaks, the weather gets bad—he just buckles down and says, "We're going to do this in spite of everything." "When the going gets tough, the tough get going," that's his attitude. I try to emulate this attitude.

It took a lot of effort to get here, and we're going to get what we came for.

Exactly, because it might be the last opportunity to get it. Very often I think a lot of fieldworkers have to work with that mentality. There are times in the field with Ian, and with other people, if conditions are such that we cannot do the main field project, then we start thinking of other things. We take advantage of what we have, our wherewithal of the moment. Often we end up with some new discoveries that way, looking at new problems.

You, more than anyone else I know as an American, have been involved with training students in marine sciences in Latin America.

First of all, I enjoy working with students very much. When I say "working with," I don't mean all the knowledge is coming from one place. Students, I feel, have a lot to contribute to professors; it's not just a one-way street. I learn from students just as they learn from me. I always try to include students in my field research efforts. In a sense, in all of my studies starting in Puerto Rico and since then, I've tried to include the students from those areas. Often they will be the next scientists to be developed from those areas. They are in their country.

And they certainly have a vested interest in what happens there.

Exactly. For this reason, I try to associate with as many good students in the host country as I can, in Puerto Rico, Panama, Costa Rica, Mexico, Colombia, Ecuador, the Galapagos Islands.

How do you locate or link up with the students?

Very often by giving seminars and talks at local universities. In many situations we need field help, so we approach local professors and ask if they know students who might like to learn more about coral reefs, and we make contacts that way. In Ecuador, the Charles Darwin Foundation likes visiting scientists to take students with them so they can learn from people from other countries doing work.

Sometimes we meet students serendipitously. For example, just three years ago, I visited an area in southwestern Mexico called La Huatulco in the Gulf of Tehuantepec about 400 miles southeast of Acapulco. I always thought there might be some coral reefs in that area because of the geomorphologic orientation of the coastline. We went out diving that day and came back to the hotel. The dive shop owner said, "Dr. Glynn, there was a group of Mexicans from the Universidad del Mar, the oceanographic institute at Puerto Angel, and they wanted to rent some tanks. I told them you were here, and they didn't believe me." I said, "I am here, and I'd like to meet them," and he said, "They'll be here at 7 o'clock to meet you." There were five or six of them, one a physical oceanographer, a geographer working with GIS [geographic information systems], and then a young student interested in coral reefs. We all got together, had dinner, and had a wild conversation about reefs and opportunities at Huatulco.

Shortly after that, I applied for a National Geographic grant for down there, and I got it, so now I'm doing a project down there. We are trying to help three or four students in the area with our expertise and with access to our library and things like that. The whole institute was devastated by Hurricane Pauline, and I've been able to locate a substantial reprint collection and back issues of journals, which I am sending down to their library. The other thing, Betsy, is that the local people know so much compared to the outside people. When I go to Huatulco, I'm just there for a few weeks. They are there all of the time. They get to explore up and down the coast, different depths, different environments. So they are really the ones with the knowledge, not I. To the extent that I can, I try to help them, and they guide me into areas that are interesting. Ultimately, too, with any knowledge that they provide me or interesting areas that they show me, I try to help them develop areas for theses and dissertations.

To ask, what are the appropriate questions here at this time?

Exactly, because for these people, it's important for them to get advanced degrees so that they can start to build up their in-house research capacity.

How has the funding picture changed in your career? Obviously it wasn't too good when you were a graduate student.

No. It got better. I would say that in the 1960s and 1970s funding was quite good. There was a lot of interest in coral reef and marine science in general. It was relatively easy to get funding in those days, NSF, ONR, and so on. As we all know, in the 1980s and 1990s, it has become extremely difficult, to the point now where the NSF says that only 25–30% of the proposals received are funded. There are excellent studies that are being received by NSF, but it just can't fund them because it doesn't have the financial base.

Another thing that is happening, Betsy, is that many of the universities in the States now like to have the professors get their own salaries. More and more, we are moving into a situation where you are hired on soft money. That is, you can work for us as long as you get your own grants and salary. We'll provide you with an office and a laboratory. In this school, the amount of laboratory space is often governed by how many grants you have and how much overhead you are providing to the school.

A problem that we often have with multiscientist, multidisciplinary projects is that they require so much money coming to so many different people from so many different institutions that often the budgets are huge. When the budgets are huge, that means the individuals in the different institutions get smaller portions of what is available and then they have harder times supporting themselves in the host institution. One of the consequences of this is that in order for a professor to support himself or herself, a person has to have four or five ongoing projects, not just one. When you have that many, you can dilute your efforts and so it's more difficult to keep up with your research scheduling, your various field activities, and so on. So people can easily spin out of control in that situation.

Where does teaching fit into this picture?

That's something that is constantly pulling me in different directions here. Even though I am a professor with tenure, I still worry about getting as much of my salary support as I can, and then in this institution, as in most institutions, you can't accept new graduate students unless you can give them full financial support. So, a lot of money has to come out of your grant proposals to support students. Every year I must get 25–30, sometimes 50, inquiries from students who want to come to study coral reefs and so on, but we can only accept a couple.

Where do you see your type of science going?

I think coral reefs are exquisite natural communities that are being threatened now because of overpopulation, especially in Third World tropical countries. There are a lot of pressures—overfishing, land management. There is a real threat of losing species in some of these coral communities. Coral reef ecosystems, after all, are extremely important to mankind in the tropical countries as a source of protein and, each year, more and more tourism. I think it is important to understand, as best we can, the fundamental processes that promote coral reef growth and health, and that's where I hope a lot of my efforts will be directed

With all of the knowledge I've gotten down through the years with my students and my colleagues, I'm trying to put together a book describing the eastern Pacific reefs, their biogeographic connections, the community structure, the things that influence their growth rates, their distributions. I'm hoping that such a book will serve as a base for continued studies of the eastern Pacific region.

You hope people don't say, "All is known that has to be known."

Oh, never! You know well enough, the more you study science, or any discipline, the more insights, the more questions you get. So that's also another beautiful aspect of science—you never know what's new around the corner.

Have you read The End of Science?

I read excerpts.

At first I thought it was tongue in cheek, but I don't think it was. He has an epilogue and said he really feels . . .

It is dead! Betsy, I don't know if you ever heard Cornelius van Niel say this, but it was one of the things he had fun with students. You know he was responsible for our understanding of the basic reactions of photosynthesis?

Right.

One time, he came into my laboratory and said, "Well, in a few years, we microbiologists will know everything there is to know about microbes, so we can all rest and feel the field has pretty well been . . ."

Exhausted?

Yes, and he would always say that with tongue in cheek, because he knew that all of his students and his postdocs were making new discoveries from one week to the next.

SUMMARY: CHANGING TIDES

This generation grew up during World War II, yet their childhood years seem to have had a special innocence. There are fond memories of natural history excursions with relatives and science clubs, shell collectors' meetings, school field trips, camps, scouting, and building radios. These activities enabled these budding scientists to communicate with the outside world and to begin to discover nature. College options increased with scholarships available for academic and athletic prowess, and many went to school far from home. Several universities seem to dominate these pages—Yale, Princeton, Cornell, Duke, the University of Chicago, Stanford, the University of California at Berkeley, Harvard, Columbia, and the California Institute of Technology—all of which had strong natural history traditions. Macintyre and Stanley each had a clear goal to work in the United States, as that country offered the best career opportunities for a scientist. As undergraduates, all of this generation had a classical training, which emphasized breadth in one's discipline and a firm grounding in the basics. Geologists took physical and historical geology, stratigraphy, mineralogy, and petrology. Even in graduate school, they were urged to broaden their knowledge through additional classwork and seminars. The biologists learned about some group of organisms in depth but often had a general background as well.

Almost everyone interviewed had extensive wilderness experiences, including traveling by fishing boats, making early ascents of peaks, serving as a hunting guide, and assisting on wilderness field expeditions. As young men, they took the time to engage in these challenging pursuits; they developed great self-reliance, quiet self-confidence, and good judgment, each led by his own inquiring mind.

In geology, in the early 1950s, the focus and the funding were pragmatic. Traditional terrestrial geology was dominated by the economic need to discover and document natural resources. The U.S. Geological Survey, a premier research institution, was mapping throughout the country; oil and mining companies were exploring for resources and supporting the education of future geologists; the military (thwarting the Soviet threat) supplied logistic support for Arctic, tropical, and deep ocean research. All of the geologists interviewed had jobs either while an undergraduate, or shortly thereafter, as field assistants to senior geologists. This was "learning the trade" through practical experience, mapping alongside experts. Marine geology developed extensively as a field in the 1950s and especially in the 1960s, due to the military's need for information about the ocean (especially that related to the threat of Soviet nuclear submarine attack). In contrast to the geologists, the biologists in this generation were not motivated (nor funded) by economic or military interests during their graduate student years. Glynn and Kohn present similar pictures of grad student financial poverty but intellectual richness. Marine stations were important for the biologists as students and for their subsequent research and teaching.

In graduate school, each of these scientists discovered an atmosphere (intellectual and funding) that encouraged the pursuit of basic research. Their advisors, including Hess at Princeton; Hutchinson, Rodgers, and Flint at Yale; Abbott at Stanford; Gubler in France, clearly recognized the balance between theory and empirical work, and stressed a regional and comparative context in which to view more detailed work. When this generation completed their Ph.D.s, the job market was somewhat restrictive, but then came the enviable period when they were sought out for specific positions because of their expertise.

All of the geologists interviewed are actively conducting field research today. Knowledge and experience gained through on-site field work enable these scientists to integrate their work with that of specialists in other disciplines, including the biological and atmospheric sciences, and to address questions of a broader scope, particularly about the role of organisms in affecting geological, physical, and chemical processes relating to the history and evolution of the Earth. While the geologists do not have formal training in biology, their field experiences have led them to collaborate with their colleagues in biology, microbiology, and oceanography in order to understand the geological systems with which they work.

Coral reef science is one of the most integrative of the field sciences that emerged in the twentieth century. Its community is small enough for the scientists to know one another personally but broad enough to include specialists in many of the natural sciences. Echoing the U.S. society at the time, coral reefs were considered "stable" systems in the 1950s. Macintyre, Glynn, Kohn, Wainwright, and colleagues of their generation began working in reef systems in the 1960s and the early 1970s. At that time, the field was dominated by the geological dogma that reefs were thin veneers on older preexisting rock outcrops and by the biological dogma that reefs were stable, self-regulating systems through time. As scuba became available as a common research tool, scientists from this generation and subsequent ones began to actually study reefs in the field—coring, measuring, identifying, and counting organisms. By describing and quantifying the patterns they found through time and space and by analyzing and interpreting these data, they challenged the prevailing dogma. Their work and that of others have contributed to a complete revision of what we can now characterize as simplistic views.

These scientists, established as successful researchers by the 1960s, noted the enormous shift in academic philosophy from a teaching-research-service balance to research as the most important element of the publish-or-perish mentality, which began to dominate academic science in the 1960s. This was seen not only in research universities, but also in research institutions, such as the Smithsonian Natural History Museum. Promotion became based almost entirely on research production. As Burchfiel points out, something had to give. In the universities, it was the time and effort expended for teaching that was curtailed. In the museums, it was the time and effort devoted to preparing exhibits. Since field courses require extra time, energy, logistics, and unusual financial commitment, they are the first to suffer. The scientists of this generation can afford to take their students into the field, to teach field skills, because they are already senior scientists (they have had tenure for a long time) but they recognize that their younger and, particularly, untenured colleagues cannot afford to do so. These men have a wealth of experience to share with the younger generation, but each is at or near retirement age. Another related issue is that in the present funding climate, when several grants are required to support one's laboratory, any scientist (even of this generation) finds his time divided by conflicting responsibilities; both teaching and research suffer as a consequence.

You may have noted that I interviewed no women in this chapter. I have encountered hardly any female scientists (let alone field scientists) from this decade, yet I know dozens of male field scientists from this era. Betty Bunce gave a hint about why there are no women of the 1950s when she pointed out that the new male president of Smith College in the 1950s eliminated 17 faculty slots from the sciences because "women didn't need that kind of education. They were going to be wives and mothers." The military was male dominated in the 1950s as well and didn't allow women to go into the field when it supplied the logistics. The military played a dominant role in direct science funding and logistical support in the 1950s and 1960s—in the Arctic, in the tropics, in the ocean. The opportunities to get field experience were extremely limited for women, and even if one were successful, job opportunities were almost nonexistent.

There is a general concern expressed in these interviews about the increasing trend toward a shift in the balance between descriptive science and theory, toward a shift in balance between teaching and research, and toward the trend for specialization early in one's career. Course requirements for both geology and biology majors show a reduction of the classical courses (including the complete absence in some curricula of organismal biology or stratigraphy, considered core courses for these disciplines by this generation), which provide a framework to relate more specialized aspects of a field. Taxonomic skills are being lost. Taxonomy is necessary not only for knowing modern organisms but for fossil organisms as well, which provide the bases for relative dating. The members of this generation ask, "Who, in the end, will be able to integrate information for the younger generations?" There is great concern that observational skills, particularly field skills, are dismissed as being old-fashioned. Yet each scientist knows precisely the value of his own field experience to any question he addresses (be it theoretical or empirical) in science.

4

The *Sputnik* Generation: The 1960s

An excellent thing to keep in mind: there are a lot of unknown mysteries out there.

Laurence Madin

I can still vividly recall the excitement I felt going out on our front lawn with my mom and brother to watch *Sputnik* pass overhead. A new era had begun, not only in U.S.–Soviet relations, but in the way science was done in the United States. My generation reaped the benefits of the national response to this technological challenge; the cry went up to increase expertise in the sciences. The scientists of this generation were naive in many ways, but most can be characterized as having a deep love for nature. Increased funding for science led to better preparation for college, supported broad undergraduate education, and provided almost unlimited opportunities to combine new technologies and ease of travel to remote parts of the globe to investigate the Earth, the oceans, and the living world. Most important, it was a time of intellectual ferment. Professors and their students discussed and argued about science with colleagues in their own and other disciplines. Groups of scientists, often at different stages in their careers, each bringing different expertise, spent time together in the field, sharing perspectives and approaches. Students were encouraged to take risks, to think of new questions, to develop new techniques, and then were given the logistical support required to test them out, to see if something "worked." It was an exciting time if one loved nature and one loved science, and that passion continues today. It is evident in each of these stories.

I'm not forceful enough to be a crusader, but my little voice in the wilderness cries out that multiple approaches need to be retained in science.

MIMI KOEHL (b. 1948)

Professor, Department of Integrative Biology
University of California, Berkeley
Interviewed in her office 17 November 1999

Mimi Koehl began college as an art major, but fulfilling a science requirement led her to the exciting idea that one can explore natural form "by learning how it works." Her approach to investigating form was refined by her courses, field experiences, and discussions with advisors during grad school. Equally important were the lessons gained through working with other kinds of scientists in the field, particularly ecologists. Learning the funding game, which I first heard about from Koehl when I was an impressionable graduate student, highlights how the selective processes for success in science are changing. Koehl illustrates the loss of generalists within academic departments by comparing her knowledge and approach to science with that of Ralph Smith, her academic great-grandfather, whom she replaced at UC Berkeley.

My mother was an artist and my father was a physics professor. They were very traditional about male and female roles, so I was raised to think that my "career goal" was to marry somebody who would take good care of me. When I got sent off to college, Gettysburg, I was an art major. Girls did that. I had to take a science distribution requirement, so I took a biology course taught by Bob Barnes. In art, I was exploring natural form, the way it looked and the way you could use it to make images. When I got into a biology class with someone like Barnes, who knew so much about the diversity of form, what I realized is: Wow, we can explore natural form by learning how it works rather than just by appreciating its beauty and using it to make images. That seemed much more satisfying to me.

Much to my parents' horror, I switched my major to biology. My father convinced me also to do the work, including student teaching, required to become an elementary school teacher, a marketable trade for a woman, in case she ended up an "old maid" with no one to support her. How things have changed since then! I loved learning from Barnes. I took invertebrate zoology, and I was just blown away by all the different forms of animal bodies. The thing he stressed was form and function. During spring break, he took us to the Duke Marine Lab. We got to see the animals where they lived, which made a lot more sense than just watching them in a dish. You could see them moving, you could see water move around them, you could feel if the sediment where they lived was squishy or hard. You could see all kinds of things, and it was fun!

I realized that I wasn't ready to stop learning about these wonderful creatures, so I applied to graduate schools, even though no one encouraged me to do so. I was horribly naive. I applied to Duke simply because I had been impressed by its marine lab. I went off for the summer to work at the Woods Hole Oceanographic Institution as a technician in a microbiology lab. People were asking if I wanted to continue the job after the summer, so I thought I'd better find out what was happening with Duke. I called the director of graduate studies in zoology, who said, "Oh. We never sent you another letter? Somebody just dropped out. Why don't you come?"

At Duke, I earned my keep initially as a TA in introductory biology, which Steve Vogel was teaching. He gave a lecture on size, on all of the profound consequences for organisms of being different sizes. I had never thought about that before. It was simple stuff, like smaller spheres have a higher surface-to-volume ratio than larger spheres, or if a cell gets above a certain size it would collapse under its own weight. I went running up after class and told Vogel, "That was the most exciting stuff I ever heard, please tell me more." He said, "That was actually STEVE WAINWRIGHT's lecture, but he is on sabbatical, so you'll have to wait until he gets back."

I didn't realize until I did a tutorial about this mechanical approach to looking at form with Wainwright that squishy things had skeletons. I'd never really thought about what a skeleton did, for example, it resists forces, it transmits loads, and it enables shortened muscles to be reextended. Or that you could do the same things with soft bodies! To learn about the skeletons of worms and sea anemones and to learn about quantitative principles of design applied to those sorts of squishy skeletons was absolutely fascinating to me. So, that's what I wanted to study.

For my dissertation project, I was looking for a group of organisms that were related and had hydrostatic skeletons, but that used them very differently mechanically. I wanted to see if there were differences at the microscopic tissue material level that related to how the whole organism functioned. I was thinking that tube feet of diverse echinoderms might be the system to study. Then Bob Paine came to visit at Duke. He gave a departmental seminar about rocky shore ecology and in passing mentioned some large sea anemones that lived in the crashing surf of the lower intertidal on rocky shores. He said that they made their living like garbage cans, waiting for animals to get knocked off the rocks and to land on them. I was blown away. All of the anemones I had seen were these little woosie things that lived in calm water. I had also read a lot about the big West Coast *Metridium*, but these were subtidal sea anemones and not subject to crashing waves. I thought, wow! Here are big hydrostatic animals, some of which live in wimpy flow and some of which are in crashing waves. Because their body design is simple, I can get big chunks of connective tissue to do my material experiments. What a great system!

In terms of my education about doing fieldwork, I must tell you that Steve Wainwright is very much a field person. I was a TA in his animal diversity class. Field trips were very important. We went out into the Duke Forest, and we talked about the form and function of whatever animals we found, and then we brought things back to the lab to study further. We also went to the marine lab. The format of the field trips was along the lines: We will go here and we will think about and discuss the organisms

Mimi Koehl (left) in the 1970s on the outer coast of Washington State, preparing to measure the water flow that the anemone *Anthopleura* actually "sees" in its lower intertidal surge channel environment. Photo by C. Birkeland.

that we happen to find. Everything someone would notice, Steve would ask, "What do you see? Is it similar to or different from anything we've seen before? In what ways? What do you see this organism doing? What do you think about that?"

I had been exposed to the approach of thinking about form and function in the field from both Wainwright and Barnes, the idea that seeing organisms in their habitat was very important. Therefore when I thought about the two different types of West Coast sea anemones that I wanted to study, I realized that I really had to get out there and look at them in nature. At that point in my studies, I had had no training in ecology, and I had no idea of how famous Bob Paine was. Therefore I wasn't timid. I just wrote him a letter along the lines of "Dear Dr. Paine, I sure thought those anemones you mentioned must have interesting skeletons. You mentioned your work on an offshore island, and it must be hard to get there, but could I please come along on a field trip there sometime?" "Sure," he replied.

We went out to Tatoosh Island in northwest Washington. Over the years, other students on those trips included Ken Sebens, Sally Woodin, Tom Suchanek, Betty Nictore, Carol Slocum, Rich Palmer, and Jim Quinn. They were wonderful to me. From Steve Wainwright, I really learned to look at the form of the organism and what it was doing. From Bob Paine, I learned about their natural history and ecology. Who recruited when onshore? How abundant were the different species at different places? Who eats whom? Who overgrows whom? There I was, a student of biomechanics, sitting around with the ecologists after the tide came in, listening to them debate different ideas about mechanisms producing the patterns we could observe in the abundance and distributions of organisms. It really expanded my horizons beyond just thinking about the individual organism. I learned that the setting of an organism was more than just the physical world around it. The crashing waves affect the anemone, but so do its neighbors, its prey, its predators—and all those effects could change with season, with age, with site. Steve Wainwright actually once came to Tatoosh. It was really fun to be in the intertidal there with him as well as Bob, because what they focused in on was different.

They looked at the intertidal with different eyes.

Yes.

Over the years, I've realized that what you want to know about the water flow and the measurements you need to make depend on the biological question you are asking. For example, for my Ph.D., working with the sea anemones *Metridium senile* and *Anthopleura xanthogrammica,* I was interested in the mechanical designs of hydrostatic skeletons that performed differently. One of the things I learned was the importance of the organism's microhabitat and that it could be drastically different from what you first thought when you looked roughly at where it lived. *Anthopleura* lives in the lower intertidal surge channels at Tatoosh, where there are crashing waves. However, what I discovered when I measured the water flow encountered by those anemones, is that they, in fact, don't "see" those crashing waves. They are very short and they have neighbors crowded around them, and the water velocities they experience are much lower than those in the free-stream flow overhead. When you see *Anthopleura* in sites that are more protected from rapid water flow, they stand taller, rather than hunkering down out of the ambient water flow. In contrast, I expected *Metridium,* which live in calmer, deeper sites, to experience wimpy flow. However, they stand up so tall that they encounter whatever flow is there, which turns out to be not all that different from what *Anthopleura* experiences. Within a habitat, exposed or protected, you have to ask what the organism sees, because it may be different from what you see.

I got interested in studying seaweed mechanics simply because they were out there when I was doing something entirely different. I was studying connective tissue mechanics of anemones. There were all these wonderful seaweeds out there, bigger than anything else in the crashing waves. I'd watch them whiplash and flail around. I began getting interested in flexibility and what that would do for the organism's ability to withstand waves or currents. With seaweeds, I started out with the idea that the big seaweeds that would be out in crashing waves were standing up there dealing with that flow. I expected that they were going to be really strong, really tough. What I found out was that, even in the crashing surf, there were some seaweeds with "good" mechanical designs and some with "lousy" mechanical designs. I am using the terms *good* and *lousy* as I would mean them if I had been an engineer designing the seaweeds not to break in waves. However, after all of those field trips with ecologists, I had learned that I needed to think about more than just the mechanical design of individuals. I needed to learn about reproduction, life history, seasonal changes, and all those sorts of things as well. What I realized, teleologically speaking, is that you can get around a really lousy mechanical design by having the right life history. An example would be that you can live in a habitat that seasonally experiences big storms and enormous waves, even if you have a crummy mechanical design, if you grow quickly and reproduce before you are torn off the rock by seasonal storms. The giant kelp *Nereocystis* are an example. They break in winter storms, but that is fine; they have already spread their spores.

When I was in the field studying *Metridium*, another thing I noticed was that, as they were gradually stretched by the current, their big, fluffy oral discs became oriented at right angles to the flow. If you watched what they were doing, they were catching zooplankton on the multitude of small tentacles on their oral discs. It was my first hands-on experience of suspension feeding. I had read about it, of course, but that does not prepare you for actually seeing an organism doing it in the field. Wow, I could watch the particles be captured! I was really turned on by that. Prior to that observation, while I was back at Duke, Vogel was teaching a fluid dynamics class for biologists. We all had to do library projects. Another student, Dan Rubenstein (who now studies animal behavior at Princeton), gave his presentation about the filtration theory that engineers used to predict the effectiveness of filters of different designs at capturing particles. He said, "These are the physical mechanisms that engineers say are involved in catching particles. I wonder if it happens in organisms?" I had that in my mind when I saw *Metridium* in the field, and I realized that I was seeing the same particle-capturing mechanisms in action that the engineers had modeled. When I got back to Duke, I bugged Dan and said, "Hey, I think this is really going on. Do you want to explore it further?" Dan and I ended up writing an *American Naturalist* paper together about all this.

I did a lot of my fieldwork at Friday Harbor Laboratories. I'd sit with someone at lunch in the dining hall and ask what he or she was studying. One of the people whose research really got me turned on was Richard Strathmann. Back then he was looking at the functional morphology of ciliated larvae, focusing on how they might be catching food particles. At the same time that Richard was using similar physical principles to understand how larvae fed, I was trying to understand how my passive suspension feeders fed. One of the reasons I was so turned on by Richard's work was that he had observed, in the larvae, mechanisms that we knew from the theoretical models ought to work.

When I wanted to do my postdoc with Richard Strathmann to test the filter-feeding model, I sent off my first proposal to NSF. I wanted to compare an active suspension feeder with a passive suspension feeder, *Metridium*. I chose brine shrimp for the active suspension feeders because I thought that they would be a good research system for this question; they have lots of identical appendages, and they are easy to rear. The rejection from NSF said, "These aren't important animals in the ocean" and "This isn't oceanography." They canned it.

What I learned from that first failure was that you have to propose to study something that is perceived as ecologically important. You cannot propose to study merely a good system to address your question, if you are trying to get funded out of the oceanography program at NSF. I suppose if I were trying to go through the biology program, that approach would have been OK, but back in those days I couldn't have gone to biology, because proposals about organisms that lived in salt water usually got sent over to oceanography. We used to refer to it as the "salinity rule."

The next time I tried a proposal to study how copepods feed. This is the story that I told you while you were a grad student in Len's lab. I was rejected again with the reply, "We already know how copepods feed. Just look in Barnes's *Invertebrate Zoology* text." [The old view of copepod feeding was based on a study in which the copepod was confined in a drop of water. The currents observed were an artifact of the minute quantity of water around the animal and not those set up by the organism in its normal environment, as Koehl eventually determined. I used her paper about this with my WIL students, discussing not only the excellent research but also the sociology of funding.]

All of my research until I became a faculty member at Berkeley in 1979 and got my first NSF grant was funded out of pocket. I had no salary on that Friday Harbor postdoc, but I had a little grant from the Graduate Women in Science, which I used to buy film and an underwater camera. I lived off my savings—and what I had saved as a grad student wasn't very much. I think things are different nowadays. I don't think that many people are able or willing to live that way.

One of the wonderful things for me about coming here to Berkeley was that I taught invertebrate zoology with Ralph Smith. I hadn't realized until I got here that he was my academic grandfather. Like EUGENE KOZLOFF, whom I had met at Friday Harbor, Ralph just knew so much natural history. He knew the names of everything. He knew under which rock at which beach you would be likely to find what kind of bryozoan and when they would be reproducing. He was this enormous encyclopedia of knowledge about the local fauna based on his experience with these organisms over years and years. I think he came to Berkeley in 1948. Having the chance to teach with Ralph and go on class field trips with him was a very special experience. Another wonderful thing about Ralph was that he was a resource for all of the students doing marine work in the department. As we were trying to figure out if a certain species would be a good system to study a particular question, I'd tell a student, "Go ask Ralph about this species because he can tell you where you can find that organism, if they can be found reliably around here, when during the year they are abundant, when they reproduce, how quickly they grow."

That generation—Smith, Abbott, Kozloff, Paul Illg—were tremendous resources for the students and other professors in the departments.

What I am realizing now, of course, is that folks of my generation are having to fill that role. That is pretty frightening. I don't have many field sites around here. I only know a fraction of the natural history lore that Ralph knew for this part of the coast.

Part of the difference between folks like Ralph and me is that I don't focus my work on one group of organisms. I am question driven and choose different species and different field locales to study, depending on the question I am trying to answer. A lot of the research I do involves looking for a pattern: Is some sort of structure or process widespread among living things? In a sense, I wouldn't have noticed the patterns if I hadn't been out there looking broadly. For example, as I studied the kelp, I realized the importance of the flexibility of an attached organism to its ability to be big in really rapidly flowing fluid. That is true of sea fans, of alcyonacians, of pines trees, of coconut palms, and so forth. Kelp is a nice system to study. They are morphologically simple, so you can analyze their mechanics more easily; they are morphologically plastic, so you can transplant them to different habitats and change their morphology; they grow more quickly than colonial cnidarians, so you can get to your answer more quickly.

I taught invertebrate zoology for 15 years, and then I passed the baton to other colleagues. In addition to the lectures, the course had two lab exercises per week because there were so many animals to see. We were, depending on tides, going on three to five field trips per semester as well. There are some other courses in our department, notably in vertebrate natural history, that also have a lot of field trips. While some of us were teaching these time-consuming courses with a lot of field and lab work, many of our colleagues were teaching courses with three lectures per week, period. Who is going to be more productive in their research? For some big service courses, like introductory biology, there are Ph.D.–level scientists in charge of the labs and staff to set them up so the professor does not have to mess with the details of preparing the lab sessions. For smaller classes, like invertebrate zoology, that is not the case at this institution. A professor in such a class must set up the aquaria, order or collect the animals, clean the microscopes, set up the demonstrations, and so on. The system doesn't reward you for teaching that way. It takes a toll on your time and energy.

We try to keep our field courses going here at Berkeley. We have a Museum of Vertebrate Zoology, and it has a course that is as old as the hills, vertebrate natural history. They go on one or two field trips a week with a class of about 100 students. Three professors teach it together. I think the Department of Integrative Biology at Berkeley may be unique in its strong tradition of field biology, and I think the vertebrate tradition will continue because the museum is endowed.

I have recently gotten a new perspective on how the field tradition is ingrained in me. One of my graduate students is interested in flying frogs. The big issue is to address the evolution of novel types of locomotion. One could study all manner of things, but flying frogs are a good system because frogs representing the stages in the evolution of gliding are alive today, and you can study them in action. But, these frogs live in the rainforest! I'm not out of that tradition. I've seen rainforest in Panama and Australia, as a tourist, but I have never tried to work in the rainforest. I don't know what to wear; I don't know when it rains; I don't know what dangers to watch out for—snakes, jaguars, spiders, bogs. I realize that I feel so at a loss. What I did in order to get that student introduced to that sort of fieldwork was to send him someplace where the rainforest tradition is strong. It is in OTS [Organization for Tropical Studies]. I said, "You need to tromp around in the field with all the people who know the folklore, the tradition." I realized how ingrained it is in me to do that, to set the stage for a student to be knowledgeable about what to expect in the field, when I realized that I couldn't do it for this one particular student.

I must stress the importance of marine laboratories and terrestrial field stations in training students about field research. Field stations, like La Selva in a rainforest in Costa Rica or Friday Harbor Labs on the shore, where people from all over the world go to learn firsthand from organisms in nature and from older folks who are experts, play such an important role in the development of new scientists. I think lab scientists have the same sort of experience at places like the Marine Biological Lab at Woods Hole or Cold Spring Harbor Lab, which offer summer lab courses. That kind of experience, no matter what the field, is extremely valuable because you can see the passion that people have for research. I know my students have colleagues around the country who are the people they met at Friday Harbor. For me, a lot of my close colleagues around the country are people I met that way.

I will be curious to read in your book about what the younger people say, about things like traipsing off to Friday Harbor for the summer. If you are going to be very efficient and crank your papers out, then you can't afford to do that. Of course, we faculty don't usually have the opportunity to teach small classes of highly motivated students like that either. I've taught two summers at Friday Harbor, and I've thought, wow, this is a rewarding experience! It is so intense. You feel as if you are really making a difference, and you are really connecting with the students. Then the summer is over, and all of the things that "the system" rewards you for, you haven't gotten done. There are the papers and grant proposals that you haven't written, and your own research that you haven't worked on. You are back to the reality of the main campus.

One other thing that I want to emphasize is the interdisciplinary nature of my research work. I am using tools from engineering as well as from ecology and physiology and trying to put them all together to try and address basic questions about how structures work in organisms. Instead of the kid with the frog in my pocket, I was the kid who took the clock apart and drew it and who looked at things and looked at things working. In that sense, I am sort of always an outsider everywhere. I hang out with marine biologists, but I'm not really one because I do biomechanics. I hang around with biomechanicians, but I'm not really one because I work in the field; I use biomechanics as a tool to address other questions.

I'm not forceful enough to be a crusader, but my little voice in the wilderness cries out that multiple approaches need to be retained in science. People who straddle fields should not be permitted to drop through the cracks. Talking across disciplines ought to be encouraged. And of course, another piece of that is that lab biologists ought to be dragged out into the field and vice versa. I remember discussions I have had with ecologists who would say, "The only thing that is important is that the organism does it; it doesn't matter how. This organism eats that organism. Who cares how?" Of course, I care how.

I must also tell you that there is an aesthetic dimension to the questions that drive me. The pattern is beautiful to me, or the actual structures are beautiful, or the way the water flows is beautiful. To understand how it works adds one more dimension to how it is beautiful.

To me good science is driven by a passion, an emotion. Some people want to take that out of science, as if it doesn't belong.

Yes. Having an emotional connection to how the answer comes out should play no part, but being passionate about wanting to understand it should be an important part. Sometimes I have a discussion with my students as they are trying to find their thesis problems. I'll say, "This is a really interesting question because of this and that reason—and this other question is also of interest for the following reasons." They'll ask, "Which shall I do?" and I'll say, "Do what excites you!" These could be good ideas, but if someone is not excited, it may not be the right project for that person, there will be no passion in the effort. Of course, if you read the *Double Helix*, [James] Watson asserts that how you choose your problems is to ask which ones will make you famous. That certainly isn't why I chose my research.

My advice to people is that they follow their hearts in what they choose to do, research, career decisions, whatever.

I feel I had more permission to do that, being a woman, when I was in school. I wasn't expected to have a career or to make anything of myself. I talked to you earlier about my rebellion, which was to just go off and do this science. My brother also loved biology, but he became a doctor. He had the pressure to have a prestigious job and provide well for his family. Who cared what I did? That gave me the freedom to pursue my passion for natural form.

When I look at scientists who study some aspect of the natural world and never, never go outdoors—I think they are missing something fundamental about the meaning and significance of what it is that they are trying to do.

Paul R. Pinet (b. 1944)

Professor of Geology
Colgate University
Interviewed in his office 12 November 1997

Paul Pinet teaches his students about the limitations of science: What are the limits of this technique, this information, this concept? He tries to make them aware of all of the assumptions every scientist makes. In this very philosophical account of his growth as a scientist and teacher, he explores the concepts of science ethics, humility, and the effort and frustration required to mature, both as a scientist and as a person. Pinet is concerned that students of today are used to being told what to do and how to do it and that people, in general, are buffered from the variability that is a reality of nature.

I grew up in French Hill, in southeastern New Hampshire, an area full of French people. It was really poor in terms of money but rich in other ways. *Sputnik* went up when I was in the ninth grade. All those people were engineers, so I initially thought I wanted to be a civil engineer. Then I went to my professor, who was a glacial geologist. He was an older guy. He was just sitting in his office surrounded by all of these photographs of glaciers, and he was reading. I saw that. Remember, this is the first time, when I was a freshman and deciding: Is this what I want to do with my life? I thought, he is being paid to do this. So, at that point I learned you needed a Ph.D. I didn't even know what that was. I applied myself to my studies. I joined the Alpine Club. I did a lot of winter climbing. That just reinforced this whole idea, marrying the outdoors to making a living. That was always in the back of my mind.

I got a master's at the University of Massachusetts, studying geology. That's when I got my chance to do work in Antarctica. That was perfect. There were two of us. We went out and spent four months examining the geology in the Dry Valleys over the Antarctic summer. It was very, very frightening. I learned tremendously. I was 21, and it was the first time in my life I really began to confront my mortality in a serious way, because we were out there alone. It was such a powerful experience. You always were alone with this one other person, basically. No matter what happened, you had to deal with it.

There was one time I remember. We were going through a windstorm in our tent.

It was very, very cold and really windy. We were afraid of the tent being sandblasted and losing the tent. Then, everything was fine, and when the storm ended, we were tired of each other. I just took off climbing for part of a day. I climbed a ridge overlooking the ice sheet. It was a cloudy day, so there was no horizon. I just remember looking into infinity. I was on the bottom of the world just looking into nowhere. It happened to be a day when there was no wind at all, so there was no sound, no sound at all. I was the only living thing there. I'm not religious at all, but it was very humbling. From that point on, I understood what humility was. I always try to use that when I teach. It was such a powerful experience.

After I got through with my master's, I wanted oceanography. I knew Bob McMaster, a professor at URI [University of Rhode Island], and I decided to go there. All that stuff was happening with sea-floor spreading and plate tectonics. You just wish all graduate students could somehow experience that, to see a field collapsing and then reemerging. You know, all the seminars and the debates, the professors arguing all the time. It was very exciting. I feel very privileged. I went into oceanography and started studying the geophysics of continental margins. It was going out to sea, but at the same time it was also getting back, dealing with all the information, and developing a story. That was important. Either one alone, for me, would not be complete. When I look at scientists who study some aspect of the natural world and never, never go outdoors—I think they are missing something fundamental about the meaning and significance of what it is that they are trying to do. It is important for a person to be out there. We forget all those emotions, all those essential things that we try not to use. I mean the smell, the taste, and the feelings you get. I really think it is critical for doing good science, exceptional science. I think it is also good for teaching exceptionally about science.

You know, so much geology now is done on the computer, with models. All those assumptions that go in those models; you try to make all these statements. To me, we are missing something about reality that's important. I've noticed that a connection between humans and the natural world is decreasing, and maybe that's what I'm worried about. Here I am at Colgate, and I teach my geology courses. I feel my contributions to our program here, which is a very good one, is that I help students understand the limitations of science. I teach everything from the standpoint of: How is this limited? this technique? this information? this concept? I try to make them understand all of the assumptions we are making.

Grounding our science in the real world is very important for geologists and biologists. The real world for me is the natural world. That's why I really believe, sometime in the next century, biology and geology are going to become "the sciences," because the real problems we are facing in the near future and in the longer term as well are these environmental problems. They are serious. I think biologists and geologists, unlike a lot of physicists and chemists, can construct stories from fragments of information. That is a real value to science. Plus, we confront reality, instead of a simplified, collapsed sense of reality. That's why we don't have those invariant laws that physics and chemistry seem to have and somehow it gives them credence. The reason we don't is because it's so complicated.

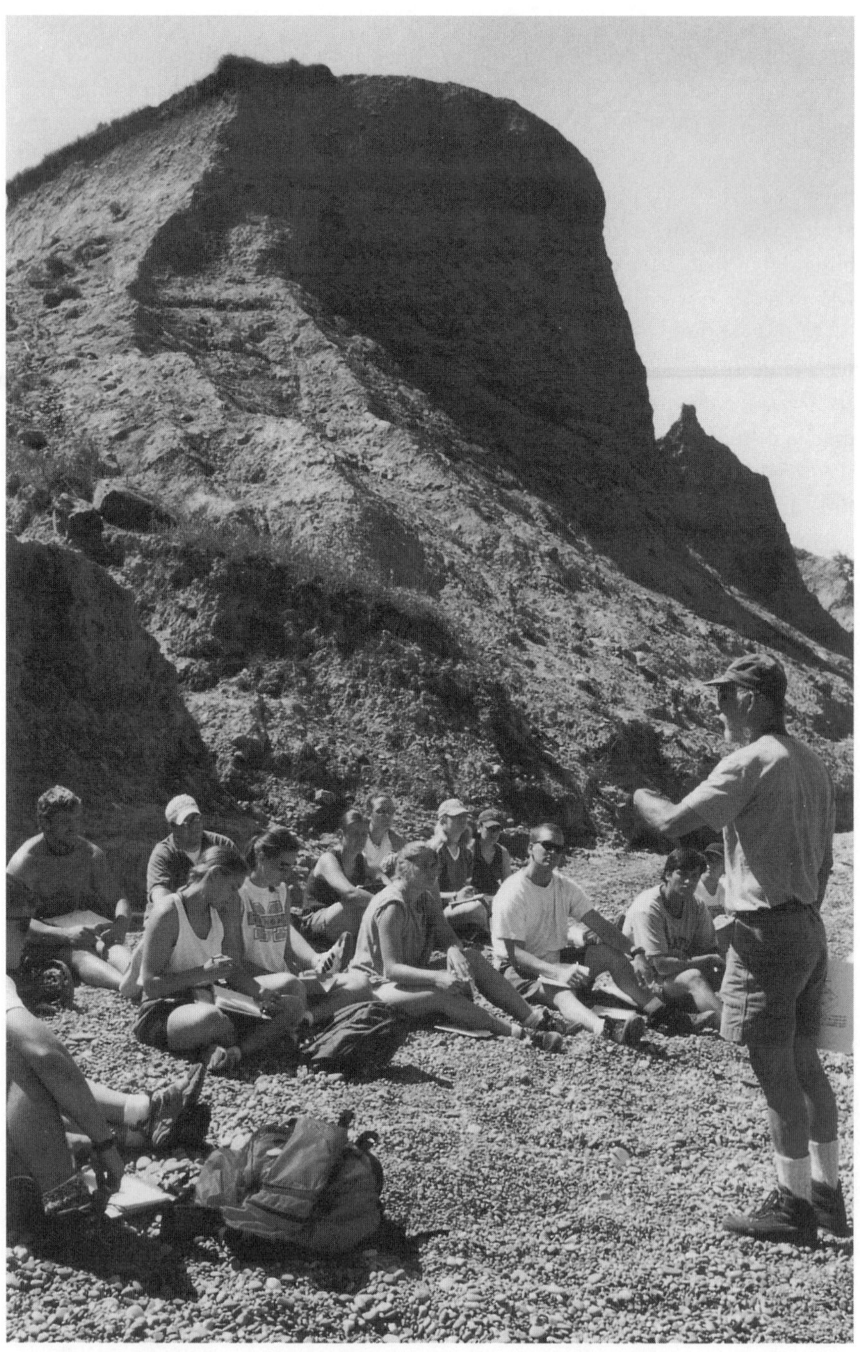

Paul Pinet challenging Colgate students in the 1990s to overcome their initial frustrations at exploring a new field site by the shores of Lake Ontario. Photo by John Hubbard, Colgate University.

Geology and biology are a little messier.

Yes, true, a little messier, but that's the appeal. To understand that, you have to be outdoors. You have to be reminded of that. You come into the laboratory. You have these numbers or whatever you have for your model, and you play around with that. You draw your conclusions, and then it is so easy to think, somehow, that it is profound. You think it is, but it isn't. You are reminded of that by going back into the real world and seeing the little thing, the statement you made, how really trivial it is compared to everything that is out there interacting on all of these different time scales, different spatial scales.

The reductionist approach to science always sees boundaries. I work on the shoreline on Lake Ontario with students. Here we can actually draw the boundary. Air, water, and solids come together in one line, but it isn't a boundary. Unless you go out there, you can't understand that box models, pieces of reality, are really incredibly incomplete if you only understand in isolation. The natural world is so complex. Yes, there are gradients where you might draw a boundary, but it isn't an impermeable boundary. At different scales, there are all sorts of things happening that are critical to whatever is happening away from that boundary. I realize we are never going to understand the natural world. There will always be questions. You have to remind yourself that our understanding is incomplete before we make all these claims of managing the world. It is very scary, very exciting.

What does it mean to do anything to the environment from an ethical standpoint? That came directly from just going back to areas where I had been periodically when I was younger and just seeing what had happened to them. But to people who are young, that has become their standard. They visit it, development is there, and they see it as a beginning point, so it is not so alarming to them. Where is all this leading? The burgeoning population but yet promoting all this ecotourism? In a sense it helps educate people, it really does, and yet it damages various things.

I never regard myself as a tourist. Rather, when we go to study the natural history of an area, we are travelers.

I like that. Why is it that I don't consider myself a tourist? I don't, because whenever I travel, I never sign up for any groups. The joy comes from the discomfort of trying to figure out what you want to do and how you are going to do it. You have that sense of freedom, but it isn't being a tourist. Of course, I'm generalizing here, but a lot of people have a lot of money and in everything they do they are being held by the hand.

Your students?

Yes, and their idea of the wilderness is exactly that. You find out a group that does more or less what you want to do and you pay them the money; they appear and they give you your equipment and they give you your food; you go have this adventure, a wilderness experience.

I'm always surprised at the reaction of students when you teach an introductory level oceanography course, and they haven't been to the ocean or they have been but they haven't really seen it. As an example, we'll just talk about a long shore current. They sort of understand it, and they can diagram it. Yet, when we visit the ocean and when they see it for themselves, it just shocks them. The fact that they see it and then when they go back and look again at a diagram, it has a whole different meaning in a very important way. It isn't just that now they have a very concrete sense of what a long shore current is, what it looks like, and what it does, but it is just that suddenly they have a sense of scale. You know, being out there and seeing the width of the surf zone and trying to imagine it during a storm. Or, if you are out there during a storm and seeing it, that's even more powerful. You get a whole new perspective. What happens is that you come back and you almost have to laugh at these elegant statements that you've made and these equations. You see all the variability out there. It helps the students really mature as scientists, in a way you just cannot do if you confine yourself to models or to the classroom.

The thing about storms is that there is so much energy there, it energizes you. All the fear and the excitement and the smells. The salt is everywhere. Everything is moving and seems chaotic. The reason it is important is because it penetrates you.

I guess I can say I'm glad I've experienced two powerful hurricanes. It's different; it's not something that you can really describe to anyone.

It's funny that you say that, because I always tell people that you probably learn more when you are uncomfortable, whether it's fear or something else. When you are comfortable, you tend not to learn that much. People at Colgate do a lot of outdoor things through groups. The tragedy there is that it is different from a real field experience. There has been so much insulation provided by the group as opposed to a real field experience. In contrast, students who go out to sea in the SEA [Sea Education Association] program experience discomfort. They, to a degree, are given a lot of responsibility for the safe conduct of the vessel, and they are doing their science. It's not like, I'm doing my science now, and now I'm sailing, and now I'm eating—it's all intertwined. That's what field experiences do. It's doing science, but at the same time that you are seeing all these things, the eating, sleeping, and all of that is occurring together. It's not split, like around here on main campus. Here you go to class, you go to lab, you go home to eat. So a field experience is much more integrated into a whole life.

When I participate in the off-campus geology field program, I have the students first. It is all about getting them to organize themselves for the camping part, getting them to cook for themselves, to set up their tents. Then we go out into the field at Crown Point. It's their first-ever attempt at actually trying to create, to make a geologic map of a very small area. The way most people used to do it is to take them and help them identify the formations. You would go to various rocks and show them what they look like at this spot. Here is one unit, here is another unit; this one is older than this one; you identify the rock from its appearance; you make some measurements to see how they've been disturbed; and then you make a map. Well, my way is very different. I just take them to an outcrop, and I don't tell them anything about it. I don't tell the name or the age. I do show them how to describe the rock and how to make simple measurements. I say, "Here is the area. OK, now, for the rest of the day, you'll divide into groups, and you need to go and map." They've had a lot of geology, so they know about it in an abstract way. They just don't know about it in a concrete way. I try to give concreteness to their notions about what these things are.

I tell them, "It is going to be very frustrating, but this is going to be one of the most important days of your life because you are going to see that frustration is eventually going to lead you to be a better geologist. You are going to wander around; you are going to be confused; you are going to see things that don't quite fit in. You are not going to make sense about everything you are seeing, but if you persist and finally get your feet on the ground with something, from that, you will start building on it." At first they get angry, but then after the first day I get them all together, at night, after we've had dinner, and I get them to talk about the sorts of things they think they saw. Collectively, I let them try to begin to understand something. I try to identify areas that are problematic for them, or areas that I know are key areas. I go with them on the third day and say, "What's this? What do you think?" There is something about the mind—maybe it is the hand holding all their lives—their potential hasn't been attained. They are so used to being told what to do and how to do it.

I see the real value of the field experience is the sensual part. I really think that is important. I get different reactions to that. Some students take it seriously, and some of them understand very well. But there is still a group that goes through the motions because I want them to, but they don't really understand it, how vital it is to their development as a scientist.

STEVE WAINWRIGHT brought up the point that field research can be at any level and involve any discipline, but it has to involve more than the eyes and the ears. Our world is so oriented to visual and audio. All the senses can and should be involved.

It is too bad. Not many people feel the way you do, particularly among scientists. Or, put it this way, not many people are willing to talk about it. It is sort of like you begin to lose credibility in a way, the way they see it. I don't. They feel that you are preoccupied with things you shouldn't be, not that you shouldn't feel them, but the last thing you should do is somehow give them importance in terms of your work.

I've often wondered, can you teach creativity?

Park S. Nobel (b. 1938)

Professor of Biology
Chairman, Department of Biology
University of California, Los Angeles
Interviewed in his office 20 July 1998

I took Park Nobel's course in physiochemical biology of plants during my first quarter at UCLA, and it changed the whole way I approach any scientific question. Nobel presents an engaging and humorous account of his transformation from being a physicist to becoming a field biologist. He notes the serious effort and considerable time needed to conduct and communicate good science, factors not always appreciated by the current generation of students. He states that he was humbled to discover, after two person-years of work developing a model to predict the ideal spacing of *Agave*, a commercially important crop in Mexico, that the result was within 2% of the spacing the Mayans had developed through experience and time centuries ago.

I was post-*Sputnik*, but very close to the *Sputnik* era. Everybody had to be an engineer. I went to Cornell, and I decided to go into what was the hardest field there, a five-year program in engineering physics. No biology whatsoever. I went to Cal Tech as a Ph.D. student. [Rudolf Ludwig] Mossbauer had just received the Nobel Prize. I was going to work at the nuclear physics–solid state physics interface. Biophysics was a word that really wasn't thought of much then—this was 1962—so I said, "I'll take this course in biology." The fantastic thing was that each week there was a faculty member who gave two or three lectures in his discipline and then three hours in his laboratory. They were all male; that's why I say "his." We went through the famous scientists at Cal Tech, one by one, into their laboratories. [Alan] Hodges was electron microscopy, Max Delbruk was there, and [Robert] Sinsheimer, a lot of people who had become quite well known. It was incredibly exciting to see these biologists, but the jargon, the vocabulary, was totally overwhelming. In the lab, the instruments were relatively simple physics: the electron microscope, centrifuges. I could talk at their level and beyond on what their measurements were, whereas I didn't have a clue what they were talking about when they had all this background in biology. I decided, hmm, this sounds interesting. I went to my advisor and said, "I want to do my minor in biophysics." He said, "Physicists don't grow sweet peas!" That was the beginning of my end at Cal Tech. I found they didn't have the flexibility to look at this interdisciplinary-type thing.

So I went up to Berkeley to visit somebody, and I had some time to kill. I talked to some faculty who encouraged me to switch from Cal Tech to Berkeley and to go into

Among the plants he loves, Park Nobel prepares to make physiological measurements on a cactus in the late 1990s. Photo from the files of P. Nobel.

biophysics. I said, "OK, fine, but you have to transfer all of my graduate courses from Cornell, from Cal Tech, the two languages that I had passed, and let me go in flying. I will take these courses without the prerequisites because I want to find out if I can do it. I don't want to spend three more years being an undergraduate again." I took the animal phys. course, which had all kinds of prereqs. I just went right into it, and it was fun. I enjoyed it. The next year I begged the instructor to let me teach the kidney, because I knew she couldn't do the kidney very well, she had difficulty with the physics of the kidney. I also did the lung and the heart.

Then it came time for a research project. I was moving from physics into biology, and we had two projects. They were both ion transport at an organelle level. The bigger one was the mitochondrial calcium transport, phosphate, etc. For that project I had to go down to the slaughterhouse at Emoryville. The guy said, "Come in here." I had my ice bucket to get a heart. He said, "OK, stand right there," and he brings in the cow. He puts the .45 right on the head of the cow, and he shoots the cow right in front of me. The poor cow falls over. They are huge when you are up close to them. He slits it and gives me the heart. I just could barely keep my food down. I took it back to Berkeley and said, "What was the other project?" Lester Packer said, "It's with chloroplasts." Same project, calcium and phosphate uptake, in chloroplasts. So, I went to the supermarket and bought some greens.

Spinach, I bet.

Yes, spinach, and it didn't cry. It was pretty easy.

I stayed as a lab scientist working on chloroplasts from my Ph.D. in 1965 until 1973, when I went to Australia on sabbatical to do a project on guard cell chloroplasts. I was making a transition between chloroplasts as a subcellular organelle and a guard cell, which was the environmental interface for letting gases into a plant. I had kept up the physics, math, and chemistry in terms of teaching and incorporated them into biology, but not too much in my own research. I felt that this transition to the environmental might allow a little more of my mathematical and physical science background to manifest itself. In Australia, the chloroplast project was impossible. Although I had a Guggenheim to do this project, it would take thousands of hours to get enough for one prep, and furthermore, the chloroplasts wouldn't be vital after that amount of time.

I had seen a wind tunnel on my second day at Australian National University at Canberra, which wasn't being used, and I had a question or two from the physiological ecology course that you took, Bio 128, about boundary layers.

So, you were teaching that course at the time?

Yes, it was another phase of moving away from being in the lab. HARRY THOMPSON was co-instructor, and then I finally was able to get enough confidence to take a class into the field. The questions the students asked were about boundary layers next to nonflat leaves. Now, for flat leaves, there is a certain boundary layer theory that makes sense with them. But, here, we were looking at what is called a bluff body. The two geometrical forms would be a cylinder and a sphere. The cylinder and sphere have completely different wind-influencing patterns compared with a flat leaf. The air is arrested or stopped and has to go around a bluff body. You have vortices, which come off the trailing edge. So it is quite different than a flat leaf, where you can actually have laminar flow over most of it and very little turbulence.

There was this wind tunnel. OK, fantastic. I took my engineering background and went into the engineering literature of Nusselt numbers, Prandtl numbers, and Reynold's numbers, where the engineers can go from one type of system to another type of system without resolving everything. They can just transfer information from one type of fluid to another type of fluid. I didn't have that at Cornell, but I could pick it up. I took that into the biologist's realm, which says, we can measure the length of something and wind speed, but we don't want to get confused by dimensionless numbers. I derived some equations, which were able to predict boundary layer thickness for some bluff body spheres and cylinders. Then I used a puff ball, a basidiomycete, as my model for a wet sphere, and I used various reeds for cylinders. I put them in the wind tunnel, I looked at the water evaporation, and I did sort of heat transfer–type analogies. It was darned interesting, and I really had a lot of fun.

I came back to UCLA in 1974. Ray Lunt, director of the Laboratory of Nuclear Medicine and Radiation Biology, and various government officials were looking for someone to do biophysics of desert plants. They had somebody else to do the biochemistry. So I said, "OK, I'll do biophysics of desert plants. What does that mean?"

Nobody seemed to know, but if I were to put in a proposal, then I could do it. They wanted someone to do desert plants, and I had the credentials of a biophysicist. I met with the chairman of the biology department and said, "I've got this urge to switch a little bit." He said, "While you were gone, we hired somebody who is a cell biologist. We want you to check, since you are a cell biologist, whether the list of equipment makes sense for her." All cell biologists had basically the same things: Sorvell centrifuges, Mettler balances, and Beckman scintillation counters. When he showed me the list she wanted, I said, "That's perfectly legitimate, that is exactly what a cell biologist would want. It is exactly what I have, and I'll sell it for 40% of its cash value." The dean was able to set her up without going to great expense. I had burned my bridge. There was no way at that stage in my career that I could go back and have the National Science Foundation, or some other agency, fund me at the level to buy the equipment that I had built up in my approximately eight years at UCLA. Now I was stuck, but we had some cash. I went and bought some instruments that enabled me to make field measurements of a bluff body, that is, some body that arrests the motion of air. I looked at the desert. There were cacti and agaves, and they are bluff bodies.

I went to the desert, and I began looking at cacti and agaves from the point of view of an engineer, a physicist, a modeler. Although a lot of people had worked as ecological physiologists in the desert, they hadn't asked the questions quite the same way as I phrased them. When I saw spines on a cactus or epidermal hairs, pubescence, I was interested in how that affected the temperature of the cactus and, from the temperature, how it affected the range or the distribution of the cactus. Or, I looked at the surface wax and asked how that affected the reflectivity and hence, the temperature. The questions I was asking were physical questions with biological importance. This happened to hit at the right time. They needed someone to study biophysics of desert plants, and this is what I called biophysics of desert plants. The Department of Energy funded that program for over 20 years, and it went many different directions.

From my initial looking at agaves and cacti as engineering, physical objects, I have to admit, I fell in love with them, and that's treacherous, because then I really wanted to work with them. I have published three books on cacti and agaves. One has recently been translated into Spanish. I love my normal book, which you have used, *Physiochemical and Environmental Plant Physiology*. It has been my main stream with my background, but my love has been agaves and cacti. It came at a very nice time, when there was sort of a vacuum of people doing ecophysiology on these plants. The instruments were readily available: growth chambers were available to use in the lab, computers had picked up speed, so we could monitor and model accurately. Total instrumentation was really developed by then, so we could measure CO_2 uptake, which you did, if you remember, with a grass called *Hilaria rigida*, which is now called *Pleuraphis rigida*.

I will tell you, you can smile, but you insisted on our measuring CO_2 in the field. I did not want to, because we had to take a very clumsy old AC-powered IRGA [infrared gas analyzer], and we also had to take calibration tanks. It was just dreadful. As soon as I said, "No, Betsy, we're not going to do it this way," you said, "Yes, we are, I insist that we do this." I thought, oh gosh!

You told us to pick the project that we wanted.

Right, I did, but I didn't want you to do that. The more I didn't want you to do it, the more you insisted on doing it; it was really very humorous.

I don't remember it that way at all.

I remember it very clearly! So, we got out there, and we put three leaves of one plant in the chamber, *Hilaria rigida*, at that stage. We had three leaves of this desert perennial bunch grass, and the reading went off scale. It read such a high rate of CO_2 uptake that it made no sense. I looked at the leaf area and I said, "No way." So we fiddled around and fiddled around until we had broken two of the leaves, and one leaf remained in the chamber.

See, I only remember that one leaf.

I remember this very distinctly, like it was yesterday. I said, "OK, go ahead, this is a student project. I can't see what's wrong, just take the numbers. I'm not going to penalize you because this is what the instrument says." It stayed on scale, and you got the readings. I couldn't find anything that we had done wrong. The pumps worked correctly, the calibration looked fine. We got a rate of 61 micromoles per square meter per second, which was the highest rate ever reported for any species at that time. I couldn't figure out anything wrong with it. So I dug up some plants and they were a little hard to get back to the laboratory.

So that's why you kept working on that, because of those data?

Yes. Those data, and it was like an inspiration to try it again under more controlled conditions and repeat it in the field. I actually got quite interested in that bunch grass, and I published a number of papers on it. But, it was your insistence and then the instrument going off scale. Nothing I could see was wrong, and I have confidence in instruments. I was confident that the measurement was correct, even though I couldn't believe that we had stumbled upon the plant, at that stage, with the highest photosynthetic rate ever published!

Unlike me, who is still nervous with instruments and figure I must be doing something wrong. It was a good thing we were both there at the same time.

Right. This is where the physics and engineering really carries through. You look at something and try to troubleshoot and really make it work. A physicist really feels the instrument can perform and is not totally baffled, even though today a lot of the instrumentation is very complex. I really understood how that instrument worked; I couldn't reason why it wasn't working properly and then carried forth.

I naively like organisms and how organisms work. I wanted to find out what the photosynthetic rate of that organism was.

Right, and I said, "Well, it's just a student project." I remember using the words *just*, *project*, and *student*, and it has since made me laugh. I tell that story many a time about the role of women in science, it impinges on that, the role of a faculty member and a student, the role of instrumentation, the role of a project, and the role of serendipity—all these things.

Back to Agave Hill. It had a trailer and that became my main research site in the desert, working with agaves, cacti, and a few other plants like a fern and this bunch grass. I began getting interested in certain computer models for predicting temperatures of plants, very complex models. A number of my students actually ended up getting Ph.D.s with me on cacti and agaves and ending up at Jet Propulsion Lab doing strictly computing. They actually beat out students with Ph.D.s in computer science, because the modeling we were doing was really quite complex.

About the same time, I wanted to test some of my models, not just in Arizona, the Sonora, and Sinaloa, in Baja California, but in South America. So I was working in the Andes, looking at a barrel cactus and a columnar cactus, looking at their highest elevation as a function of latitude and then using models to reduce the variation in stem diameter, in pubescence, in spines, so that I could say where it should be. We were able to predict top elevations of populations within 20 meters, which is astounding.

In the Andes, you go up very quickly from the Pacific Ocean. You can go up every valley, there is a trail on one side or the other, and you get to high elevations very easily, on all the little rivers flowing west in Chile. Coming down one day from one of these measurements at a high elevation site, I saw these cacti growing in a field and asked, "What's that for?" They told me they were growing cacti for the fruit; they are grown for fruit in a lot of different countries. This was about 1981. I began working on *Opuntia ficus-indica*, a prickly pear, now generally called a cactus pear. That became interesting and exciting because now there was an economic motivation for the modeling. We were looking at how to space the plants to maximize growth per plant and then growth per unit ground area, and they are quite different. For maximization of growth per plant, you put the plants far apart. To maximize biomass productivity per ground area, you have to have an optimal spacing, and so it is ideal for computing.

I also worked on *Agave fourcrogder* (henequen), which is in the Yucatan peninsula of Mexico, and spent about two person-years developing a model for a plant with 160 leaves, a very complex model, putting the reflection of one leaf onto another. Then I developed a prediction of optimum spacing of the plants to get the maximum productivity. I was humbled, after this two-year effort, to find the Maya had developed the ideal spacing within 2% of what the computer model predicted. It was a nice sort of interchange between modeling and then having the time the Maya had to develop henequen as a crop.

Presumably by trial and error?

Trial and error and lots and lots of labor. Some of the work with *Opuntia ficus-indica*, the modeling work there, has led to a change in the agricultural practices in terms of the spacing of the plants, where they are getting a much higher yield per hectare per year than they had with the traditional spacing. That's very interesting. I gave a talk at a congress about ten years ago and a person who was basically a businessman-farmer listened and took it more seriously than my own colleagues did. He began planting his farm using this model of the orientation of the plants and the spacing of the plants, and he is doing very well financially. I like that.

I began as a lab scientist, moved to sort of environmental things, became interested in environmental interactions at an ecological sense, and now have as much interest in the agronomic aspects of the environment out there, still with the fun and excitement of bringing a physical approach as much as possible into biological questions. With lab work, interestingly, before I went to Australia in 1973, I was delegating my lab work to technicians. That was, to me, a very bad sign, that I was losing some enthusiasm for it.

When I came back to UCLA, I realized the only way I could direct students in the field, where I wasn't trained, was that I had to go out into the field myself. So I went out every other weekend for two years. Fifty trips I made to Agave Hill in a two-year period, to look at plants for two whole years, in all the seasons, in all the changes. I saw some phenomenal things. I was lucky to have an interest in instruments, so I could measure things right away, but there were some really interesting, almost classical studies that I did at that stage. I basically became a graduate student, at the same time that I was a tenured faculty member.

When I have sent people on fieldwork, some days I get itchy to see what they really see. You go out into the field, the plants aren't in straight rows, they aren't of the same age. You have to make a lot of judgments. A lot of time is spent in the selection of plants that you are going to measure. You walk around to get a sense. Is this slope really normal? Why is this slope different from that slope? You look at the substrate. There are a lot of things you have to look at before you begin measuring. Whereas, in the lab, you have uniformity. You have material that may be the same one day to the next day to the next week to the next year from now. In the field, you are running into complexities, which are much more exciting but are very hard to anticipate.

So, how does somebody get that kind of experience?

It's hard. Back in 1975, I made some measurements that took about eight minutes to make, then I had a two-minute break, then eight minutes of measurements again, a two-minute break. I did it for 24 hours without sleeping. I began realizing that I was seeing the response of net CO_2 uptake to light for a crassulacean acid metabolism plant for the first time in the field over 24 hours. I was seeing that the dark side of the cactus really was not doing too well in terms of CO_2 uptake, while the sunlit side was. But I felt that, after 24 hours, with never more than two minutes from the next measurement, I really wasn't quite where I wanted to be. So, I stayed up for another 24 hours without sleeping, making measurements for eight minutes, then two minutes off for good behavior, then more measurements. After that, I had to get into a car and drive back to Los Angeles, a little over a three-hour drive. I was exhausted before I got into the car, but I decided I wanted to get back and work, because I wanted to analyze the data more. Instead of going to sleep after that 48 hours, I stayed up and plotted the curve. That, now, is one of the first light-response curves for a CAM plant. It shows, indeed, that a cactus or an agave is not saturated even under the high light levels of a desert. That became the underpinning evidence that was used in the computer models for plant spacing.

What I find now is that a lot of students don't like the idea of spending 12 hours making measurements continuously. Nobody really wants to spend 24, let alone 48. [I don't know] whether it is a generational thing or just personal; perhaps it is probably a bit of both.

I've often wondered, can you teach creativity? My feeling is that you can encourage creativity, but it is really very hard to teach. To be original, you have to be really disciplined; you have to be really dedicated. Certain measurements are hard. You have to be willing to shut other things off, to focus in on things. Science is getting harder. With the uncertainty of job markets for recently graduated students, the difficulty at a university in publishing, getting grant support, getting tenure—it is harder. Personally, I look back at my own background in these quantitative fields, and then the shifts and the growing interest in plants, and then the excitement with plants and combining this excitement, wanting to be outside, wanting to be in the field, wanting to do new things, wanting to be creative with the tools, the discipline—that was a fortuitous combination that I could not have predicted. It wasn't the shortest path, by any means. But you can't tell when you are moving forward. You can't tell whether taking this course in chemical engineering, taking this course in computing, taking a course in a foreign language is really going to benefit you. I think the bigger your arsenal, the better the chances that you are going to hit your target. So, [Louis] Pasteur's comment—chance favors only the prepared mind—has been my motto. It's a comment I feel very strongly about.

Ultimately, there are individual minds that have creative ideas and that's what should be supported.

PETER F. SALE (b. 1941)

Professor of Biological Sciences
University of Windsor, Canada
Interviewed in St. Croix 28 August 1997

In contrast to the confident and polished professor I have known as a close colleague for many years, Peter Sale portrays himself as naively stumbling onto opportunities along his career path. He points out that factors which are of vital importance to organisms in some settings may be completely different from those of importance elsewhere. He is a strong advocate of the necessity of fieldwork in behavioral and particularly ecological studies of fish, a truth he learned through experience. He laments the fact that due to today's tight funding, graduate students are often forced into more of a technical role in research projects, rather than given the opportunity to display (and thus to develop) their own creativity, a necessary element in the production of good science.

I didn't know what herpetologists did, but I thought that is what I'd become. I actually tossed up between architecture and biology, but I didn't know what kind of jobs biologists had. I thought they only worked in museums and taught high school biology. Looking back, my naiveté was incredible. I had no idea about biological research as a field. At the University of Toronto, I had a very conventional first-year course. They made a real effort to employ students in the summertime. They phoned me up and said that there was a possibility of a job, which would have involved me in setting out live traps for rodents along the highways. The problem with that was I didn't have my driver's license yet. I'm probably the only person of my generation who didn't have a driver's license by the time I got to college! And so, they gave that job to somebody else. Instead, I got a job helping a graduate student do some temperature tolerance experiments on fish. Running lethal temperature tolerance experiments required 24-hour-a-day observation to detect when the fish died. The lab the student was working in was within easy access to my home, which made it another plus. I got to see that there were people who did research for a living, and in fact, during my first year as an undergraduate, I began to realize there were a whole bunch of careers beyond teaching high school or working in a zoo.

The first-year zoology course had a lab every week, and the people who were teaching that lab seemed to be very interested in the teaching of it. I was just taking these courses because I was learning. I certainly didn't have the kind of focus the premed students had: I'm going to get through this stuff and get an A+. I didn't do that.

I was terribly interested in whatever they were telling me about biology. It just suddenly dawned on me that there was a whole world of science out there that I knew nothing about. I was totally unprepared for it. I don't know if I ever talked to a high school guidance counselor, but if I did I didn't get anything useful out of it. I don't know how many of today's students are as naive as I was, but I am very conscious of the fact that first-year students can be quite unaware of what the possibilities are for them. They go into biology for all kinds of reasons. Some of them are absolutely committed to being doctors, but many of them go into it because they like biology, but they don't know what this offers them. I think one of our jobs is to try to educate them about that.

One of the consequences of my first summer job was that the subsequent winter I was hired in that lab on a part-time basis. It was an experimental trout hatchery, so there were lots of trout being reared. They hired me to be there on the weekends, to make sure the water was running, and nobody was dying. Following that, I was employed at that lab, not working for a graduate student but working for his major professor, the following summer, and in fact, for the rest of my time at the University of Toronto. The professor was one of the outstanding fish physiologists in Canada, Fred Fry. I ended up doing a master's degree with him, so it was very serendipitous in the sense that I had had no particular thought of working with fish when I entered college. I got into fish purely by accident. If I had had a driver's license, I probably would have become a mammalian ecologist. That's one of the things that has actually always fascinated me. When you start talking to people about how they got to do what they do, there are tremendous numbers of forks in the road where you take the fork without even realizing you are at a choice point. I guess some people really plan ahead, but I certainly never did.

The summer after my second year, Fred Fry gave me a task to run some temperature tolerance experiments with a race of trout that came from some lakes in northern Ontario. There was some controversy about the taxonomic affinity of these trout. This work was going to help decide what they were. He put it entirely in my hands. He was a very busy man, not in the lab most of the time. He used to come in, talk to people, and then disappear for a week and be off doing other things. I had to plan these experiments. There were other people in the lab who did a lot of helping. In particular, I remember a postdoc who helped me do regression and covariance analysis, back in the days when doing that with ten data points in the sample was a real achievement and took about a day and a half. There was mentoring from people who were graduate students or postdocs or other people who were fully fledged Ph.D.s and were working in the lab. My first paper came out of that summer's work.

What I've always recognized about Fred Fry, as a mentor, was that he gave me the project; he taught me how to do it; he steered me when I needed steering; he helped me write it up. In fact, he took a major role in writing it up and making sure it was acceptable before it was sent off. It was published under my name alone. I think that had an enormous impact on me as a scientist in terms of how to deal with my own students. Then, that grew into a master's degree with him because with that same species of fish he said, "How about doing some fieldwork on it?" I did the fieldwork in northern Ontario on my master's degree after my third year as an undergraduate and then after I graduated, so I put two summers in before I officially enrolled as a master's student. Then I just stayed on one year to write up the master's. During that year, I also took a genetics course and a required course in the history and philosophy of biology. Those were the only courses I took. That was the typical Canadian system. They didn't offer courses for graduate students.

Actually, the master's degree was a real problem because the fish was practically extinct in its home lakes. I spent a lot of time fishing and caught hardly any. Mainly I tried to do the biology of the thing. Just as an aside, one of the things I did was to sample the water. In Canada, you automatically look at the nature of the environment in which the fish is living. You take water samples, you send them away, and you find out what is in them. I looked at oxygen, pH, and all those sorts of things. Well, these lakes were acid. They were about pH 4.85, and I duly noted all of this. The lakes had no mollusks in them at all, which is a good sign of their acidity.

Too acid for the mollusks to deposit shells?

Right, they can't make shells, and what was happening is that these fish were disappearing because of acid rain.

Ah, but of course, that wasn't a big deal at the time.

Nobody knew about acid rain, and I didn't pick up on it. I saw that the lake was acid, and I assumed it was an acid lake. I didn't assume it was becoming more acid. In fact, many years later, somebody reported on the fact that this particular story—of how this fish became rare in these lakes during the 1940s, 1950s, 1960s—was one of the first pieces of evidence that industrial air pollution was causing an acid-rain problem in Ontario lakes. So I blew it! I could have become famous for acid rain, but I didn't know anything about it. It never dawned on me that something like that could be happening.

Following that, I wanted to get a Ph.D. I was now at the stage where universities were paying me, sort of, to go to school so I wasn't a burden on my parents. I was, incidentally, the first person in my extended family to go to university. My parents encouraged me to continue. I had this feeling in my mind that I wanted to go to a place where I could continue to work on fish, but I wanted to work in a system that was different than northern Ontario lakes. I thought I probably wanted to work in a marine system. I was very interested in animal behavior, which came from some lectures that Fry gave. I asked Fred Fry for some ideas. I had the audacity to make an appointment with Miles Keensleyside to seek further advice. I made a special journey (by this time I had my driver's license) and drove from Toronto to London, to the University of Western Ontario. I don't know if you've ever run into Miles, he's one of the nicest men around. Miles was a fish behaviorist, now retired. He did some very early work using scuba watching juvenile Atlantic salmon in streams in New Brunswick, then he went off to Europe to all of the ethologists, and then came back to Canada, got a job, and spent time looking at cichlid behavior. He'd never met me; I didn't know him from Adam. I found out he was there, and I made an appointment to go and see him.

I walked into his office, this brash young graduate student—and I was very young looking, so I was really brash and young—and I said that I came to talk to him about places where it might be worth going for a Ph.D. in animal behavior. I made it quite clear that I wasn't interested in working with him! I've thought about it afterward; he handled it very well. I think I'd have thrown me out of the office, but he made some suggestions of particular people. He confirmed that the universities that Fry had suggested were appropriate (the University of Miami, the University of Michigan, UCLA, the University of British Columbia).

But the moment I got a positive response from Hawaii, I lost all interest in other places. I got out my *National Geographic* map of Hawaii, I learned how to pronounce the place names, and that was it. I was going to Hawaii! There was a coral reef and that was what I wanted to do. I didn't go to Hawaii to work with any particular person. In fact, again, there was some incredible naiveté in the way I approached the whole thing. Once I got there, I started figuring out what I could do. I would have been a total failure in today's much more competitive world.

The first year I was in Hawaii, I had to relearn everything I had learned in Canada. I spent a lot of time wandering around rock pools measuring temperature and salinity, and I was even measuring oxygen content. I was going to study the behavioral cues and behavioral responses that got young fish up into the tide pools. In fact, I was going to find out how they used the environment differently, so they could all coexist on a coral reef. The first year was a complete waste because the fish that live in rock pools in the tropics can tolerate just about any salinity the pool can throw at them. It throws different salinities and temperatures at them daily. What I'm saying is that I had to learn to completely reorient from the Canadian perspective, where temperature preferences and other responses to the physical environment were tremendously important in determining what fish could do and where they could live, [and change] to the view that these factors were tremendously unimportant in the tropics.

I did my Ph.D. with a very loose kind of supervision. I had a committee, but there were some interesting problems. I had an ichthyologist on the committee, Bill Gosline, who didn't believe in animal behavior, and a behaviorist, Ernie Reese, who, at that time, didn't believe in people working on fish behavior; none of his students had worked on anything except crustaceans. I just bumbled along with my committee. It wasn't clear who my major professor was going to become, but that was solved when Ernie took sabbatical the year I graduated and so he was off my committee. He didn't even sign my thesis, even though he had tremendous influence on some of the things in there. I became Bill Gosline's student. I had, at the time, considerable frustration and yet, afterward, enormous respect for the man. We used to get into amazing arguments. I remember showing him the chapter of the thesis that I thought was the absolutely central heart of the thesis, and he thought it should be an appendix.

I would identify Bill Gosline and Fred Fry as probably my two most important mentors. Both of them let me do my own thing. I worry that that is one of the things that we've lost with all the emphasis on having to have publications in order to get grants. Today, there is tremendous lead time needed to get a grant and a total lack of any other kind of money to do anything. I think graduate students really lose out when they're kind of shoehorned into a grant that is already written and they're told, "Do this bit of it." If they start straying, they are reprimanded because we have to have these papers come out so we can get the renewal of the grant. I don't think that's the best way to educate our future scientists.

In Hawaii, because of the Canadian influence, I think, I wanted to do experiments. Probably two-thirds of my Ph.D. work was done in aquaria, and only a small part was done looking at fish out in the field. I recognized, as I did it, some of the disadvantages of working with fishes in aquaria, but I persisted. It was when I got to Australia, where the Heron Island Research Station had no adequate facilities, that I suddenly realized that you had to work in the field. Now I've reached the point where I'm very unsupportive of a student who wants to do something in an aquarium with a few rare exceptions. It depends. If you're looking at details of behavioral movements, they'll probably move the same way in an aquarium. If you are looking at ecological questions, the aquarium is just far too artificial. Fishes become conditioned to the way the water circulates, when they get fed, how they get fed; they become different animals. Fish that you never see behaving in a hierarchical way in the field become very much a dominance hierarchy when placed in an aquarium. Space is just not there. But I learned that fact only because the facilities were not there at Heron Island to enable me to continue working in the aquaria. It's amazing how blind we can be to the borders of our paradigm. My view when I went to Australia was that you did experiments in aquaria because you could do them in a more controlled way. Now, I think you do it in the field; it's more difficult but more realistic.

I'm concerned that the process of producing field scientists has drastically changed. Not only are there fewer people getting the opportunity, but even those people are coming in through a different set of mechanisms. You know, a lot of students get their training on these team projects. Team projects are only as good as the members of the team. It is very rare that a group of individuals will work synergistically in such a way that they will produce something that is better than could have been produced if they talked to each another occasionally and went about doing their own stuff. There is this huge myth out there about interdisciplinary science being the way to go, and it is not true. It is so firmly inbred in the thinking of the system that increasingly you just have to play along with it. I'm very skeptical about it.

Ultimately, there are individual minds that have creative ideas, and that's what should be supported. The training of young people really requires that you try and identify if they've got an inquiring mind. If they do have an inquiring mind, you put them in a situation where they can use it. It doesn't mean you hire them on your project and tell them what to do, because then there is no inquiry. As a person who has been department head for a while, I've seen the range of variation in the way people handle students. Some people treat them as serfs. That's not training. If they do go into academia, they end up not thinking for themselves because they never had the opportunity.

If you want to talk more about how times are changing, because of the pressure of funding, the bandwagon scenario is becoming much more important. I've seen people switch into more molecular approaches to what they were doing simply because that was the only way to get funding. I don't think that that is going to make for necessarily good molecular science. There are lots of interesting questions out there that don't have to be approached at the molecular level. On the other hand, there are a lot of people out there who seem to think they should be funded to do natural history without any focus, and the system can't afford to sustain that.

How long did you stay in Australia?

Nineteen and a half years. When I finally left, the only reason I left was that there were more jobs to apply for in North America than there were in Australia. I got job offers in the States before anything came up in Australia, and I was interested in leaving the University of Sydney. I wasn't interested, particularly, in leaving Australia. At the time I was leaving Australia, I was invited to join the council of the Australian Institute of Marine Science, effectively the board of directors of AIMS. That was the last administrative post I was offered, and I had to turn it down. I was very actively involved in the administration of Australian marine science. I served on some other national bodies. The people at the University of Sydney were getting to take me for granted, because I had been there since I was little, and the usual story—you get to thinking that it's time to move on.

What's going on in Australia in coral reef science now compared to when I first arrived is terribly exciting. When I arrived there, the first time I applied for a grant to do some work up on the reef it came back rejected with the argument that you could do this with some fish in Sydney Harbor, you don't need to go to Queensland to do it. There was nobody producing graduate students on coral reef biology in Australia. The people who were becoming coral reef biologists were doing it because they were very independent souls, like Ross Robertson and Howard Choat. They both got their degrees at the University of Queensland. There was no one on the faculty at the University of Queensland who really knew very much about coral reef fish ecology. Howard got inspired by Joe Connell, and in fact some of my inspiration in Australia was Joe Connell. I met Joe for the first time at Heron Island. I think Ross Robertson was strongly influenced by Jiro Kikkawa or others of the behavioral biologists at Queensland at the time (none working on reef fish). Now the Australians recognize, as a country, that understanding coral reefs is important to them, and they fund it. They fund it bravely, also. They don't reject anything because it is a little risky, like NSF does. There is this huge group of people that's doing some very exciting things. You can see that the growth and development is really a plus.

For undergraduates, I think it is important to foster the knowledge that you can discover interesting things without a lot of sophisticated equipment.

JAMES G. MORIN (b. 1942)

Director, Shoals Marine Laboratory
Professor of Ecology and Systematics
Cornell University
Interviewed at his home 13 November 1997

I know Jim Morin is an excellent teacher as I was his TA three times at UCLA. He admits that he never realized how he himself got good grades until he began to teach. This self-deprecating approach belies his interest, enthusiasm, intelligence, and hard work, which his mentors in junior college, university, and grad school recognized and rewarded with opportunities to do research and teach. He describes how innovations in biotechnology, leading to fortunes, can have small, humble beginnings in basic science. The value he places on undergraduate teaching is an important asset to his current position as director of the Shoals Marine Laboratory, founded by JACK KINGSBURY with undergraduates foremost in mind.

I always wanted to be a biologist from when I was Ben's age [his son, about five years old], and I was in Minnesota. I always had frogs and snakes. I *really* wanted to be a herpetologist. Then we moved to California. That's when I discovered the ocean. I said, "Oh, forget about the snakes." My parents never pushed education. I went to Hartnell Junior College in Steinbeck's Salinas. There I ran into a fellow, Howie Feder, who had gone to Stanford's Hopkins Marine Station for his doctoral work with Don Abbott. He subsequently went on to the University of Alaska. He didn't get very many good kids at Hartnell so when he had someone like me come along and show some interest, he put in the effort. He was doing the early work on the effects of the starfish *Pisaster* on limpets, the mushrooming and running away. I merely did the photographs for his paper on it, published in 1960. My appearance in press was for photo credits, and that was a big deal to me.

Then I was asking, "Where am I going to go as an undergraduate?" I didn't know beans about where I should go. I had some friends who lived in Santa Barbara, and I'd never been on my own very much, so I went and stayed with them. It was on the beach and that was great! First of all, I grew up as a fundamentalist. That was a really big problem at Santa Barbara. I was a boarder in the worst fraternity; it was really bad, but I really liked UCSB otherwise. One of the first things I managed to be able to do was take the inverts course, and then I was encouraged to become a scuba diver. This is when there wasn't much diving except at Scripps [Institution of

Oceanography]. I became UC-certified, and I became a collector for the department, so I was initiated into the fraternity in a sense.

There was another development that was really important to me. I was just a country bumpkin, I didn't know what opportunities were around, but somebody told me I should apply for a particular program. It was a Ford Foundation grant called EPIC, Experimental Program Instructors for College. The idea was to provide incentives for undergraduates to see if it would make a difference in terms of the outcome. They provided money for books, and so I bought books! I got the whole Hyman invertebrate series then. I was also allowed to be a TA as an undergraduate and because of that I'm a firm believer in having undergraduates as TAs.

But then I got really involved in research, first with Demerast Davenport and, even more important, with Jim Case. He sort of took me under his wing. What I found really impressive about him is that he saw enough in me that he was willing to take me, instead of some graduate students, to Enewetak to study hermit crab chemoreception. That made a big difference in my life because it got me off to an exotic place and to a coral reef. This was early on, before I scuba dived. Boy, I'll never forget because it was the first time I was ever on my own. I had $50 of my own money to my name and instead of going back with Case and his crew of graduate students, at Kwajelein, I had the opportunity to get on an old flying goose that flew once a week to Ponape. I didn't even have a guarantee of a plane back because Micronesians had priority. If the plane filled up, I'd have to wait another week for the next flight. I'd have one meal a day, a bowl of rice and a fish. But it was an amazing experience.

I worked a lot with Jim Case, so that was really a formative time. I was working on *Renilla* but on feeding behavior. It gave me a lot of opportunity to work independently. It was valuable experience; it gave me the opportunity to branch out. I didn't have any confidence at all. I could never understand how I got As in classes at Santa Barbara. It wasn't until I became a professor that I realized.

That there are a lot of dumb people around.

Exactly! When I went to Harvard, I had never been east of Chicago. I put all of my belongings on the top of my VW bug and went through the end of a hurricane in Tennessee. Arriving in Cambridge, I remember driving around looking for Harvard only to discover that I was driving back and forth through the Harvard campus for an hour at night before I realized, really, the campus was part of the city. At Harvard I had to decide, who am I going to work with and what am I going to do? I had my own NSF grant, and I had the option of working with E. O. Wilson, Ian Cooke, who was a neurobiologist, or with Barry Fell, who worked on echinoderms. I figured that neurobiology was technically the hardest thing, so that was probably the way to go. In retrospect, it paid off. What is ironic about it was that since then I've done a lot of behavior and a lot of systematics.

I got started working with Woody Hastings. He's a cell biologist who works with circadian rhythms, dinoflagellates, and bacterial luminescence. After I discovered the *Obelia* system, Ian Cooke, it turned out, had very little to do with my dissertation except for providing space and some equipment. The irony is that when I met Ian some years later in his position in the Von Bekesy Lab in Hawaii, he had forgotten that I was his student! On the other hand, at this same time, Woody had forgotten that I wasn't his student. So, over time, I've changed advisors. Woody and I have continued to publish papers together for years now. We had a celebration of his 70th birthday at Woods Hole in June 1997 with all of his former students. It turned out I was one of the early ones who had worked in his lab. That was a lot of fun. So that's my history of the things that mattered in academic development. The main things were my driving curiosity—and not having that squashed—and being given some opportunities.

How did you discover the Obelia *system?*

Two things, fortuitously, came together. During my first year at Harvard, I knew I wanted to work on primitive nervous systems and how nervous systems control light emission. I started working on the neurobiology of *Hydra*. There were a lot of people working in that area—Max Passano, Bob Josephson, Adrian Horridge, Norm Rushforth—and some of them were at Woods Hole. I was there and got to meet them. But, *Hydra* is not luminescent, and since I also wanted to work on luminescence, there was *Renilla,* which I knew from my Santa Barbara days. I spent one whole wretched summer sitting in the darkroom in Woods Hole with imported *Renilla* from Santa Barbara trying to record nervous activity, and I couldn't do it. Subsequently, it's been done by others, with better techniques than I had at that time.

Near the end of that summer, one evening I was out on the Woods Hole Yacht Club dock with my wife. I reached down and rubbed my hand along the side of the dock, and there were all these waves of luminescence! It was *Obelia*. It was like—this is it! I didn't know at the time that *Obelia* was luminescent so I took it into the lab.

You actually first saw it for yourself in the field?

That's right. I remember it was one of the few times in my life that my ex-wife was ever in the lab with me. I used the apparatus I had set up for my *Renilla* work. I had a photomultiplier hooked up to an oscilloscope and a stimulating electrode. I was watching through the microscope as my (ex-)wife watched the oscilloscope. I saw pulses of light travel along the colony. All my (ex-)wife saw were vertical lines as the flashes went off the scale. What an exhilarating moment that was! I had been trying, without success, to record electrical activity from *Renilla*. I now tried the set-up on *Obelia* and voila! Beautiful, big potentials. I now had a wonderful colonial nervous system that coupled to luminescence to study.

Then there was also the luck of the draw that Woody came to Harvard in 1966 from Northwestern, just after I had arrived. That second summer, he invited me to work on *Obelia* with him at Woods Hole. At that time the photoproteins were being discovered, and he said, "Let's do some work on this," so he offered me a place in his lab. It was during that summer and the next that we discovered the green fluorescent protein that has now become a multimillion-dollar business as a reporter gene, as a reporter protein because the gene is known. I discovered GFP in *Obelia*. I had no idea what its importance would be and that subsequently it would be used as a reporter for all kinds of things. GFP and fluorescence microscopy allowed me to see the luminescent cells. The first time I saw these cells in the microscope was another of those remarkable experiences. I had put several pieces of information together to lead to a giant step forward in my dissertation. Merely shining UV light on *Obelia* allowed me, noninvasively, to visualize the light-emitting cells. I could now ask questions about individual cells.

There was just an international symposium on GFP. The GFP gene has been cloned and can be attached to other genes people want to know about. When these genes are expressed, so is GFP. Using fluorescence microscopy and looking for GFP, you can now see where and when the gene in question is being activated. A miracle! GFP was discovered in bits and pieces. I put it together—and it's all because of one experiment Woody said that I shouldn't bother with. I had the in vitro emission spectrum of the *Obelia* photoprotein. I wondered if the in vivo spectrum would be the same. It was a very difficult experiment to do because we didn't have the fast methods for analyzing spectra that are available today. I got a very different spectrum, which indicated that it was absorbing blue light that was re-emitted as green light. Thus, you should be able to shine blue light on it and have it come out as green. I vividly remember putting *Obelia* under the microscope and seeing all these green cells! From very small beginnings. Anyway, that's how it all happened.

In terms of my career, I've never had a lot of money. I've followed my curiosity rather than immediately fundable routes, and I've been quite happy.

But, isn't this a problem with modern academia? The people who have the most technicians and thus bring the institution the most overhead money get selected for?

That's one of the attractions for me about a place like the Isles of Shoals. The faculty who teach there are people who use their brains to learn about nature and not something stuck between their brains and the organism, that is, technology. They are doing things in a way so they can make direct discoveries. They can get kids to learn to use their brains to generate interesting data with very little equipment. For undergraduates, I think it is important to foster the knowledge that you can discover interesting things without a lot of sophisticated equipment.

If you keep learning things, you never get to the point where you think you know it all. Basically, after a while, you begin to realize how much you don't know.

Ian Stirling (b. 1941)

Senior Research Scientist, Canadian Wildlife Service
Edmonton, Alberta

Interviewed at the Arctic Seas Conference, Connecticut College, 20 October 1998

I remember sitting in a cookhouse in British Columbia in 1960, a young girl riveted by Ian Stirling's tale of his encounter with a grizzly bear. Meeting Ian again, almost 40 years later, I again found myself absorbed—this time, in the story of his career. His field sense was developed as a boy, wandering on his own in the wilderness, but his scientific field skills and research skills were honed at the University of British Columbia, through the lively academic atmosphere and the tradition of excellence in wildlife biology. For his Ph.D., he chose to spend time in the Antarctic to provide a comparative perspective for his career work, which he intended to do in the Arctic. Stirling discusses many important aspects of the education of a scientist; among his points, he says that he counsels his students to take the time to appreciate the nature around them.

Probably the thing that influenced me the most initially was growing up in a small mining town in the mountains in southeastern British Columbia. It was the 1950s, just after the war. There was no television or other sorts of distractions so we pretty much made up our own entertainment. I lived about five minutes' walk from the bush, which seemed to run forever, hundreds of miles of primitive wilderness. I was attracted to that just like iron filings to a magnet. Almost everyday after school I either went simply walking or hunting for grouse or fishing for trout or cross-country skiing in the woods. In the course of doing that, of course, I started to notice a lot of things, even as a kid. You'd find tracks of this or that sort of animal at different times or different places. Various species of birds came and went. Places that were swamps or little lakes dried up. There would be a series of rainy years and a series of dry years. It just got me continuously thinking. Without really realizing it, I wanted to become a biologist. I think that is interesting in retrospect, because the principal kinds of role models for success, in those days in most small towns, were the various professional people. But it was living things that really excited me, right from the beginning, for reasons which I couldn't explain precisely. They just did.

I arrived at the University of British Columbia [UBC] at a very important time. There was a man there, Ian McTaggart Cowan, who was one of the fathers of wildlife biology in North America. Certainly at that time, virtually anybody who was in-

Combining the need to do research that is not only good and interesting science but that also has some relevance to conservation has motivated Ian Stirling in his polar ecology studies for almost 40 years. Photo by Andrew Derocher.

volved with conservation and wildlife biology in Canada had been one of his students or had gone to UBC in Vancouver. That was a very lively place. Even as an undergraduate I used to go to a lot of the faculty seminars because there were just so many interesting things going on. I realized what I wanted to do: I changed my aspirations to wanting to do research on wildlife. Nevertheless, my old interests in management and in doing research that is not only good and interesting science but has some relevance to conservation and some of the problems of the world have also remained with me as a combined driving force.

I couldn't say why I've been so interested in polar regions as opposed to the prairie or the mountains. I got my bachelor's degree in 1963. In those days, there were so many opportunities you almost didn't have to apply for grad school. People were approaching you. When I finished my master's degree, I was literally offered six jobs, doing everything from working on caribou in the northwest territories to working on game birds in the mountains of British Columbia to working on water fowl. This was quite a junction in my life. My priority, my greatest interest, was in polar regions. I had an offer to go the Antarctic to work with the New Zealand Antarctic Research Program and study Weddell seals. At the same time, through a friend of mine who had gone to Africa, I had a possibility to go to Tanzania and study elephants. I was sort of stewing about this. My heart was sort of in the Antarctic and polar regions when suddenly, unpredictably and very sadly, the person with whom I would have worked, had I gone to Africa, was killed in a plane crash while surveying elephants. So that was the end of that. I went to the Antarctic, and I've stayed in polar regions ever since.

So, my first field experiences? It wasn't until I actually finished my third year of university that I got my first real break and got a good job as a field assistant. What I did previous years was to go and get jobs where I could make more money in the summer than I would need the following winter. That allowed me to bank some against when I got my first job in biology. I knew I couldn't go back to school with what I would earn as a biological field assistant. I got my first break through a class I took in mammalogy and ornithology. I was one of the highest graded students. Certainly, if there is a message in trying to be successful, you really do have to get good grades in whatever it is that you are interested in. We had a lot of field assignments. The lab instructor had noticed my interest. When a friend of his called him and said someone had canceled out as an assistant for a field expedition for the summer studying and collecting birds, he thought of me. The person had to be able to shoot, to find one's way around the woods, and so on. He came down and told me this job was available. It was not advertised. I never would have found out about it had it not been for this particular instructor.

It was wonderful. I spent three months in northern B.C. and the southern Yukon in some of the most beautiful country I've ever been in my life. It was one of those museum expeditions, a basic survey. If it moved, you shot it; if it didn't, you cut it down and put it in a plant press. We weren't collecting large mammals. Everyone knew there were moose there and so on, but the distribution of some of the smaller species, such as the mice, were not well known. This is where some of my growing up really came in useful. When I started taking the course in ornithology and mammalogy, I came to appreciate that I had already learned and come to understand a certain amount about animals. For instance, you'd find certain birds in savannas, certain kinds in spruce forests, and so forth. I had already taken this in without realizing it. I had it in a mental structure that I hadn't formalized. There were four of us on that first summer field trip. One of the kids was a graduate student from eastern Canada, and he had not had a wilderness background. He literally couldn't find anything, and he couldn't do what he was supposed to do. He was so frustrated. He became overwhelmed by the vastness of the area. Simply put, he wasn't used to being in wilderness; it frightened him. Eventually, the party chief had to send him home, which was a bit sad really.

I know you are interested in the value of field experience. I often think back on that, and I realize that of initial benefit to me was the amount of time that I actually spent outdoors with an interest in birds and mammals, looking for them or looking at them. I developed a sense that you find loons or scaup or mallards in very different sorts of circumstances. You can read about animals 'til you are blue in the face, but it is actually going and looking at them. The Germans have a word—they talk about having a gestalt with something—a sense of understanding or a oneness with whatever process or situation is present. Let the environment and animals tell you what is important. Follow your own curiosity. You see things happening. As you develop a sense of watching anything, from a ground squirrel to a mountain lion to a polar bear, you see they do things under certain circumstances or not. Where you find them in different seasons and weather, these observations naturally bring the questions out of you. I don't mean to downplay the value of doing a lot of reading. Learn as much from it as you can because when you take that factual learning into the environment, it can help you understand what you see. It can help you to be a little bit more sensitive to questions that crop up. But there is no substitute for the real thing. You've got to see it! You've got to be there. There is just no substitute for that.

Many people can look at the same thing and see completely different things when they are looking ostensibly at the same thing. This was really brought home to me vividly once when my oldest daughter was 11. I took her up to the Arctic with me for a week or so. One evening, we were about 80 or 100 miles offshore in the helicopter, just the two of us plus the pilot. It was in the late spring, when you get these long sunsets and lovely light. We were standing by the edge of an open lead, and there were polar bears walking along, there were seals breathing in the breathing holes, there were bear tracks and signs of life all over. For some reason or another, rather than being somewhat unimaginative and saying, "How do you like those bears?" I don't know why I did this, but I asked her, "What do you see when you look out there?" There was not a comment about an animal. It was all about the texture of the light on the snow, the beauty of the color of the water with the light on it like that. And of course, polar waters in the spring are absolutely crystal clear. It is a blue that you just don't see in oceans anywhere else. There are fantastic pressure ridges from crushed ice and so on. She was enjoying herself every bit as much as I was, but she wasn't seeing the animals at all! In contrast, I was totally focused on the animals. The eiders had started to arrive, there were murres out there on the water in the leads. It was just teeming with Arctic life. It was a spectacular moment. But it was also a spectacular moment for her, for just as valid reasons, but quite different. I've never forgotten that.

One of the summer jobs I had was working for the warden service in the National Park Service in Kooteny National Park. I had noticed that a small population of mountain goats lived on a particular mountain and also that they migrated vertically with the seasons. I thought they'd be very interesting to study. When I was looking for a master's degree, I initially started out seeing if I could get someone to support me to go and do that. I realized with what I know now, that I could have found that support. But fortunately, I had a couple of other opportunities. It was put to me, and I think, with the benefit of hindsight, quite rightly, that if I went out by myself for two years to study the goats on the slopes of Mt. Wardle, I would learn a lot about goats, and I would probably learn a bit about myself, but I would not learn very much about science, or how to conduct research, or how to recognize the other kinds of questions that might be popping up right in front of you.

Instead, a prof named Jim Bendell suggested I do a master's degree with him. He had a group of grad students working on blue grouse. I worked on the birds' behavior and vocalizations. Most of them were captive birds, but I also took things that I learned from the animals in captivity, went out, and did experiments in the field. Jim and his students also had a big field study going on. They had an area in which they knew a lot about the birds that were there, including where they were. This provided me with a wonderful field situation in which I could do some simple experiments and interpret the results.

The other thing which Jim did with his students was to have an evening seminar, every two weeks or so, where one of the grad students would give a talk on his thesis work and his progress to date, and everybody else critiqued it. It was a very interactive environment during my master's degree that I certainly wouldn't have had if I had been living by myself in a tent on the slopes of Mt. Wardle. I participate now in a similar seminar program with some professors from the University of Alberta and our current grad students.

The experience I gained during my M.Sc. was absolutely crucial for me in terms of the development of my career. It is probably one of the most important things that happened because I learned so much about doing research through the interactions with other people, the value of group discussions, and different ways of looking at things.

When I went to New Zealand to do my Ph.D., it was under the British system. That is, essentially, they throw you into the deep end of the pool, and if you are able to get down to the shallow end and crawl out, they'll give you your degree when you dry off. If you drown in the process, well, nobody worries about it very much. When I got to New Zealand, there were people in the department who had been working on their theses for several years who were still working on them when I left, basically because they didn't know where they were going. They weren't getting much direction.

They weren't having this interactive feedback that you had had during those two years.

That was how I saw it, yes. I felt I had learned so much more about how to do research, how to ask questions, how to use the libraries, not to be afraid to interact and go and ask people and initiate discussions. All of this was learned without knowing it. Just like I didn't know what I had learned as a kid walking around in the woods. I didn't really realize until I came to do my Ph.D. what I had learned in my master's degree. It was interesting that the three people I interacted most with as a Ph.D. student were all people who had had a prior research experience similar to mine. One was a New Zealander who had done his master's degree in Australia; he turned out to be one of the world's leading scientists in his particular field. Another was a woman who had gone to Hawaii to the East-West Center and done her master's degree, again in an interactive environment. The third was a fellow from New York who had gone to Cornell for his master's and was down in New Zealand on a Fulbright fellowship. So the four of us sort of interacted like electrical circuits, all the time.

I really think in terms of a master's degree as a research-training degree. It is where you should learn the tools of the trade. At the same time, hopefully, you will do something good enough to be able to write it up, maybe to produce something that is useful for management. You want to get training that will be useful for doing a Ph.D., for which a great deal more rigor is going to be expected.

I was very lucky in my career in that I've had two or three really great mentors from whom I've learned a great deal. Possibly the most important was Robert Carrick, a Scot who lived in Australia for most of his working career. I first met him when he came to the Antarctic as the guest of an American scientist. I knew he was coming, and I had invited him to come out on my field study to my field camp with me for a week in McMurdo Sound. I don't think I've ever been out in the field with someone who had such an absolutely clear and intuitive way of asking the question, peeling off anything that might distract you to what was really the central issue. I would sit and talk with this guy and I would sometimes almost feel like I was so stupid that I should be taken out and shot, because I hadn't thought of things that seemed so logical. The other thing that was really remarkable about it was that he didn't point things out to you as if telling you. Through discussion, he led you to it. Robert was incredible, he would bring you through something and suddenly you would think, ah, yes, that's what it is. That is what this particular discussion is all about! I had my first experiences with him at about the middle of my Ph.D. period. That was one of the most important things that happened at that stage of my life because it helped me focus questions a lot.

Having a week with Robert was just like having a ray of sunshine suddenly come in. It helped me clarify all sorts of things I was trying to understand. So, when I finished my Ph.D., I went and did a postdoc with him as well. I got a grant on my own from the Nuffield Foundation to study in his lab in Adelaide, South Australia. I worked on some questions related to fur seals and seal lions. I was in the field literally 50% of the time I was in Australia. But, almost everyday when I was at the lab, Robert and I would have a sandwich lunch together, along with the three or four other students who were there. Again, it shows the value of having three or four people at a similar level and then someone brilliant to really guide us and lead us along. That was the most stimulating year and a half of my life. It provided some of the focus for me for the next 30 years—not just ideas, but also ways of approaching ideas. Again, it comes back to this feeling I have that after you learn things, you don't even know you've learned them until you realize that you are applying them. Suddenly, you stop and think back, do you realize where that came from and why you are doing that?

This raises the enormously important role of modeling, which goes through everything from parenting to politics to science. I was lucky to have known Robert Carrick and Ian Cowan as role models. I've taken as much as I can from them and tried to inject it into the way I train my own graduate students. Two of my former students are here at this conference giving papers. It's kind of fun watching the next generation come along. I am pleased that, so far, all of my students, including the women, which is sometimes a bit more difficult, have gotten jobs, and all are still in the field. I think at last count there were about five women marine mammal biologists in the Canadian Department of Fisheries and Oceans, and three of them are my former students. One is here to present her research.

Part of what I get out of having students is simply the enjoyment of being around them. I enjoy watching the enthusiasm of people who enjoy things. But they also ask me questions. They think about things that I've maybe been looking at for years but hadn't thought about in terms of a question. Could it be this or that that is happening? There are so many questions all around you. The world is so complex, you couldn't possibly think of everything yourself. I don't think we are ever going to know all we'd like to know about animals in the wild and about how ecosystems and environments work and how they interact.

Students do a couple of particularly important things. In the field, they see things because they want to learn. The first thing is that good students are a little bit impatient because they want to learn quickly, right away. They have tons of energy, and they wear you out with endless questions and a desire to do everything immediately. I try not to get frustrated because that's how they are going to learn, but sometimes it can be tiring. Second, because I am really busy with a lot of other things, I always tell them that they are responsible for keeping track and making sure that I am aware of papers they think are relevant to what we do. So I learn a lot from them. For example, what I am going to give my paper on here—long-term changes in population ecology in polar bears in relation to environmental change—is something that I've been working on for a long time with several objectives. Through some references and contacts that the students initially found, we've been able to figure out what was happening, put those things together, and get a plausible explanation. That doesn't, of course, mean it's right, but our explanation is logical.

A problem with large-scale ecosystem research is that it is very difficult to test hypotheses in a manipulative way. The standard experimental control-type thing just doesn't work. You just can't go out and modify half the Arctic, so you have to use what I call natural experiments. Sometimes, this approach is hard to rationalize to molecular biologists, who really don't understand ecology and how variable things are because they've got control of everything in a test tube. You can't even go out to a field across from the lab and control everything in the environment. The kind of work that I do is what you might call quantitative natural history. There are a number of processes and questions which we are interested in. We try to set up our data collection in a structured fashion so that if we've asked the wrong question or we want to come back ten years later and ask a different question, we hope the baseline data were collected in a similar enough fashion to be useful in perpetuity.

I thought it might be a good strategy to try to do my Ph.D. down in Antarctica. It wasn't easy to get in down there, but I was lucky again. I wrote to a fellow named Bernard Stonehouse at the right time, and I got accepted virtually by return mail. I also wanted to work in the Antarctic for my Ph.D. because I knew I wanted to live in Canada, so this seemed like the right time to go somewhere else, work for a few years, hopefully learn something, and hopefully contribute something useful. People aren't going to be upset with you if you leave after a few years of being a grad student, whereas in most places, if you take a job, you are not really that much use until you've been there three to five years. So I thought, I'm going to do my Ph.D. in Antarctica and maybe I'll do something else in the Canadian Arctic afterward. As it turned out, I spent five years in New Zealand and Australia before coming back. But

I specifically went to the Antarctic because I wanted to do that first, then come back, and work in the Arctic. I wanted to have experience in both regions.

I was lucky because I was accepted to study the population ecology of Weddell seals. It was up to me to decide what it was that I wanted to try to find out. Again, the British system. I decided to focus on factors which affect their distribution, abundance, and reproductive success. At that time, no one had developed a way to capture and tag these animals. There you are, faced with an 800- to 1,000-pound animal you want to tag! You can drug them, but that is very expensive; it involves risk to the animal; it takes a lot of time. I wanted to tag hundreds of animals. Quite frankly, people laughed at me when I said I wanted to do that. Well, we lucked out. We invented a way of catching the animals and restraining them temporarily within about two or three days. We ended up marking about 3,000 seals over the next four years. I was able to look at seals in individual colonies, the way that ecological factors affected distribution, why they were in some places and not in others, or at least some of the aspects related to that. I thought I knew a lot more then than I realize now that I did, but that is part of the learning process. If you keep learning things, you never get to the point where you think you know it all. Basically, after a while, you begin to realize how much you don't know.

My field circumstances did not permit manipulative experiments, so I had to use observation and tagging. I had to quantify my observations in ways that enabled me to ask questions that made predictions that I could test in different ways. For example, the effect of the distribution of ice cracks in relation to the differences in the thickness of sea ice in different years caused animals to be distributed in different ways. You didn't see very many subadult animals, so why was that? You'd see pups that were born, and you'd see adult breeders. One important aspect appeared to be access to breathing holes. I found, just as an example, during the breeding season when adult females were lying around on the ice with their pups, they kept other animals at a greater distance from them than they would after their pups were weaned. As soon as the pups were weaned, all of a sudden the size of these little colonies would increase severalfold. So, obviously, there were seals hanging around under the ice or in adjacent areas that were not being allowed in.

While I was in New Zealand, I worked on Weddell seals in the Austral summer in Antarctica. But during the winter, I got really interested in the fur seals around the New Zealand coast. I started reading about them. I noted a very peculiar thing. It is an example of what one can get from reading. I think I read pretty well everything that was written in English about fur seals and quite a bit that was written in French, although those papers took me quite a bit longer to read, but I did it. I found that there seemed to be two different kinds of descriptions of the social behavior of fur seals. Either one group of scientists was recording their observations totally correctly, while the other group of scientists was recording things incorrectly, or different things were happening in different areas. It was all kind of confused. I thought, I wonder if there are different behavioral forms of the seals, which was a totally new concept. Some of the people with whom I raised the hypothesis were not very interested in it.

However, when I told Robert Carrick this idea when he was visiting me at my field camp in McMurdo Sound when I was a Ph.D. student, he was very interested, and we talked about it quite a bit. I said, "Well, what do you think about that idea? Everyone else so far thinks I'm crazy." He simply replied, "Well, how would anyone know that? Have you talked with anybody who has looked at the two types?" I said, "No," and he said, "You might be right." So that was when I wrote to the Nuffield Foundation and got a fellowship to go and look. My reading included six or seven studies of fur seals, which had been done in different places, and a couple of them seemed rather different from the others.

Recent references?

They varied. Some were recent, and others went back 20–25 years. Although in the last 15–20 years, there has been practically a nuclear explosion in the literature on marine mammals, in those days it was a lot more limited. There were probably only a few hundred really important references. I think I had probably read practically anything that had been written, in English at least, and at that time I had a good memory, and I could remember almost all of that material.

I finished my Ph.D. and went to Australia. After doing my first season of fieldwork on the fur seals in South Australia, they seemed to be doing much the same thing as the seals in New Zealand and several other places. I arranged with Bob Warneke in Melbourne to visit his study site, where one of the groups of fur seals which seemed, according to the literature, to be different, lived. To make a long story short. I went over to Bob's study area with the Tasmanian fur seal; I walked out in the colony, and within ten seconds I had my answer. It was simply so clear. Their vocalizations were different when you looked at them, their behavior was different from the way they laid out the colony to the way they interacted with each another. It took me some work to quantify the differences, but the really exciting thing was that within ten seconds, I had my answer. It was like sitting under the apple tree. There it is. It was one of the most exciting little experiences of my life! Nobody had compared the behavior of the two forms.

To come back to one of your original questions, that experience illustrates why it is so important to be in the field with the animals, because I could never have figured that out just through reading the literature, even if I'd memorized and recited every paper that had ever been written about fur seals, and yet . . .

Something was nagging you?

In ten seconds, it was there. However, I still had to also have had the prior field experience with fur seals and knowledge from the literature to be able to suddenly interpret what was there in front of me. So, if I hadn't done my homework, I couldn't have drawn the conclusions I did.

Through a long chain of events, I was offered a job with the Canadian Wildlife Service working on polar bears. Initially my interest in polar bears was as a predator of the seal, because at this time my research focus was on the Arctic seals. I was interested in the broader question of how an animal evolves into its environment. In my case, what I was interested in was polar environments, and so I'd worked in the

Antarctic where there are no terrestrial predators. So you can walk up to penguins or seals, and they don't run away. People refer to that as being tame, but it is not; there has been no need to develop an escape response from a terrestrial predator. Whereas in the Arctic, there is a very obvious need to develop such a response. There are wolves, bears, humans more recently, foxes, and so on. Animals need to behave very differently.

If they hadn't evolved that?

They wouldn't be there. There is an evolutionary aspect of life in the presence of a very effective terrestrial predator, superimposed on top of the ecological things that are associated with patterns of ice break-up and freeze-up, currents, and the distribution and abundance of amphipods, fish, and whatever else they live on. I hadn't really thought of it consciously at that time, but of course then I came to realize that the leopard seal is basically the Antarctic equivalent of the polar bear. I went back to the Antarctic again with a colleague named Don Siniff from Minnesota in the late 1970s to work specifically on crab eaters and leopards [seals]. That helped me to get much more insight into what was going on in that ecosystem so that I could compare the evolution of seals in these two systems in the presence or absence of predators, either surface or subsurface.

That sort of led into another aspect that I had been interested in for a while. It is a truism in ecology that animals aren't usually distributed evenly over a habitat. You find them in some places and not in others. Why is that? When it comes to polar bears and seals, it seems to me that there is quite an association with areas where there is water, because of course the seals need to be able to breathe. Ringed seals have solved the problem of access to air in the winter by scratching the ice with their claws to make breathing holes. They stay in fast ice areas, just like Weddell seals in the Antarctic. The other species of seals move around a certain amount to stay with places where the ice cracks and remains open in winter. Open water and the absence of ice also allows more light into the water, which stimulates biological productivity, so maybe areas which are more broken and have more leads have a higher level of productivity. I got really interested in the role of *polynyas* [a Russian word referring to the ice cracks] and shore lead systems in relation to distribution and abundance of seals and bears. This gets back to the old theme of distribution and abundance in the Antarctic, although I hadn't thought about polynyas and shore leads much then. I was sort of led into recognizing the biological importance, the trophic implications of those areas in the Arctic.

Through a series of systematic surveys of ringed seals, bearded seals, and walruses, we found very good association between their distribution and abundance with certain types of habitat, which also related to bear distribution. We have been able to use that information to identify critical areas for protection or modification of offshore practices for exploration and production of hydrocarbons, so activities can take place with minimal disturbance.

When we started working in western Hudson's Bay, I was particularly interested in the possible effects of long-term change. All of the bears in Hudson's Bay come ashore when the ice melts in the summer, and they fast for about four months. They live off their stored fat. We also know that pregnant females go into dens, and they don't feed for eight months from the time they come ashore in late July through to the following March or so. It is often referred to as hibernation-type physiology. But pregnant females that come ashore during the open water period can't go back and feed because they are about to give birth soon after the freeze-up occurs in November. So we got to studying how these animals are really able to do that.

Through contacts made by one of my students, we started working with Ralph Nelson from the Carle Foundation at the University of Illinois. He has been working with black bears and the way they are able to metabolize fat. My student was quite interested in fat metabolism as well, so he, to a large degree, was responsible for bringing Ralph and me together. Again, an example of a student getting into interesting ideas through reading the literature and bringing me in. That led to a long study of how bears use fat, which several students of my former students were involved in. Ralph was interested in how bears metabolize fat because it has enormous applications for the treatment of obesity, anorexia nervosa, and kidney disease. It has huge human health applications. A hibernating bear lives almost entirely off its fat. It recycles its wastes. It gets its own water from the fat and just uses the fat and loses only a small amount of its lean muscle mass. If you or I stopped eating and drinking, we'd be dead in a week or ten days. A bear just goes on indefinitely until it eventually runs out of fat. If you put on enough fat, a physiologist friend of mine projected that a really fat adult male might not need to eat for a year. Pregnant female polar bears are not only looking after themselves, but they also give birth to a couple of cubs, which weigh about 600 grams, and nurse them up to 11 or 12 kilograms or so when they leave the den. Pretty amazing.

Then we got to thinking that something more important might be happening. If you take a black bear and take food away from it in the summer, it will starve to death. It can go into a hibernationlike physiology all right, but it can only do so seasonally. If you take food away from a black bear a couple of weeks before it is likely to go into its den anyway, you can advance the onset of the physiological state. But you can't do it in the summer. We found from some feeding experiments on a captive polar bear that they can do this at any time of the year. In other words, polar bears have taken this ursine characteristic and modified it for the Arctic, so they can eat when food is available and can fast anytime during the year when food is not available. So a polar bear can fast in the middle of winter in cold and inclement weather or when it can't hunt very well, and it can slow down its metabolism and live on its stored fat. They take on most of the fat for the year between April and July by feeding on fat, weaned seal pups. But if there is an opportunity to kill something or a whale washes up on the beach in August, they go right back in and gorge themselves. Then, ten days later, they return to the fasting mode, if there is nothing else to eat.

There we were, looking at it as such a wonderful adaptation for life in the Arctic! Ralph was just as excited as we were, but for totally different reasons. It was like when my daughter and I looked at the sea ice. Ralph thought, if we can find out how the bear does this, we might be able to help a person suffering from obesity. Maybe they can metabolize their fat without losing their lean body mass, like a bear. It doesn't eat or drink, and it can recycle its own water. What if there is a way biochemically you can do that with humans? You might also be able to treat kidney diseases without dialysis, which is a horrendous process, I'm told.

That whole question of energetics led to thinking about the relationship between numbers of predator and prey. We have pretty good estimates on the numbers of polar bears in different areas, because we have been doing population assessments, which we need to do for conservation, for setting sustainable harvest levels for hunters. I had been doing a lot of behavioral studies on bears because we needed to know what bears were doing when they were not running away from hunters or biologists in helicopters. How many seals do they kill? How do they use their habitat?

How do you figure that out if you are not in a helicopter looking at them?

Exactly! Well, you have to find a place where there are a number of bears active in different seasons of the year, in a location where you can watch them undisturbed, which basically means the top of a cliff. I have two little camps perched at the edge of a cliff in areas where there are a lot of bears. We just sit there with telescopes, and with the 24-hour daylight in the Arctic spring and summer, we can keep track of individual bears continuously unless they walk around the corner into the next bay or the fog comes in. The longest we've been able to keep track of the same bears is two weeks, 24 hours a day. From these observations, we can learn a lot about their energetics.

One of the things which really struck me from our aerial surveys of seals, to put it quite simply, was that there didn't seem to be enough seals to be supporting all of the bears that we knew were there. Something wasn't adding up, but I couldn't really see why. So we spent a lot of time working on our behavioral information. We then had our physiological information coming in from the work that we did with Ralph.

I then went back to work again with a Norwegian ecophysiologist, Nils Oritsland, with whom I had worked periodically since the early 1970s. I knew he was doing similar things with reindeer, and you hit a certain chemistry with people. He was a scientist with whom I just clicked scientifically, and we were very good friends on a personal level. So we sat down with all the information, and it took us five or six years to write a paper and figure out what in fact was going on, partly because of our geographic separation and partly because we were working on other things as well. At any rate, it turns out that there are probably enough seals there, if you can interpret what is going on ecologically. We found that although polar bears are capable of killing adult seals, most of what they are eating is the young of the year after they are weaned up until the time of the break-up. These seal pups are 50% fat by wet weight. The digestibility of fat by polar bears is something like 97 or 98%! Also, because the mortality rate of seals through their first year of life is much higher than adults, the polar bears can kill a much higher proportion of the weaned pups without harming the adult population.

We worked on it independently to compare results. He went at it from a metabolic view. A few years ago, I had sampled seals at different stages and run them through an agricultural grinder to determine the caloric value of animals at different ages. So we had that information. We also had quite a bit of information from over the years about the ages of seals which had been killed by bears, and about 80% were the young of the year. To make a long story short, when we worked out how many animals a year they'd have to be killing to support themselves, using behavioral data or physiological data, the results came out surprisingly close together. It worked out to be about 43 seals per year per bear. Obviously, big animals take more than little animals, but on average, that's what we found.

So that's sort of where we are now. We are continuing the long-term monitoring in Hudson Bay that I talked about, we've gotten into other sorts of big picture questions in one sense. We are looking at denning habitat. The polar bears do some unique things in Hudson Bay, such as using dens in the frozen peat. Suitable habitat for this is limited, so it should be protected or at least disturbance to the bears in their dens should be controlled if we want to have bears into the future. In the high Arctic, we've gone full circle. We were there ten years ago looking more specifically at polynyas through the eyes of walruses which are wintering there. Now we are involved with part of a large-scale multidisciplinary project on polynyas in a place called North Water in northern Baffin Bay. Exactly where all of the next questions are going to lead, I don't know, but my experience throughout life has been that there has never been a shortage of interesting questions cropping up, as long as you stay on top of the system and continue to read and think.

One of the things I always tell my students that I want them to do everyday in the field is to spend at least once a day, 10 or 15 minutes, just appreciating where they are and things in general. "Take a power break," in the modern lingo. I used to say, "Go vibe with the environment." Appreciate where you are. People pay tens of thousands of dollars to come up to the Arctic just to see what you spend a month or two doing. I want you to do the science, of course, but I also want you to have an appreciation of the area. I tell them to do that because I really do want them to have an appreciation of just what's around them, but also by stopping, thinking, and looking around, you are subconsciously taking in information, and that may lead to the next question. Just like when I was a kid and would go to the woods after school. It's not much different.

A great deal of what happens in one's life is, I believe, a matter of opportunity and timing: who you meet, when you meet them, and whether you seize the opportunities or not.

Yossi Loya (b. 1942)

Professor of Marine Biology
Raynor Chair for Environmental Conservation Research
Department of Zoology, Faculty of Life Sciences
Tel Aviv University, Israel

Interviewed at the Society for Integrative and Comparative Biology Meeting, Boston, 6 January 1998

From his escape from the Nazis as a baby to his position as dean of the faculty of life sciences at Tel Aviv University, Yossi Loya engages us in his life history. Through humorous but touching stories, Loya illustrates how he was greatly influenced by two mentors, Larry Slobodkin and Tom Goreau. His detailed quantitative studies of reefs in the Red Sea have enabled him and others to make use of natural and manmade disturbances to learn more about how reefs are organized, but he credits comparative work on reefs in other oceans for broadening his perspective. Loya and others of his generation brought not only their own expertise in science, but also an American academic philosophy back with them to Israel to change the way science is done. Loya has been a close colleague of mine for almost 25 years.

I was born in Plovdiv, Bulgaria, in 1942. One morning in September 1944, when my parents were accompanying friends to the train station, they heard over the loudspeakers that this would be the last train to leave Bulgaria until the war was over. My father was about to be taken to a forced labor camp. On the spot, they decided to flee, leaving everything behind. They had barely 20 minutes. They rushed home. At that time, during the war, we always had a suitcase packed for the shelters. They grabbed me, a baby; they grabbed my grandmother. Then, with just this small suitcase with a few clothes, they rushed back to the station, where the train was already moving. They had no official documents, only some papers that "appeared official," bearing the stamp of the youth movement they were members of ["Hashomer Hatzair"] a Zionist youth movement, well known for establishing Kibbutzim in Israel—no visas, no tickets, no money—nothing. But these were terrible times we are talking about.

Their friends bribed the Turkish border policeman to turn a blind eye, and so they managed to enter Turkey. In Turkey, the Jewish community, sympathetic with our escape from the Nazis, received us warmly, giving us clothing, food, and milk for me, the baby. We crossed the Bosphorus by boat and then continued by train through Syria, Lebanon, until we finally arrived in Haifa in October 1944. The whole trip took about a month with many frightening moments on the way (as my mother later told me). I was then two and a half years old. My parents had nothing but the clothes they stood up in; no money, no belongings, just me, a sickly infant with an eye infection, and an elderly grandmother.

It is quite a dramatic story, how the state of Israel was established in 1948. There are many heroic stories of that period. Jews were fleeing from all over Europe to Palestine, ruled at that time by the British mandatory regime. The British were preventing people from entering the country and deporting them to closed camps in Cyprus. In Haifa, we lived for a few months in tents provided by the provisional government. My parents then moved to Jaffa, which at that time was mostly inhabited by Arabs. They spent their first night in an apartment they had rented with the small amount of money they were given by the government. That night all their clothes were stolen and they were left with nothing. That's the way life started out in Palestine—with nothing, everything left behind. That is also how the state of Israel was largely created; by those who had escaped from—or managed to survive—the Holocaust.

I went to elementary school in a small village, Givat Shmuel. I then attended high school in Tel Aviv, followed by compulsory army service. In Israel, you are inducted into the army at 18 and serve for three years. I was a communications officer in the army, and until the age of 43 I continued to serve 20–30 days every year in the reserves. I took part in the [1967] Six-Day War, fighting in the Sinai desert, and in the 1973 Yom Kippur War, in Jerusalem and the Sinai again.

I earned my B.Sc. in biology and M.Sc. in zoology at Tel Aviv University.

Why biology?

Well, my parents, of course, wanted me to be a doctor, a physician; doesn't every Jewish mother want her son to become a doctor, an engineer, or a lawyer? But I loved biology. Animals always fascinated me, so that's where I headed. I began my M.Sc. in 1965 with Lev Fishelson at the zoology department. My research involved growing *Tilapia* [Saint Peter's fish] in two small experimental ponds filled with highly saline water, about half the salinity of sea water. *Tilapia* fish, and especially *Tilapia* hybrids, were known to withstand such high salinities. During my first year as a master's student, Larry Slobodkin came over to Israel from the United States on sabbatical leave. You probably know Slobodkin—the famous ecologist. I'll tell you a little about him because he is an important figure in my life. At the time he gave a course at Tel Aviv University on population ecology, based on his first famous book.

Growth and Regulation of Animal Populations, *right?*

Yes. I was fascinated by his personality. He insisted on lecturing in Hebrew, largely gained, however, from the Bible, an English-Hebrew dictionary, and some past experience in Israel. As you can imagine, his translation of ecological terms could become quite amusing at times. My basic grounding in ecology is from him.

Larry joined one of my field trips to Neot Hakikar, which was a small settlement

On the shores of the Sinai peninsula in 1969, Yossi Loya (*left*) and Tom Goreau (*second from left*) examine a recently caught shark with two colleagues. Photo by Amikam Shoob.

in the Arava Valley near the Dead Sea, where I was working. The study area itself looked like the Wild West, desolate other than for desert shrubs and venomous snakes—and these two experimental ponds situated in the lowest place in the world, close to the Dead Sea. Everything had to be done by hand, and everybody pitched in—my family, neighbors, colleagues, and friends. Oh, the things you do when you are young! Larry viewed this as a major Zionist endeavor—"a young Israeli biologist blossoming the desert!" There was wonderful chemistry between us. He invited me to do a Ph.D. with him at the University of Michigan after completion of my M.Sc. degree. Finally, I went from being a fish biologist—essentially a fisherman, if you like—to a coral biologist, as I shall shortly tell you.

The Six-Day War broke out on June 6, 1967, toward the end of my M.Sc. studies. That's also when I met my wife-to-be, Shoshana [Shosh]. Two months later, we decided to get married. I had been fighting in the Sinai desert and, like many of my friends, had to remain there on reserve duty for two months. I completed my master's degree four months later.

In your master's were you basically doing everything you needed to understand how Tilapia *survived in its environment?*

Not exactly. The purpose then was mainly to perform a study of economic value. It's necessary to understand the infancy phase that ecology was in in Israel in 1965. What we wanted to find out was whether we could grow fish in the desert without needing to invest any money other than in digging ponds to be filled with the otherwise useless highly saline water that flows into the Dead Sea. The idea was to grow *Tilapia* hybrids obtained from a cross between *T. nilotica* and *T. aurea*, resulting in about 100% males, which are almost double the size and weight of females. The experimental design was such that one pond received additional food, as practiced by the fish growers in the north, while the other pond had only the natural phytoplankton developing in the pond. Every two weeks I measured the growth of the fish, performed water analysis, nutrients and oxygen analysis, those sorts of things—the whole chemistry of the water, with the procedures we knew then. Within three months we had fantastic fish that grew from about 20 grams to 600 grams! We then emptied the ponds completely, measured and weighed large fish samples, and restocked the ponds with fingerlings. There was no significant difference between fish growth in the pond that got added food and the one that didn't. In terms of commercial value to fisheries, this was extremely successful compared to the fish yield in similar ponds in the northern part of the country. The economic benefit of using the waters near the Dead Sea is obvious: besides being free [their high salinity makes them useless for agriculture], since they remain very warm throughout the year [25°–30°C], which is ideal for *Tilapia* growth, it enables up to three to four fish stockings per year. In the northern part of the country, which has a relatively cold winter, the growth rate is much slower, permitting only one stocking of the ponds per year.

We arrived at Ann Arbor, Michigan, in November 1967. It was snowing, the first time I'd seen snow in my life, and also the first time I'd seen television. Coming directly from the hot Israeli summer, we were terribly cold. I remember telling Shosh, "I'm not going out. Look it's all white outside; it is snowing!" Of course I had to go out eventually. In the coming months I took the required basic courses in the ecology program. I remember especially Nelson Hairston's excellent laboratory course in community ecology of soil mites, where I first exercised the concepts of species diversity. At that time Slobodkin was invited to chair the ecology and evolution department at the State University of New York [SUNY] at Stony Brook. He gave me the choice of either staying in Michigan or going with him to New York. We went with him. He inspired me greatly with his unusual and warm personality, wisdom, intellect, ingenuity, novel ideas, and ecological philosophy. Later on, this also affected my own approach to coral reef studies.

So, why reefs and corals?

This was all quite by chance. While working on my M.Sc. I had also helped other students who were working at the Coral Nature Reserve at Eilat. From the moment I saw the coral reef at Eilat I was overwhelmed. It was incredible. I was captivated by its fantastic richness and beauty. In the back of my mind I instantly knew that this was the environment I'd like to work in in the future. I had gone to study with Slobodkin on a research grant he had at that time to study coral reefs with Lev Fishelson at Tel Aviv University. I had a perfect arrangement: obtaining my formal education in the United States and doing coral reef fieldwork in Eilat.

Larry, it turned out, brought me to the United States not only because I had struck him as "a nice fellow" or because he thought I would be a good graduate student, but because he also wanted to take on an Israeli student, train him in the United States, and send him back to pass on his knowledge in Israel. Slobodkin's first Israeli student, Uzi Ritte (who helped us greatly in our first steps in Ann Arbor), was about to finish his Ph.D. and return to Israel to join the Hebrew University in Jerusalem. At that time, the Israeli academic system was greatly influenced by the German system, with traditionally rigid teaching approaches. The American philosophy of education was very different, as you undoubtedly are well aware. This demands a book of its own, so I'll not go into it here. However, I do I believe that my generation brought about the big revolution in Israeli university life, because those students returning from the United States brought back with them the American idea of how science should be approached and university life managed. I was therefore lucky indeed to meet the Slobodkins, who became later like family, and to have the opportunity to study in the United States. A great deal of what happens in one's life is, I believe, a matter of opportunity and timing: who you meet, when you meet them, and whether you seize the opportunities or not.

We arrived as students in the United States with almost no money. Shosh intended to complete her B.Sc. studies in biology. Our entire income was $220 per month from my fellowship. While in Ann Arbor we had managed somehow to survive by renting a tiny inexpensive university apartment. However, at Stony Brook there was no way we could survive at such a low level of income. Ever the caring advisor, Larry discovered, after some inquiries, a big deserted house owned by the university, which was offered to us for free. It was a sort of huge mansion three stories high, reminiscent of those seen in *Gone with the Wind*.

One of the big old estates?

Yes, indeed, it was the Child's Estate! The original owner even had a private train from his own courtyard to the beach. Having been vacant for many years, the house needed "a little work" to make it livable. After fixing up two of the rooms, we finally moved in. Our first night in the house was a nightmare. It was a stormy night, and we went out to a movie, *Rosemary's Baby*. Have you seen it?

Yes!

Scary, isn't it? So back we went to sleep in our new home that first night. We did not yet have shades for the windows and the house was close to the beach, with a lighthouse nearby. The nearest other house was half a mile away. We were surrounded by woods, an incredible place. In the middle of the night we woke up to hear screams and scratches on the ceiling. The beacon was flashing on the window, and the howling wind was rattling the shutters, which started to collapse. We felt like the whole place was about to blow apart. Of course we got no sleep that night. In the morning, somewhat shaking, we hurried to talk to the person at the university responsible for the property, but he merely noted laconically: "Sure, didn't you know? The place is full of raccoons. The screams you heard were the raccoons!" We lived there happily, with the raccoons, for three years. Our daughter Yael was born in May 1971, a few months before I finished my Ph.D. and a week before Shosh completed her master's. Superb timing. Baby Yael was very content in the "raccoon house," except for screaming from time to time, in the middle of the night, sympathizing with the raccoons. That's how we managed to survive economically at Stony Brook. It was a joyful and wonderful period.

I spent the summer of 1968 in Woods Hole, at the Marine Biological Laboratory. Larry taught the marine ecology course there. What a stimulating and exciting summer it was. Then came the opportunity to go back home and work on corals. At that time, questions like, "What generates and maintains a high diversity of organisms?" were a hot issue as they are today in fact. I was struck by the concept of information theory and enthused by quantitative indices developed to describe communities. Where could I use it? For coral reefs, of course! That's what I wanted. My Ph.D. research proposal was to study species diversity of corals at Eilat.

Very little was known about community ecology of corals at that time, and almost nothing was known about the corals of the Red Sea. The proposal was to perform line transects for a quantitative study of the coral community structure at Eilat. I was lucky because I had the opportunity both to work with Slobodkin and to get funded by a Smithsonian grant. Such grants were being awarded to developing countries, like Israel at that time. Today, we have the Binational Science Foundation, established by the two countries, but at that time the United States was helping to advance Israeli science by directing funds for this purpose.

I was looking forward to returning to Israel. I had passed all my preliminary exams and completed the remaining formal courses required in the ecology and evolution program at SUNY Stony Brook. For two summers, I carried out my fieldwork in Eilat and during the year took courses and analyzed the data back at Stony Brook. Remember, at that time, computers were in their infancy. We studied and wrote programs in Fortran IV, which was the "state-of-the-art" at that time. I sometimes wonder what we would do today without computers, so taken for granted.

The work was mainly to study stony corals and their zonation in Eilat. We knew nothing about their biology; we hardly knew the genera. I started from scratch, performing the line transects and measuring every coral beneath the line. I had to sample a small piece of almost every coral and put it in a bag, number it, and give it a tentative name of my own, according to whatever it looked like to me. Since I was not familiar with coral taxonomy, in order to be on the "safe side," I ended up with a huge collection of coral samples totaling about 200 "species" (as I described them). Working later on the collection at Cornell University with John Wells, to whom I owe my taxonomic training, we reduced the number to about 115. This coral collection is now exhibited in the Natural History Museum of Tel Aviv University.

Tom Goreau came into my life in the summer of 1969. It was Tom, in fact, who connected me to John Wells. As you know, Tom Goreau was the founder and director of the Discovery Bay Marine Laboratory in Jamaica. He was the "dream-mentor" of any student who wanted to study coral biology at that time. Tom came to Eilat for a short visit to work in the Red Sea on the "crown-of-thorns" starfish *Acanthaster*.

I remember Tom coming to Eilat. He was very enthusiastic about my work, enabling quantification of coral communities. At that time, all the research on coral community ecology was qualitative. He was so excited about the methodology of quantifying coral communities that he became "my assistant." I was all of 26 years old, and here was the great Tom Goreau coming to work with me—my diving buddy! He would hold the small nylon bags, give them to me, and I would put in a little piece of coral sample. Tom was my diving partner in Eilat for a week, just doing these transects. Then we went south to the Sinai. This was shortly after the Six-Day War, so the area was still mined. Today, the trip only takes three hours to drive from Eilat to Sharm-el-Sheikh. But then there were no roads, so we drove along the beach, accompanied by armed soldiers, who had volunteered for the job, in armored vehicles; the place was still aflame and dangerous, lots of field mines. It took us about a week to get there. We dived at different spots along the stunning reefs of Sinai, doing transects. In retrospect, when you think of it, we must have been mad indeed.

After returning to Stony Brook Tom invited me to come to Jamaica, to quantitatively study the community structure of the Discovery Bay corals. This was another first for me—after my first trip to the United States, the first time encountering television, and the first time feeling the delight of snow. Being invited by Tom Goreau, "the guru of coral reef research," was a great honor. I arrived at his lab to find a whole gang of enthusiastic, hardworking graduate students, among them Judy Lang, Jeremy Jackson, Dave Barnes, Henry Reiswieg, and Bob Kinzie. Today, they are all famous coral reef biologists. All were Tom's students, or descendants, if you like.

On my first day at the lab, Tom already decided to teach me the deep corals of Jamaica. It was quite stormy that day, with rough seas. I can't stand rough seas! White and pale from snowy New York at Christmas, I arrived straight into the tropical Jamaican heat. Tom dove, dropping like a stone to 60 meters, and I followed headlong. Diving was fanatical there at that time. Well, perhaps I exaggerate—maybe it was 50 meters—but still very deep. I tried to keep up with him, looking up from time to time toward the distant light. Would I ever surface again? I was fine in the water, but seasick immediately after we got back in the boat. He drove me hard, to the limits. We surveyed these line transects exactly as we had in Eilat, at different depths, diving twice a day for two weeks. Each time we returned to the lab, Tom would complain about terrible stomach pain, go to his room, and close himself in. We believed he had a stomach ulcer. I thought at the time that the reason for his pain was his eating habits: In order to regain his energy after diving he used to eat those terrible fish in cans—what do you call them?

Sardines?

Yes, oily sardines and hardboiled eggs was what he was always eating. We continued to work hard every day. Tom felt wonderful underwater, no pain, very happy. Out of water he suffered terribly.

Three months after returning to Stony Brook, Tom called me from New York to tell me that he had been hospitalized and asked if I could come to visit him. He wanted to talk about our work in Jamaica. I went to see him and was told that he had stomach cancer.

A week later, he was gone.

Tom was to have given a paper on the coral reefs of the Sinai at the meeting that David Stoddart and Sir Maurice Yonge organized in London in May 1970, "Regional Variation in Indian Ocean Coral Reefs." While still in hospital, Tom asked me to go instead. I said, "Of course!" He requested that Stoddart invite me in his place to talk about the coral reefs of Eilat. This was my first international symposium, which resulted in my first paper on coral reefs. Sir Maurice Yonge paid a tribute to Tom's memory, summarizing his huge contribution to coral reef science, and the entire conference was dedicated to Tom Goreau's memory.

My first publication was thus on coral species diversity in Eilat. The coral survey methodology used in this study became what people now sometimes call "Loya transects." I am still amazed to see that in many reef monitoring programs these very simple but highly informative and efficient methods are still being used with the latest photography and video techniques.

After completing my Ph.D. in the summer of 1971, as the first Ph.D. student to graduate from the ecology and evolution department at Stony Brook, I obtained a postdoc position at Woods Hole Oceanographic Institution with Howard Sanders and Fred Grassle. From diving in warm tropical waters, I went on to diving in the cold waters of Buzzard's Bay with Fred as my buddy. As you know, Betsy, January in Woods Hole is not the most pleasant time for diving. It was bitterly cold underwater. We studied species diversity of polychaetes settling in mud trays polluted by oil, compared with clean mud trays. The experience and knowledge I gained from this study was to help me greatly in my studies on oil pollution effects on corals. In October 1972, I began my work at the zoology department of Tel Aviv University as a lecturer, "closing the circle" in the same place that I had started my academic career, and where I still continue to work up to the present.

To return to the detailed data set from Eilat. I was now able to use these data in different studies, as baseline information for comparison with what had happened to the reefs over the next 20–30 years. The initial data were obtained in 1968 and 1969. In 1970, there was an extreme low tide in Eilat, totally unpredictable, never having happened before. Water dropped to 40 centimeters below the reef flat, exposing the reefs to 40°C for four days in a row, between 11 A.M. and 3 P.M. The result was 90% mortality of the shallow reef corals along the Sinai beaches, as could be ascertained from comparison with my earlier records.

Then another catastrophe occurred, of manmade origin. Many oil-transporting supertankers arrived in Eilat between 1972 and 1975 (after the Yom-Kippur War), mostly from Iran, at that time still under the rule of the shah. From Eilat, the oil was pipelined to the Mediterranean, and from there it was transported to other countries. Two or three large oil spills occurred every month, for five or six years. Unfortunately, the Nature Reserve is situated very close to these oil terminals, and during such oil slicks it was covered black by crude oil. It was disastrous.

What does oil do to corals?

At that time there was little information available. What could be found in the literature, in brief, was something like: "I visited a coral reef near an oil terminal in Saudi

Arabia, where I saw lush coral growth. Conclusion: Oil does not harm corals." The author's address: Shell Oil Company, Houston, Texas. This reminded me of Sanders's oil pollution studies in Buzzard's Bay, where oil companies tried to re-examine his studies in an [unsuccessful] attempt to prove that no harm had been done to the environment.

Since so little was known at that time about the life history of corals, we had no direct evidence to prove that oil harms corals. Oil may be spilled above coral heads and not appear to affect them. However, although it does not kill them instantaneously, it does harm what goes on inside them. At that time we were also studying coral reproduction. In the textbooks we read that most corals reproduce by brooding, releasing planulae larvae. We knew so little then on the mode of reproduction of stony corals. We began to study the reproduction of *Stylophora pistillata*, which is the most abundant coral in Eilat. We tried to capture the planulae by putting plankton nets over coral heads. I remember well the first time we saw under a stereoscope what the *Stylophora* planula looks like. It was one night in January 1974, definitely an exciting moment. It seems trivial today, but then it was, in fact, a major drive in our research. At that time I already had many graduate students who did the main work. Buki Rinkevich, for example, worked on oil pollution effects on *Stylophora*.

He was one of your first graduate students, wasn't he?

Yes, both Buki and Hudi Benayahu, working on soft corals, were my first graduate students. In the coming years many more joined the crew.

Since we already had data on growth rate and knowledge on the reproductive biology of *Stylophora*, we did most of the experiments on *Stylophora*. Our findings gave us many reasons to worry about the fate of our coral reefs. In a series of field and laboratory experiments, we showed that crude oil was harming corals by affecting different stages in their life history, mainly their reproductive system. We found that fewer eggs, sperm, and planulae were produced per coral head in areas affected by oil pollution, compared with colonies in clean areas. Moreover, during the reproduction season, colonies aborted immature planulae, which failed to settle. To make a long story short, even today, over 25 years after the oil spills and the fatal low tide, the Nature Reserve of Eilat has still not returned to its former coral community structure. In 1969, we had in the Nature Reserve of Eilat high diversity of 15 different coral species per 10-meter transect in shallow water. Today, there are only about seven species.

These chronic oil spills that occurred every month for years caused a community phase shift, and the shallow reefs of Eilat's Nature Reserve were taken over by algae. The situation was and still is totally different, however, in other places along the Sinai coasts, where corals suffered similar mortality during the fatal low tide but were free from oil pollution. Within three years, these clean reefs had more colonies and more species, as recorded in the same 10-meter transects. With time, four to six years later, the coral community in the nonpolluted reefs recovered and gained its original composition. It took, however, an additional 20 years for the nonpolluted area to regain its original coral living cover. In fact, today, coral living cover in the pollution-free, shallow reefs of the Sinai is even higher than before the devastating low tide. I regard this example of a long-term field study as my most important contribution to coral reef field studies.

Stylophora became the major species to be studied in my lab for years to come. We worked on many aspects of its life history and physiology, including population dynamics, reproduction strategy, growth rates, colony integration during regeneration, pollution effects, predation, competitive networks, histocompatability, spatial heterogeneity and fish diversity, relationship with ectosymbionts, bioerosion by boring organisms, relationship with its endosymbiotic zooxanthellae, energy budgets (autotrophy vs. heterotrophy), calcification and productivity, hydrodynamics (diffusive boundary layers), Heat Shock Proteins, *in situ* colony stress assessment, bleaching effects, and others. There is no way I can describe to you now the results that came out from all this research. I hope to be able to summarize one day, perhaps in a book, the significance of *Stylophora pistillata* in reef research. I cannot refrain from saying that this is definitely my favorite coral species, since it fed my graduate students for many years.

For my first sabbatical in 1979, I was privileged to win a Queen Elizabeth Senior Fellowship to work at the Australian Institute of Marine Science [AIMS] and at James Cook University. Michel Pichon and Carden Wallace were there at that time, leading an enthusiastic group of graduate students including Bette Willis, Peter Harrison, Russ Babcok, Vicki Harriott, Andrew Hayward, Jamie Oliver, Gordon Bull, and others, many of whom were studying coral reproduction. Today, most of them are internationally recognized as leading coral scientists.

I had a very enjoyable and stimulating time in Australia, learning from the field experience and frequent constructive interactions with students and faculty. One of my publications from that period deals with tumors in corals. In a way, it bears some relation to what I am studying today. I participated in several cruises led by my "mate," Charlie Veron, together with good friends like Terry Done, Carden Wallace, and many others, cruising all over the Great Barrier Reef. I also spent a lot of time in the field with Roger Bradbury and Russ Reichelt [who became the director of AIMS 15 years later], doing line transects for a study of coral community structure in several GBR reefs. These studies yielded two joint publications. It was an incredible year.

Later, John Bunt, then director of AIMS, invited me to visit the institute several more times. At that time, AIMS was focusing on a large-scale study of coral communities along an inshore offshore transect. This resulted in a very productive and successful project, due to all those enthusiastic young scientists recruited at the time by AIMS, all working within the same reef region, on a wide variety of exciting aspects of coral reef research. Since then, AIMS and James Cook University have grown rapidly and become a world focus for coral reef research.

In science, my major contribution is to the field of biology and ecology of Red Sea stony corals. To date, I have published more than 130 papers on corals from the area, including a book on coral reefs (in Hebrew, now also translated into German and Dutch, together with Ramy Klein, my graduate student then and very special friend and colleague today). A great deal of this publication record is largely due to the excellent group of students that joined my laboratory throughout the years. It was working at the Great Barrier Reef, however, that gave me a wider perspective of the dimensions, complexity, immense diversity, and significance of coral reefs. I had visited and worked in many reefs around the world, but none, I believe, matches the GBR, with its colossal 2,300 kilometers of reef systems. Charlie Veron describes it as "the biggest structure ever made by living organisms on Earth."

One of the most amazing discoveries to excite many coral reef researchers was the spectacular annual event of coral mass spawning in the GBR. This is one of the most dramatic events on the reef's calendar. More than 150 species of corals release their gametes after dark, five days after the full moon in late spring. This incredible phenomenon was first observed only in 1982 by the "James Cook reproduction gang," that fantastic group of students I mentioned before. During one of my visits to Australia I heard about this from them. I was astonished not only by their amazing discovery, but also because our own studies on reproduction in Red Sea corals had indicated exactly the opposite pattern.

Although in the Red Sea we have the same coral species as in the GBR, "our" coral species reproduce in different seasons, different months, or different moon phases within the same month. That is, they show clear temporal reproductive isolation. Most of the Red Sea coral species reproduce during summer, starting in May and continuing through September with almost no overlap in their reproductive timing. This was the Ph.D. dissertation subject of my student Yechiam Schlesinger. It is easy now to report the facts, but it was much harder to determine them, as you can imagine.

When I met up again with my Australian friends and we exchanged our coral sex stories, their story was somehow more . . .

Dramatic?

Yes, definitely; I'd even say "sexier"; really stunning! A good story for *Science* magazine, which is, indeed, where they published it in 1984. We decided to wait with our contrasting story until after their publication had come out. To tell the truth, I was not sure that our own study would even get into *Science*, if our story was simply that Red Sea corals exhibit temporal reproductive isolation. Such a story could be interesting but was certainly not so sensational and would probably have a better chance of acceptance if the GBR story made it first. Temporal reproductive isolation makes sense evolutionarily and ecologically—less interspecific competition, higher chances of successful fertilization, less hybridization; but I'll not go into that now. So, it is not "earth-shattering" compared with the GBR astounding tale of the "coral sex orgy." We all know that the editors of *Science* like "contradictory stories," to provoke discussion and stimulate further and broader research. In short, we succeeded, and our story came out in *Science* in 1985. I am sure that you are familiar with both these publications, which I believe stimulated many additional studies on coral reproduction. Since then, coral reproduction research has increased exponentially in many areas, as has our knowledge of the subject.

Coral reproduction continues to be a major subject in my lab, not only of Red Sea corals but also of Mediterranean corals. For example, we made the cover story in a recent edition of *Nature*. I'll tell you about it, briefly.

Probably one of the reasons why corals are among the most successful organisms in the marine tropics is due to the high plasticity of their reproductive patterns. In a general and simplistic way of describing it, they exhibit either hermaphroditism [where both sexes occur in one coral] or gonochorism [separate sexes]. They may be hermaphroditic or gonochoric brooders [with internal fertilization], or hermaphroditic or gonochoric broadcasters [releasing their gametes into the water, where external fertilization takes place]. Within these general patterns there are a wide variety of different patterns that we'll not go into here. Corals, however, also exhibit asexual reproduction, such as fragmentation, and polyp bailout.

Our cover story in *Nature* [May 1997] reports on yet another asexual means of reproduction in corals, which we termed polyp expulsion. This finding was first observed by two of my graduate students, Esti Winter in the Red Sea and Maoz Fine in the Mediterranean. We noticed that in extreme and unfavorable reef areas with a lot of sedimentation and wave action, some corals like *Favia* in the Red Sea and the encrusting coral *Oculina* in the Mediterranean are quite abundant, while other corals are scarce. We wondered how and why these specific corals managed to occupy such areas, especially because they don't fragment like branching fragile corals. One could expect that asexual reproduction would have an advantage over sexual reproduction in such a harsh environment that induces wastage of gametes.

We discovered that in harsh environments corals that do not fragment are able to expel whole polyps from the colony. In these corals, some polyps, including their calices, rise up on elongated calcareous stalks before detaching and "taking off." [Loya shows me a picture from the article.]

I've seen that!

You've seen it, of course! I've seen it too many times but thought it was some sort of a tumor or an abnormal growth. I had never given any thought to the significance of this phenomenon: why, when, and how it takes place. That's what I wanted to show you, Betsy, in this cover photo: a polyp just taking off from the coral surface.

Wow!

The lesion left on the surface of the colony regenerates within two weeks. You can clearly see the hole left in the mother colony, where the polyp had resided. The expelled polyp usually falls near the mother colony and grows into a new colony.

That is incredible!

Yes, I think so, too. Polyp expulsion differs from the polyp "bail out" process [described by Sammarco], where dying corals release whole polyps [without skeleton] from their calices, some with planulae. Polyp expulsion occurs in healthy corals, and whole polyps, including their calices, are expelled. You need to have wonderfully "crazy" graduate students like Maoz, who spends many hours underwater, to be able to be there and witness the very moment an event such as polyp expulsion occurs.

Another subject that I have been pursuing in my lab is the use of stony corals as proxy indicators of regional climate change. This was the Ph.D. thesis subject of Ramy Klein, who worked on the sclerochronology of contemporary and fossil Red Sea corals. As you know, skeletal bands of alternating high and low density in *Porites* coral colonies have been widely used to record their growth history. Ramy found periodic sequences of yellow-green fluorescent bands in fossil corals he collected from Quaternary reef terraces in southern Sinai. Collaborative work on the fossil coral samples with Peter Isdale from AIMS yielded another *Nature* cover story in May 1990. In this study we provide evidence that during the late Quaternary reef-forming peaks, the climate in the Sinai was wetter than today's extreme desert conditions, with possible summer rainfall regime. Incidentally, the beautiful cover photo accompanying this article of the Sinai fossil terraces at Sharm-el Sheik, as well as the "polyp expulsion" cover photo [and the photo of Loya and Goreau in this volume], were taken by our departmental photographer and my good friend Amikam Shoob. Amikam proves indeed that "a good photo is sometimes worth more than a thousand words."

After becoming a full professor [1985], I was asked to take my turn as chairman of the zoology department. All academic-administrative posts in Israeli universities are based on a rotation system. During this period, we also needed to elect a new dean for our faculty of life sciences [every five years]. Two members of the Department of Plant Sciences were running for the post. The tenured professors of the faculty elect the dean.

Three weeks before the elections, one of the contestants retired from the race. At that time [1990] our faculty was undergoing a similar process to that which was taking place in many universities around the world, where molecular and cellular biology were taking over the organismic biology sciences. It was a critical point for the fate of organismic biology studies in our faculty. Various members approached me to enter the race and save the "homeland," as they said.

I had no plans whatsoever to become dean, which usually takes a year of political canvassing. I had a very successful lab, very interesting and fulfilling research, and felt still too young to run for such a position and take on the vast commitment. Our faculty numbers 125 members and about 150 technicians and administrative staff, incorporating five departments (plant sciences, zoology, molecular microbiology and biotechnology, cellular biology, biochemistry and neurobiology), and we have around 200 M.Sc. students and 130 Ph.D. students. In a family consultation my wife said, "I have never stopped you from doing anything, why should I stop you now?" My graduate students then [Micha Ilan, Ramy Klein, Avigdor Abelson, and Ofer Mokady] said, "Go for it; we are behind you." In the end, I gave way to popular pressure and decided to raise the "organismic biology" flag. The long and short of it was—I was elected by a vast majority. Within two weeks, I had suddenly become dean of a very large faculty, something I had never even considered before. I remember the following days, waking up in the morning, holding my head and asking myself, "What do I need this for?"

Today, three years after completion of my deanship, I would still do it again. I believe that I achieved the major goals I had worked for. This is not the place, of course, to recount this most interesting period in my life. I would, however, like to mention the most important achievement of this period. During my post as dean, I managed to obtain from the government 25 permanent fellowships directed to a new M.Sc. program of ecology and environmental studies.

But you must have had to do some politicking—how did you convince nonscientists that this was important?

Oh, it was a very long and complicated process. Starting within our faculty, I had to convince first our departments to approve the establishment of a new academic program in ecology. This was not easy, because it stood in conflict with their immediate interests in the thin university "budgetary cake." Then I had to compete with other programs, proposed by my rival deans representing other faculties; to convince the university leadership that our program was unique, timely, and had the best chances of being approved by the National Council of Higher Education [NCHE]; to go through many other academic procedures, which I'll not bore you with; and, finally, if successful in the internal university race, to compete with other academic institutions in the NCHE.

In the Israeli system the government channels funds to the NCHE, which allocates it to the universities. Each university gets its share in accordance with its number of students, novel teaching programs it proposes, and many other academic criteria. Clearly, there is fierce competition between academic institutions on these funds, which are their major source of support. My main task, therefore, was to convince the NCHE that Tel Aviv University [TAU] was the best university in the country for ecological studies [which is true]. Needless to say, they were convinced. We were generously granted with additional faculty positions, student fellowships, equipment, and more. It took me almost three years of hard work to complete the whole mission.

We now have a very strong program in ecology and environmental studies, including a new M.Sc. degree in that field. Currently, there are more than 50 M.Sc. graduate students in this program. Many of the outstanding students graduating from this program continue their Ph.D. studies in related fields dealing with environmental problems. I feel that this was my greatest achievement and contribution to the advancement of ecological and environmental studies at TAU and to a large extent in Israel as a whole.

Coral reefs had been the most important thing in my "scientific" life, but now I began to learn that there are other things too.

I know for you your family is important.

Of course it is. I have three children.

I've met the "little urchin" who has eyes like you when I visited Israel in 1990; he must have been six or seven.

Yes, his name is Assaf; he is now 14 and fills our lives with joy. We call him our own "grandson" because of the big age gap [12 years] between him and our adult children, who at times play the role of acting parents. Our daughter Yael was born in New York. She is 26, an ambitious, very busy, hardworking lawyer. Our middle son, Shay, is 25, a pianist, a real scholar, studying musicology at the Music Academy of TAU. My wife Shosh earned her Ph.D. in microbiology at TAU and works at the medical school there on AIDS research. We are both collaborating now in a large consortium comprised of many scientists, from different biological, chemical, and medical fields, in a large-scale project dealing with novel bioactive compounds derived from coral reef organisms. A great deal of whatever I have been able to achieve in my academic life is due to Shosh, who has supported me all the way.

Certainly, your scientific family is important too—just the way you've talked about your graduate students.

Absolutely. I'm very proud of my present and past graduate students, four of whom have joined our faculty at TAU, after completing their postdoc training at U.S. universities. Others obtained positions in other Israeli universities, research institutions, government agencies, and NGOs dealing with environmental protection and conservation. As a matter of fact, to date, 25 M.Sc. and 14 Ph.D. students have graduated under my supervision. It is very gratifying to watch my past graduate students developing their own successful scientific careers, gaining international reputations, and leading busy labs with crowds of graduate students, most of whom significantly contribute to coral reef research.

I'm proud to have been a student of Larry Slobodkin and a descendant of the Hutchinson School. You are probably familiar with the Hutchinson "scientific family tree," which includes many famous ecologists like [E. S.] Deevy, [R. H.] MacArthur, [G. A.] Riley, [H. L.] Sanders, [L. B.] Slobodkin, [W. T.] Edmondson, [T. F.] Goreau, [H. T.] Odum, [R. T.] Paine, [A. J.] KOHN, [F. E.] Smith, [P. H.] Klopfer, [W. D.] Hartman, to mention a few. I appear on the Hutchinson tree as a tiny leaf branching from the Slobodkin trunk.

You're not a leaf now; you are a branch with lots of leaves on it!

Yes, today, 30 years after my first steps in science, it is wonderful to observe how the Hutchinson tree has grown to produce so many generations of ecologists all over the world. As to my "scientific family," I must say that I am very happy to already have "grandchildren" in coral reef science: Could we call them "branching corals"?

Don Abbott had just retired and I thought, oh, my God, the whole generation is gone now, beginning to die and retire.

Daphne Fautin (b. 1946)

Professor of Biology
Curator, Natural History Museum
University of Kansas
Interviewed at the Society for Integrative and Comparative Biology Meeting, Boston, 6 January 1998

Growing up literally "in the field," helping her dad collect mammals, provided Daphne Fautin with rare skills and knowledge she is just beginning to appreciate. She is the world's expert on sea anemone systematics and also is an ecologist, focusing particularly on anemone–clown fish relationships. She describes an NSF program, PEET (Partnerships for Enhancing Expertise in Taxonomy), which is designed to help save and restore the enormously important but disappearing art of taxonomy by establishing apprenticeships. Emphasizing that personal interactions and friendships among scientists are a crucial element in the advancement of science, Fautin traces the beginnings of coral reef science as a separate, recognized discipline in the early 1970s.

I'm a firm believer that much of education is through personal friendships, and there is a lot of transfer of information from generation to generation that isn't just factual information.

Well, that's exactly what this PEET grant program is for. When it was first being considered, the guy who was the director of the KU Natural History Museum at the time, who has since retired, referred to it as a geriatric brain dump. Although all of the knowledge that all of the anemone systematists of the past have put down on paper still exists on paper, the actual, how do you *really* do a nematocyst smear and make these distinctions? and all of those kind of things are not formalized. The knowledge is really very dry and sort of trivial when it's reduced to paper, but when it's active and when you are apprenticed, it's very different. The PEET project is for people who have expertise in taxa for which the expertise is rapidly disappearing. If we don't rescue it, it is sort of like Bob Johannes's tale of the boat builders of Micronesia. You've got to get those skills while somebody is still around to teach them.

Somebody convinced NSF that this was important?

Yes. The very interesting and very scary aspect of it is that the first summer of my grant, my brand-new Ph.D. student and I toured ten natural history museums of Eu-

Suspended above the reef, Daphne Fautin gathers data on clown fish–anemone interactions in the Red Sea in the 1990s. Photo from the files of D. Fautin.

rope. Everybody at these museums knew about this PEET program, and they were all having the same problems.

People I've already interviewed said, "I hope that the Europeans are continuing to do this."

No, they're not. So they were saying, "It's really good, and if it's a success we can convince our governments to do something similar." There is real pressure from NSF on us, as a cohort, to produce systematics that the community will find useful. There was this feeling in the old days that what we did was nineteenth-century, dry, and trivial, basically stamp collecting. But systematists all over the world know about this new program, and so there is tremendous pressure on us. All of these students can recite the cell cycle, but they don't know the context of the cell.

They don't realize that there are lots of organisms out there, made of cells. The idea of biodiversity is that there are lots of organisms, but they are not always easy to tell apart, and there are systematic ways to do this.

Even worse! I was very impressed when I was still in San Francisco and Bill Hamner sent me some corallimorpharians from the Great Barrier Reef. He sent them air freight, which meant they came in through customs at San Francisco Airport. Fish and Wildlife is there. They have this form to fill out, and it says "genus" and "species." I said, "I don't know." "Well, we can't let it in. Where can we go to get it identified?" I replied, "That's why it is being sent. I'm the expert." It turned out to be a new, "genus" and "species." They were astonished! These are Fish and Wildlife employees, presumably bachelor's level biologists.

Where are you from?

Actually, I was born in Urbana, Illinois, but as I've told everyone, I had the good sense to move to Wyoming when I was three months old. My dad [Reed W. Fautin] had gotten his Ph.D. just before World War II broke out, at the University of Illinois in ecology. He was in the last cohort of Victor Shelford, so intellectually my roots, my partial biological and academic genealogy, go back to the origins of ecology in the United States. He went off to the Pacific for four years and, of course, he couldn't apply for jobs from there. So, when he came back, they gave him an instructorship at Illinois for a year, until he found a job. During that year, he got the job at the University of Wyoming, and I was born. He went out and started teaching summer field camp, which he taught every summer of his career. After he retired, they kept him on for an extra summer to teach the field camp, so those are the bookends of his career.

Mom and I took the train out at the end of that first summer. Because he was a field biologist, we went camping for weeks at a time during the summer. Dad collected small mammals for the University of Wyoming Natural History Museum. Among my earliest memories are running the trap line with him. If what he caught was damaged or a duplicate, something he didn't need, I played with it. Those were my dolls.

Among my best friends, when I was growing up, were my father's graduate students. I assisted them in their research. My father had a woman grad student—I must have been about nine years old when she came to the University of Wyoming—and her project was a bird census. Laramie, Wyoming, gets about 10 inches of rain a year, a real desert. The only place you could do a real census, where all the birds stopped, was where there were trees, and the only place there were trees was the cemetery. Every Saturday morning, we'd get up at about 5 o'clock, and we'd spend about three hours walking through the cemetery. She had this map, and she'd mark down what birds were there.

When I was three years old, if a ground squirrel would run across the road, I would say, "Oh, look, there goes *Citellus lateralis*." I knew the animals only by their scientific names—not the birds, but the mammals—which is what my dad worked on. One of my students gets really upset with me for asking her to learn these kinds of things: "Well, you grew up with these things, you know about them. It's natural to you and it's not to me." But that's not the point of the story; it's that these aren't hard. A three-year-old can learn them! I grew up in a scientific, academic environment. Somehow—I guess it was both genetic and because of this early exposure—I just always knew I was going to be a biologist. When the University of Wyoming began its Ph.D. program in zoology, my father had two students, one male, one female. His cohort at Illinois, whether this had anything to do with it, was two men and one woman. My mom was a biologist. I never got any messages that I couldn't do it, it was unladylike, or anything. It took me a very long time to appreciate that I grew up in a very unusual culture, unlike a lot of girls. I just thought they were wimps and making excuses. Of course, girls can do this.

I wanted to be an ornithologist. Dad had gotten his master's in ornithology, and there had been this graduate student whom I'd befriended and with whom I'm still friends. I always was a birdwatcher, wherever I went. Just at the time I was making my decisions about college, the first edition of Carl Welty's *Life of Birds* came out. He was a professor at Beloit College in Wisconsin. All of us, when we became teenagers, worked at the University of Wyoming summer field camp, where dad taught every summer. The majority of students were geologists, but there was a botany course and a zoology course, too, that dad taught. One summer, the geology course was just full of students from Beloit. I had never heard of Beloit before. I had been thinking of Carleton, and just then, Carl Welty's book on birds came out. I thought, this will be great, to study with a great man. He became a lifelong friend. He was still alive when I was teaching at Beloit in the early 1980s, and he died right after I left.

When dad was a graduate student at the University of Illinois, his professor Victor Shelford had a long-standing relationship with Friday Harbor Labs; he went there most summers. And so, he required all of his Ph.D. students to spend a summer at a marine lab.

It used to be that way at Berkeley in the 1960s.

I think a lot of places used to have summer field station requirements, and the collapse of those caused the collapse of a lot of the stations. Dad went to Hopkins [Marine Station] and talked about working with Ed Ricketts, collecting for Pacific Biological Supply in the intertidal. But he was too far along in his research at that time to change, although years later, he said he wished he had had that experience sooner, because his life might have been very different. At the University of Wyoming, he couldn't require people to have that experience, but for his serious zoology students, he suggested that they spend a summer at a marine lab. He had a lot of positive feedback from students who had gone to the University of Oregon Marine Lab at Charleston, so I spent a summer at Charleston after my second year at college. It was at my father's suggestion. In many ways, he was my mentor and my guide in career development.

When you got there, did you know right away?—gee, I want to do more of this.

Oh, yes. That was it. First it had been ornithology and then veterinary science, but I had found it! I had grown up in the mountains and just loved the mountains, and I couldn't understand all this fascination with the sea. Something just spoke to me. I had been to the ocean before; it wasn't like it was my very first experience. We'd actually traveled around the world. When I was 11 years old, we lived in Afghanistan for two years. The University of Wyoming had an exchange program, which provided technical expertise to the University of Kabul, because the climates of Afghanistan and Wyoming are very similar. So, I wasn't just a rube from Wyoming!

But you never had a formal introduction to the rich flora and fauna of the West Coast?

Nope, and that was wonderful, just wonderful. The first day at OIMB [Oregon Institute of Marine Biology], we went out into the field. People returned with things. Of course, collecting was what everybody did in those days. There were no concerns about overcollecting or anything like that. We were each issued a bucket. People returned with buckets of these spiny, purple things. I literally could not figure out where they had gotten them. I had no search image. The next week, when we went back there, and I saw all these *Strongylocentrotus purpuratus*, I thought, how did I miss them!?!

I had had that experience living overseas at a very formative time of my life. The minute President Kennedy announced the Peace Corps, the thing that propelled me through college in three years was to get into the Peace Corps. I really wanted to go back to Afghanistan, but the timing was bad. They were starting to train people at the beginning of the summer, and I was going to summer school to graduate. I wasn't going to be available until the end of the summer. So who knows what my life would have been had this not happened? I went to Malaysia in the Peace Corps because they started training in September.

I trained in Hawaii. Then I was in this coastal situation in Malaysia and went to work on coral reefs. I was teaching biology, but the coral reefs were right out there. I befriended fishermen, who would go out to the reefs. I learned to snorkel in Hawaii during training and I had my snorkeling gear. These rich white people would come from Singapore and charter boats. I would work as a boatman and just take the time while they were picnicking to snorkel around. Now, it's a major tourist area. I was back there about three years ago and augh! but then, it was just renting a fishing boat. So that's how all that happened.

I met my first husband [Frederick L. Dunn] in Malaysia. He was doing tropical disease research. He is retired now, but he is still a professor at the University of California, San Francisco. I had actually already made arrangements to stay at the University of Singapore, after my Peace Corps two years were up, to do a master's degree there and stay on those reefs, off the town I was in. He persuaded me to go back with him. He owned a home in Berkeley, so I applied to Berkeley, to go to graduate school there. It was totally happenstance.

My professor at OIMB in 1965 was a fresh Cadet Hand Ph.D. I had heard Cadet's name. It is distinctive and memorable. People ask me how I got involved with sea anemones, and again, it was total happenstance, serendipity. My first husband was actually very formative for me in professional development. He's exceptionally rigid and doctrinaire, I realize in retrospect, but it was just what I needed then. He is a stuffy easterner, went to Harvard, and so forth. But he showed me how to use the Berkeley library. He read all of my manuscripts. He instructed me in writing professional correspondence. It was wonderful to have this kind of tutoring. Of course, the problem—it was a great Pygmalion thing, but when I became a professional in my own right, he couldn't stand it. But it was great while I needed a professor in the house.

He told me, while we were in Malaysia, that really the right thing would be to write to one of the professors in the department. I got the literature from Berkeley. I read it, and I chose to write to Ralph Smith. Ralph was on sabbatical, but he answered me, even though he was in Finland. Fred was very impressed that this professor had taken the time to write to me from sabbatical and in longhand! Of course, neither of us knew Ralph then. So Fred's interpretation of that, being fairly self-centered and having an attitude, I think, that there are proper ways and there are payoffs for doing things differently, was that he must be doing that because I was in the Peace Corps.

Fred figured that Ralph probably had a child in the Peace Corps and so he felt a duty. Because Fred couldn't understand otherwise why a Berkeley professor would write longhand from sabbatical.

When I got to Berkeley in the middle of the year, Ralph was still on sabbatical, and I met Cadet. He started talking about *Epiactis prolifera* [a sea anemone], which I remembered from Oregon. He'd been trying to get a student to work on this. It was a beautiful little project, and so I said, "Sure." By the time Ralph returned, I was already launched, and the rest is history, as they say. Funny thing is, years later, I said something to Ralph about that initial letter. I was very close to Ralph. In fact, I arranged retirement parties for both Cadet and Ralph, when they retired from Berkeley. I told Ralph of Fred's reaction, because also, when I got to Berkeley, I discovered that he *did* have a son in the Peace Corps. In fact, he still lives in Micronesia, in Majuro, I think. So, years later, I said, "Well, that must have been why you wrote to me." Ralph was so offended, because he didn't need any ulterior motive. He was kind; that's why he did it. He wasn't the intimidating guy that he has sometimes been painted. I really felt bad.

We have a professor at KU who was an undergraduate at Berkeley, and boy, she just, "Oh, Ralph Smith, he was [so aloof]." I said, "Look, Linda, you were an undergraduate, don't say those kinds of things about him."

She probably called him Professor Smith.

Oh, yes. The thing is, we had what I think is really a very nice convention at Berkeley. When you passed your Ph.D. orals, you could call professors by their first names. Of course, now everybody uses first names. It was a rite of passage then; you became a colleague. It was hard to call him Ralph.

I'm sure. I don't think I called Don and Izzie [Abbott] by their first names until they came to West Indies Lab in the late 1970s, not while Bill was a graduate student and certainly not when I was an undergraduate.

I still have Don's plant, a mother-in-law's tongue. It was doing poorly for a while and I thought, augh, I have to keep this thing alive. When they were leaving California, I was there talking to him one day, and he said, "We're trying to get rid of our house plants, take this one." I carried it with me to Kansas.

Isn't that nice to have? [Don died about ten years ago.]

Yes. That's my Don and Izzie right there.

How did you formulate or how were decisions made for you on a career path after that?

I've been really lucky, I think. I guess most ecologists wouldn't consider me an ecologist any more. I even purposely entitled my thesis "natural history." Cadet told me, "You'll never get a job doing that, but you're married to a rich doctor, so you don't really need a job." Hmm! But anyhow, I really am interested in the animals, have tried to keep them foremost, and have tried to transmit that to students. One reason I left

Cal Academy is so I could have students. I would teach courses and give lectures occasionally, but it's not like following someone through a career and helping professional development.

It's interesting. Between Ralph and Cadet, they divided up the students so that the ones who were in need of really hands-on supervision were Ralph's. I think it also had to do with discipline. But I think that they actually thought about the degree of attention a student was going to require. Because at that time, Cadet was director of Bodega, and he was on the Atomic Energy Commission's Public Review Panel and traveling all the time. He could see my thesis area from his office, and he practically never came out to it. Sometimes, when students come into my office for our weekly meeting, and they ask a question, I tell them, "Wait for a week and think about it." The first time, some of them get really upset. I say, "I remember distinctly calling Cadet from Berkeley on the tie line with a question, and his response was 'Think about it.'" I could handle that, and maybe the ones who couldn't sort of fell by the wayside. But I think it was also that I was one who was seen as being able to handle that.

I find that I would like to be there if people need it, more with some and less with others. I really like developing in students, in cultural terms, a perspective and a way of looking at things, not making them fit into a narrow hole but taking advantage of their strengths and weaknesses.

Not telling them what they have to do.

Right. It's the STEVE WAINWRIGHT sort of questioning: "Well, explain yourself." One student I had regarded that as me saying, "You are wrong," and no amount of persuasion could convince her that I was just asking her to explain her answer to me. I'm not saying, "It's wrong," just "Can you explain it?" And she didn't, and she left. So, there is some selection going on in the process.

Steve Wainwright was telling me about his relationship with Len Muscatine [my Ph.D. advisor] and how they were good as graduate students together, because they were so critical of one another—but in a positive way. "Explain yourself!" "Well, what do you mean by that?" I remember that Len was like that with us, his grad students.

You have to learn what works. I really like that interaction and the strength of character it shows or it builds. Science doesn't need to be unkind, but it has to be rigorous. One can fall into the other, unfortunately, but I think it's not a fixed thing. But it's clear thinking that I'm trying to develop.

I teach scientific writing because I think that communication is another skill that is not being taught well. I regard that as an investment in the future, multiplying my science. I can help them to clarify. I regard editorial duties similarly. And then you've had an effect on producing science that's going to be, if not important, at least clear and well received, a lot more than I could have personally done.

I think one of the positive things, perhaps, is because of the disposable income now, lots of students seek field-type experiences. They can travel to the Great Barrier Reef.

My graduate student, while in high school, spent a month or two traveling in Africa. I think there is that element. People scuba dive as a sport now. I learned to scuba dive as a postdoc when there was almost no sport-diving industry. I wrote a paper on the history of coral reef research in which I said, "I really do think the growth of research on coral reefs has been greatly aided by the development of sport diving and the development of lodges and destinations that air links have opened up." Places to go have opened up because sport divers have to have the newest, most beautiful coral reef, and the few scientists who would go there couldn't support a place to stay or an airline link, so it's made for better places to work. I've led dive trips for the last three years for a group that likes to go with scientists to help in coral reef research, for example. So that, I think, is very good. I think the bad part is that most of this has so little depth.

When I get serious undergraduate students who want to be marine biologists and they are in Kansas, I say, "For your sake, so that you don't get to graduate school and then discover, well, this isn't what I want to do, for the credibility of your application, so that they know it's not some kid from Kansas who decided to be a marine biologist by watching Jacques Cousteau on TV, you need to have a summer at a marine station." They come in with these glossy brochures of clearly for-profit operations, with a person whose credential is that he studied marine biology at the University of Miami. You don't know whether that means he took one course or he has a Ph.D. They say, "College credit may be available for this." I say, "Probably, if you take that, KU is going to send you over to me to assess whether I think they should give you credit." Why don't you go instead to Oregon Institute of Marine Biology, to Friday Harbor, to Duke, where we know what the standards are, where it is academic? Yes, the brochure may be black and white but . . .

You are going to meet lots of other people with your same interests.

And we know what the qualifications of the instructors are and so forth. It's not that I'm being snobbish. It is just that a lot of people are getting information that either isn't very well filtered or it's overstated. I think they are being cheated in some way. Most of them are buffered too much, away from the natural world.

There are, in addition to academic rigor, also the social interactions with your peers and with people of older generations, which I think is so important at academic marine labs.

That is exactly the point I was trying to bring up the other day. I was presiding over the invertebrate zoology business meeting. Donna Walcott was just stepping down as secretary of invertebrate zoology. She was saying something about colleagues or a cohort or something. I said, "Donna and I were graduate students together at Berkeley, and then we went our separate ways on opposite coasts. Yet here we are, we are interacting here, and you keep seeing the same people over and over again." I was saying that for the benefit of the graduate students.

In fact, that was actually my motivation for doing this history of coral reef research at the International History of Oceanography meetings. We were sitting after the Miami workshop a few years ago, talking to Chris D'Elia. Bob Kinzie talked about the Caribbean and Jamaica, the Discovery Bay days. Many of us were reminiscing about the ship the *Marco Polo* and the Second International Coral Reef Symposium in 1973. The symposium was held on that ship, cruising the Great Barrier Reef for ten days. That formed a cohort, like the group from Jamaica. It wasn't of students, but it was a very analogous situation because all of us were on this ship together, and you couldn't get away. Chris said that one of the great regrets of his life was that he wasn't on that ship. He really missed that sort of bonding. Don Abbott had just retired and I thought, oh, my God, the whole generation is gone now, beginning to die and retire. It's not like this meeting today where Stephen J. Gould comes in and talks for an hour and goes away; you were on the ship together. Usually the grand old men are the people who zip in and out of these meetings, but the other thing, of course, is that that was in the early days of coral reef science, and there weren't many grand old men. C. M. Yonge would have been a malacologist, Harry Ladd was a geologist. They were all something else. So, I mark the birth of coral reef science from that; there was a group identity.

Man wants to understand his surroundings. In that sense, science is worthwhile.

Rolf P. M. Bak (b. 1942)

Professor dr
Netherlands Institute for Sea Research
University of Amsterdam
Interviewed at the Society for Integrative and Comparative Biology Meeting, Boston, 4 January 1998

Rolf Bak left the cold and darkness of the Netherlands to begin work in Curacao with the vague direction to "study the coral reef." However, as no one knew which corals were there, he first had to figure out the species, not an easy task in those days, as not many descriptions were available. To add to the confusion, the taxonomic status of many species was unclear. From those beginnings, he spent more than 15 years there, diving twice a day, making many contributions to our knowledge of the ecology of Caribbean reefs. I have known him as a friend and colleague for almost 25 years. His cheerful account clearly demonstrates his continued love for the challenges posed by coral reef ecology.

I grew up in a not very attractive city in the Netherlands, but it was right by the sea. Close to our house was the outer dike, confronting the sea and built of large stones. As a kid I would jump from stone to stone and sometimes I'd slip. I always used to come home with wet feet. My mother would say, "Why did you do that again?" I tried to hide it, but she always found out. Playing there, I noticed crabs scuttling around in small pools and other interesting things. With the small boys of the neighboring families, I organized all sorts of exhibitions. We had empty jellypots, filled with sea water with crabs or with mussels and with seaweed, *Fucus* mainly. Then we put up a couple of boards and had an exhibition, and we tried to charge money. But nobody would come. It was not a success. When I went with my father to the dike, he wanted to walk, not get wet feet. Then we both would have . . .

Gotten in trouble with your mother?

Yes, right! Around us, it was just flat land and cattle in a very dull Dutch landscape. When I was 14, a friend and I took charge of the aquarium in the museum—our own seawater aquarium!—and it was primitive. Screwing in new lamps and getting terrible shocks because everything was wet! We had to get animals for the aquarium. We could get animals at the fishermen's harbor. Every Friday, the boats came in. The last thing the fishermen did was to clean their decks, and if they didn't do it too carefully, we could find some strange crabs or mollusks, things you couldn't find at the dike or the beach because these came from farther away in the North Sea. It was exciting. I must confess, I even had a whole collection of shells that weren't brought in for little boys. Sometimes we caught things ourselves at the dike. Those were very often the most exciting. I remember one time we found what we thought was a new species for all of Holland. God, I remember that. It was a little fish that is common in the channel, but for some reason it appeared higher up north. Another thing that always thrilled us were great storms. We had these enormous southwesterly storms. Algae growing on the coasts of Normandy and the channel would be swept up on our beaches, and all of the shells and weird things we could find in there were very interesting.

The connection with my studies? I could have become something else. For a time, I had vague thoughts of becoming a preacher. It had a certain lure, I don't know why. I found the language of the Bible impressive. Those old English texts, I thought, wow, this is fantastic.

I can't imagine that lasted very long?

Well, no. I went to Amsterdam to university. I must say that during large parts of my studies, I was neither a diligent nor an interested student. In the city, nature lost its allure. Contact with nature was limited to excursions. We had to learn all of the plant species of the Netherlands, so we had excursions for three summers to learn the plants of different parts of the country. My behavior was not that of a dedicated biologist, even though I knew those species, of course. Once again, in those days, things were fairly strict.

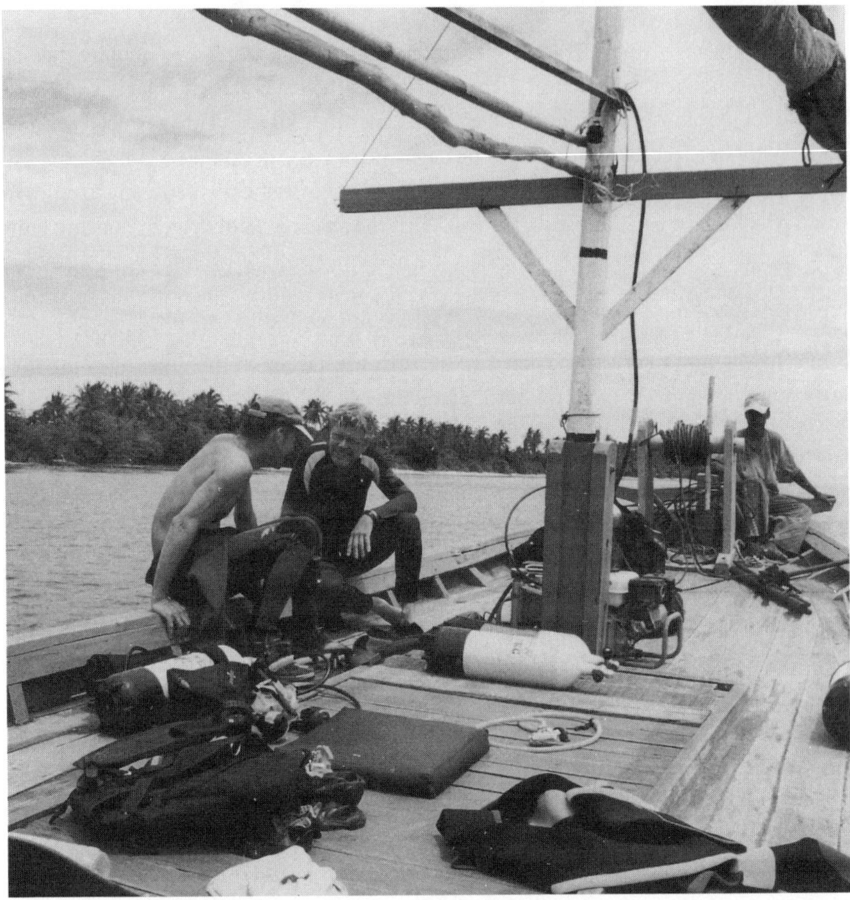

Rolf Bak and student Erik Meesters compare notes between dives from a native craft off the island of Pulau Tunda, Indonesia, in the mid-1990s. Photo by Gerard Nieuwland.

I had a friend at that time, from my school days. He was studying biology also, but at Leiden, and I used to visit him sometimes. We used to talk about what are interesting organisms, what we were going to work on. We had a discussion about the advantages of working with animals which don't run away when you approach them. Somewhere here is a connection with coelenterates. I knew coelenterates from the dike when I was a little boy.

Like sea anemones?

Right. I think I was browsing in the library, and I came across the Odums' [Howard T. and Eugene P.] paper.

Odum and Odum 1955?

Yes, that must have been one of the starting points which got me interested in reefs. I wanted to get out of Amsterdam. I was nearly finished with my formal education there. I was attracted by the tropics because at university I had worked on the biography of tropical terns, *Anous*, the noddy terns. Oh, they were fantastic. At my university, you couldn't do anything on the tropics, but there was one professor who was an authority on birds. I went to talk with him, and I explained that I was interested in tropical biology. He mentioned that he had the notes made by the captain of a merchant vessel, who was a birdwatcher, and during his trips all over the world this captain had been taking notes on the seabirds. The professor asked if I would be interested in working on the notes and especially the noddies. There was a species problem, there being possibly two or three species, but the difference between *minutus* and *tenuirostris* was debated. Nobody knew where they occurred, breeding seasons, etc. So, I was going to put that all together using the literature and these notes of the captain, and I did. It took an awfully long time because it was not always inspiring, I must say, but in the end I wrote a report that was appreciated later by others who studied seabirds. At the time, I had never seen a living noddy, of course. Later, after university, when I was in Curacao working at the marine lab there, one day a lady came with a box and said, "We caught this." I looked inside, and there was a live noddy. Wow! The first live one I ever saw. It was fantastic. I still think, whenever I see noddies now, they are fantastic birds. They are so beautiful. If you sail or are on a boat in a lagoon.

Like we saw in Tikihau?

Yes, right, on that lagoon, fantastic, those black, white-capped noddies around the boat and flying over the blue lagoon.

What intrigued you about the tropics?

I don't know. I began to read about corals, and I began to look for ways to get out there. At the time they were rejuvenating the marine lab at Curacao. It had been a one-man operation, and it had barely been surviving. A visiting scientific committee decided that, to become a real operating marine lab, more staff was needed. Instead of just one director and no one else, there were going to be the director and two other staff members. So I applied for that, and I got that job.

This is before you had a Ph.D.?

Right. But meanwhile, still in Amsterdam, I had discovered the first papers of Tom Goreau. Oh, that was exciting! The first set was on descriptive reef zonation. There was a strange photograph of spur-and-groove formation, but they were descriptive papers with all sorts of drawings and diagrams of the reef. I was living in an apartment, second floor, and I was reading this, looking outside, measuring the depth of reef zones on the houses on the other side of the road, and wondering how deep I probably would have to dive. All sorts of considerations you have when you have never dived and never seen a diver.

To get to Curacao, we could either fly or take a ship. The ships going to Curacao were merchant vessels. For the first time in my life, I had some money. So my wife, Birthe [Bak-Gade], and I had a hut with beds, chairs, and portholes to look out. It was fantastic. There were few passengers, and I was allowed to get on the bridge and look at birds. The ship had troubles, now and then, with the engine. In the Sargasso Sea, the motor had to be overhauled. We stopped and floated around for two days. The change to the good weather and being out, looking overboard, and seeing there is *Sargassum* algae out there, what a thrill. It was like I learned in the books, you know. I said, "Wow, we have to do something about this." I talked with the boatswain, and with chicken wire and rope we constructed some kind of net and started fishing this *Sargassum* algae. There were these weird little fishes with all of these strange projections and crabs, you know, straight from the books about how these animals are adapted to mimic their seaweed environment. That was great.

The real feeling of the tropics came actually in the first harbor. At night we arrived in Kingston, Jamaica, and in the morning, we opened the curtain for the porthole and there on a piling was a pelican. Wow! That was great. Behind that, in the distance, we saw the shore with the palms. God, that was good. That was fantastic. Then, of course, getting onto the island. The tropics, the smell and the feel. The tropics have something special. The vegetation. The smell is so intense, you know. I love it.

Curacao. Not very much to tell at first. I took some time before I got into a real groove. I was appointed as marine biologist. My duty was vague; I was to study the coral reef. To start with, I didn't know where to work or what to do. The first thing, of course, was to learn the corals. Coral identification at the time was a mess. There were these old weird books in the library from Cuba and other places with contradictory identifications. There was nobody really I could ask. There were these vague rumors that in Jamaica they knew much more, all sorts of knowledge about new species. The only Dutchman who had been working in Curacao was in the Netherlands, and I wrote to him about the determination of corals. He said I had to undo the Gordian knot myself. There was a slight problem in that he had published two papers, but they were not without mistakes. It took me some time to realize this authority before me had made some mistakes and even longer before I figured out how things really worked.

I went to Jamaica at the time. The original appointment had been made with Tom Goreau, and I didn't know he had died just before I came there. There was nobody in charge at the lab. After a couple of days, I met Judy Lang and a couple of other people, and I saw all those new exciting coral species. They were mostly dried skeletons in cupboards that I was allowed to see. At the time, I had decided that I wanted to study the growth of corals. I wanted to get an idea of the growth of the reef as a whole, not realizing of course, that to get an idea of that, coring would have been a much better idea. My geology background didn't focus on that.

So, first, I wanted to measure growth underwater. In discussions with other people, I tried to invent an underwater weighing apparatus. We talked about that a lot. The suggestion was to neutralize a float with weights. When you knew the upward pressure of your float underwater, it could be expressed in weight. Then, you can put on a coral and the number of weights needed for equilibrium would change when the coral would grow, that is, increase in weight. So we constructed a weighing apparatus and put it out on the reef at about 18 meters. That depth is where I did much of the work for my thesis, sitting there for hours running into decompression problems, finding out the weight increase of corals at different depths, and having controls to account for the boring of excavating organisms and the accretion of crustose corallines.

I was a marine biologist, rather happy-go-lucky, living from day to day, not thinking much about formal things such as academic degrees. Then our lab was reviewed by a couple of professors from the Netherlands, much feared by everyone, because they were God and his right-hand man. Actually, maybe, two devils combined; that's how we viewed them. At that time I had found some sort of groove. I was working at the reef, I was feeling certain about the things I was doing, and I was liking it, not realizing that later it would feel even better—much, much better. I was looking at the zonation of the reef, where organisms were found. Because I was looking at the weight increase of these coral colonies, we discovered *Diadema* [a sea urchin] was eating live corals, so we started night diving and counting *Diadema*, to see what they were doing to the corals.

One thing led to the other?

Yes, but the whole thing centered on the growth/survival of the coral colonies and some of the influences on that.

During the much-feared review, it turned out the reviewers judged my work to be the best at the institute. By that time there was a whole group of Ph.D. students working at the lab. When the reviewers talked to them and to the other staff, people would say, "Rolf Bak says this, or Rolf thinks that." That rather enhanced my status at the institute, which had been very low until then. One of those visiting professors said, "One of your fellow students from your year has now graduated [with a Ph.D.]. Isn't it time you should?" That was the first sharp realization that, indeed, I should use the job to get a degree. Initially my position was a three-year contract. That was renewed for three years, but I was told that was final, no way could it be renewed. Meanwhile, I got my Ph.D. Because of my increased status, I got into a position in which I could initiate projects. People would turn to me, and I could get students to work with me.

I was daily on the reef, of course. I dove, day in and day out, except for some weekends, twice a day. I was sitting there underwater at this buoyant weighing apparatus. I spent a lot of time underwater. You look at things and you think. At the time, I had a lot of questions and things that I wanted to know more about. For example, at that time it was still thought that corals were very stable, stationary things. You didn't think of corals as disappearing, gone. Corals were there, and the reef was virtually unchangeable, because everything was growing very slowly, and the reef was very stable and the organisms very specialized. But where did small juvenile corals come from? That was one of the questions. We worked on that with great enthusiasm. That became one of the papers I worked on with Sabine Engel, maybe my most quoted paper. The good thing was, that made me very happy and it still makes me happy, when findings emerge and form a pattern.

All the small corals we found! The smallest were two to three millimeters and, contrary to what everybody believes, you can find them if you look carefully. Most of them were *Agaricia*; at the time, we thought they were all *Agaricia agaricites*, now we know *A. humulis* is also important.

Very strange to find that. *Agaricia,* such a dominant coral, has such a high input of juveniles. Earlier, we had marked permanent quadrats at the reef bottom, and after years, when we did the first analysis, much to my joy, we found that larger *Agaricia* were the most vagrant, mobile corals. They came and they went, while other coral species were very stable.

We had also started to look at regeneration. I had data on growth, I had data on recruits, I had data on the behavior and mortality of the larger colonies, and we were doing work on regeneration. We found that *Agaricia* was having a high turnover rate. All these things started falling into place. At the time, to discover that corals were so diversified, had such diverse life history strategies, were doing these special things, and to see how these things were coherent and make sense—that was fantastic! That is one of the best things for me in science, in the work that I do. I have to be very careful. If I say "make sense," I do not mean this in a metaphysical way. I look at the existence and see no sense; in essence, there is chaos. But, on the scale of the reef, you see a pattern, and it makes sense. Not that it is part of a large design; that isn't the way I think about the world. Did we talk about that, the caveman going out of his cave, licking his finger, holding it up, and feeling which way the wind blows? That is sort of the basic feeling, I think. Man wants to understand his surroundings. In that sense, science is worthwhile. But I don't believe in science and progress in the sense that it will make the world better.

What he taught us was: find a system we know nothing about and start to manipulate it.

Donald C. Potts (b. 1942)

Professor of Biology
University of California, Santa Cruz
Interviewed at the Society for Integrative and Comparative Biology Meeting, Boston, 4 January 1998

He explored the Australian bush and mangrove coast as a boy, but the global tradition of his father's family's seafaring background and his mother's family's British foreign service background provided Don Potts with a broad perspective as he began a classical university training, which emphasized practical work, independence, and a collegial atmosphere. He discusses the period when Atlantic reef workers felt reefs were stable ecologically, while Pacific researchers held a contradictory view. Potts points out a disturbing trend in university curricula today: separating the lecture portion of a course from the practical portion (laboratory or field).

I was a solitary child, very shy, and I had few friends. I had a bike, which I won when I was ten in a competition where I prepared a frog skeleton. Once I had the bicycle, I was unconstrained as to where I went. Growing up in the outskirts of Brisbane [Australia] in those days, the middle 1950s, I would go up to 50 miles away along the coast and explore mangrove swamps. I was interested in anything that was sea, land, bush; my brother was entirely into things that had engines. My interest in biology developed as I traveled around more and more, but quite independently of school. I remember very little of school, except being bored out of my mind for years on end.

My father was a sea captain. He'd spent his life at sea, beginning when he was seven years old, in 1905, when his father abducted him from school, to go to Valparaiso from Liverpool via Cape Horn. My grandfather was captain and owner of a clipper ship. The family tradition is that there had been master mariners at least back to the Elizabethan days. I was born in Scotland. My mother got her degree at Cambridge in the days before they actually gave formal degrees to women. Her titular degree states in Latin, "If she had been a man, she would have had a degree." The scientific interest came through my mother; her background was chemistry and math. The things marine are from my father's tradition. To do things biological was my own.

When I went to university, I had never done any biology or field science, but I had a lot of physics, chemistry, and mathematics behind me. My first year at university, I took zoology, botany, geology. There were hundreds of pre-meds and others taking zoology. At the end of the first undergraduate year, eight of us were invited to go on a two-week field trip the following January; I have no idea how they selected us. We went out to Heron Island. That was the first time I had ever been on a coral reef! The first week I was wandering around with my eyeballs dropping out, and it was just an incredible experience. My recollections are still the variety and diversity. Botany and geology also had a lot of field trips. All the science classes automatically had labs, three lectures a week and three to six hours of lab each week. In our final year as undergraduates [third year], we were given keys to the building which included access to the biology library. The assumption was that, as senior undergraduate students, you'd be reading widely.

The other side of it, why I became a biologist, is that I hovered for two years between biology and geology. Why I became a biologist was simply approach, because geology, at that time, was taught in a very traditional descriptive way. There were reef people around, like John Jell and Dorothy Hill. These were the times when modern ideas about plate tectonics were just beginning. When I was a third-year undergraduate, the zoology department invited people like [Samuel] Carey and [Stanley Keith] Runcorn, the paleogeomagnetism people, to come and give seminars. I was taking a geology course at that same time, and they didn't even bother to announce these seminars in geology. We had a situation where the biologists were all getting very excited and talking about the implications of sea-floor spreading, plate tectonics, and the geologists were largely ignoring it. This was also the time when [Philip Jackson] Darlington was coming out with his book *Biogeography of the Southern End of the*

World, remember? It was the ultimate, invoking land bridges in a static world. Now, we were getting another viewpoint, which the biologists were picking up on, but the geologists didn't seem to be. I was attracted to the process orientation of biology, dealing with things dynamic, especially thinking of things today as a consequence of a long time in the past. That is what excited me so much to become a biologist because the geologists I encountered weren't thinking that way.

I went on to do honors (which is basically a research-oriented master's degree) on the settlement of fouling communities on plastic plates, using an experimental set-up, caged and uncaged. This was in the early 1960s, when Joe Connell's experimental work was just coming out. He turned up in my first year doing honors. I vaguely knew there was an American on sabbatical somewhere in the department, but I had never seen him. One day, for no obvious reason, I used the elevator in the building, which I never did on principle because I liked the exercise of running up and down the stairs. At any rate, I happened to step into it with a person I had never seen before, who said, with an American accent, "Hey, do you want to go up and work in the rainforest with me?" I said, "Yes." On the next floor, he stepped out, and I didn't see him again for six or eight weeks. Then he popped up again and said, "I'm going up at the end of the week. Are you ready?" and I said, "Fine." That was Joe Connell. I heard later that he had already asked a number of graduate students, in fact every graduate student he ran into, if they would go, and everyone else had reasons not to. My principle was that I would take any opportunity that comes along.

Coming to the United States was my first exposure to formal ecology. I had six lectures on ecology in all my undergraduate years. When I got to the United States, the ecologists at Santa Barbara were excited about [H. G.] Andrewartha and [L. C.] Birch's 1954 book, *Distribution and Abundance of Animals*. We called it the "Big Red Bible" at UCSB. Without knowing it, I had been imbued with a lot of that kind of thinking, but as a student, I had never heard of that book per se. I had no idea then that Australia had generations of ecologists—[V. A.] Bailey, [A. J.] Nicholson, Andrewartha, Birch, [Thomas] Browning—and how different their thinking was from the ecological thought of America and Europe.

The equilibrium view of the northern temperate world versus the "boom or bust" view of the subtropical of Australia.

Yes. Equilibrium thinking versus nonequilibrium, the physical environment versus competition. That was all very new for me, coming to graduate school in the United States.

One of Joe Connell's principles about his students was that he was not interested in cloning himself. He did not encourage his students to work on his systems, but rather he encouraged his students to find their own systems. As I recall at that period, we all ended up going off and doing things he had tended to discourage. His idea was that he could tell us about good ways to do ecology, using experimental approaches in the field, but he left us to find our own system. This was a very different tradition from what I heard about the other experimental schools at that time. I ended up working on introduced land snails, *Helix aspersa.* Coming home late at night from the bar after a late seminar and parking my car in odd places, I realized that the snails were active at night if there was dew on the ground. This was in midsummer and in a semi-arid desert. Connell's view was that it was impossible to do ecological work on land snails because Santa Barbara was a desert, so the snails would be inactive most of the year. He said that in so many words. Whereas I promptly went ahead and did my thesis on the comparative demography, including field experiments on the introduced land snails, in different habitats, some favorable, some unfavorable.

At that time, I shared a house with Chris Onuf, another student, to whom Joe Connell said, "It is impossible to do experiments on a surf beach." So he worked on *Olivella,* a snail that lives on surf beaches, and he did experiments which actually worked. Laurel [Fox], my wife, one of Bill Murdoch's students, started working on streams. Joe Connell's reaction was that one couldn't work in streams because they were too dynamic a habitat in which to do experiments. She developed theory and techniques for doing field experiments in streams, by using pools as experimental units.

BOB WARNER *said that when he got to Santa Barbara, Joe said, "You can't answer evolutionary questions by doing field experiments." Do you think that is part of Joe's personality and how he brings out creativity in people, or do you think he really believes what he says?*

I think Joe thinks ecologically. He is very open about saying this. He thinks on ecological scales, not evolutionary scales. He thinks on practical terms. Joe's approach is: you don't spend a lot of time designing the perfect experiment before you go out and do it. That is where he breaks from the [Tony] Underwood and the [Bob] Paine tradition. His idea was and what he taught us was: find a system we know nothing about and start to manipulate it. Take a bucket of snails, move things around, and see what happens. He encouraged us to go out, look, and see how things responded to changes you might introduce—and the questions would come out of that.

In my second year in the United States, Bob Morris from the University of Oregon called and asked if I wanted to go to the Antarctic. I said, "Yes." "Can you leave in a month?" "Yes!" Then I went to Joe and said, "I've just agreed to go to the Antarctic, is that OK?" So, after barely a year of graduate school, I took a year off. I went to the Antarctic for four months and then took another five months to wander back to Santa Barbara. In Brisbane during my return, there was a cable from Joe Connell, asking me to help him in the rainforests. That was the first time I went to New Guinea. Ten days after standing at the South Pole, I was working in the hot, humid coastal rainforest near Lae! That was the biggest climatic shock I had ever had in my life. I came back from that break totally refreshed. My thinking was totally different. When I came back to Santa Barbara, I had switched from thinking about a purely mechanistic kind of ecology to starting to think of these space-and-time kind of relationships. I was aware of the vastness of the systems in ways that I hadn't been before. When I went back to working with my snails, I didn't do any of the things I had been planning to do the year before. I went toward the comparative approach, rather than taking one population and going into it in tremendous detail. I started moving away from that reductionist approach. I started looking for the most different kinds of environments I could find and comparing the populations in them.

When I went back to Australia on a postdoc, it was with the question I had raised in my thesis about natural selection operating within the extant generation shaping ecological properties. Even though the snails I had worked on were pretty sedentary, I wanted something where I could do the genetics in the field, something I could clone, and something sessile, so I could find them all again. In Canberra, I did experimental manipulations of reciprocal transplanted material among habitats. I set up two systems, one on the reef working with *Acropora* and the other—same experiments, same questions—working with a herbaceous rose, *Acaena*, that grows from the top of Australia's highest mountains to coastal sand dunes. In each case I used a single species. I looked for five habitats that were as different from one another as you could imagine, in which the species was present. Then, I did reciprocal transplants of cloned material among them. I continued both sets of experiments for two years.

At the end of that period, it was very obvious that the terrestrial system was not working well. In every manipulation I did, there were artifacts being introduced, changing of soil properties. Even the little holes I dug to put the seedlings in had effects. In one case, after two years, you could see that square on the hillside from two miles away, because the whole texture of the soil was changing. This happened at several sites in several ways. In fact, at one site, which I visited ten years later, there was a 10-by-10-meter square in which the vegetation was totally different from its surroundings. All I had done to it was to make holes several centimeters in diameter and plant seedlings in them. Just trampling around on a semiarid hillside compacts the soil, alters the erosion, and that scoured out the whole area when it rained. In the coral system, as far as I could tell, there were no artifacts of my manipulations. With the corals, the data were coming in from the beginning. They were clear, consistent, no obvious artifacts, and they were easily accessible. I've worked with corals ever since.

The positions Laurel and I had at Canberra were five-year nonrenewable positions, so we had to move on. In the last two years in Australia, there were no positions advertised in Australia that either of us could have applied for. The late 1970s was one of the low periods. We were offered a split appointment at UC Santa Cruz. When I started analyzing my data from Canberra, I realized that many of these morphologies, characteristic of different habitats and which are clearly genetically determined, had been described as different species. Then I started thinking, if you can get such intense selection in one generation, then if there are isolated reef systems, of which there are many, with different arrays of habitats, you'll see speciation like mad. I thought, all I have to do is find these reef systems with different environments, and I shall see speciation in action. So I brushed up on the literature, and I found what, to me, was an extraordinary paradox. For most reef organisms, the more isolated the reef system, the more you have endemic species, geographic races. This is true of mollusks, crustaceans, fishes, but it is not true of the corals. In general, the Indo-Pacific morphological species of corals had pantropic distributions. The morphological variation which had been described locally on one reef was about equal to the variation which had been described globally for that species, which is a total contrast from most other organisms found on reefs.

That started me thinking about what is different about the corals. Trained in comparative demography, I thought about life histories and what it means being a long-lived organism in a spatially and temporally variable environment. That's where I came up with the idea that the key thing is not longevity per se. The main fact is that they have indeterminate growth. The reproductive output of a given genotype is going to increase exponentially with size and with age, without limit potentially. The other component is that there are overlapping generations. Once you get a few very large genotypes present in the population, they are essentially going to dominate the gamete pool for the rest of their lives. That says to me that the fact that intense selection is going on locally—I can certainly measure it in local habitats 50 meters apart—has no effect in the larger scale. Because, as long as these large, old genotypes are present, the gene pool is not changing. Barring catastrophic events, once you get over a certain threshold size, the probability of the oldest genotypes dominating the gene pool will continue to rise. I come back to this idea of periodic disturbances. Some recruits get established and survive, and they are going to be the dominant ones until you get another disturbance big enough to kill a lot of them off.

In the 1970s, before the first big recent Atlantic hurricane to affect well-studied reefs, Hurricane Allen in Jamaica in 1980, most Caribbean researchers were thinking that reefs are stable systems. Whereas in Australia, there was a very different way of thinking, at that same period. We didn't really understand the Caribbean literature, which was so dominated by competition, stability, equilibria, that kind of thinking. But, when I came to the United States, I became very aware of what seemed to be an incredible dichotomy between the reef workers on the East Coast, who were working in the Caribbean, and the much smaller number on the West Coast, who tended to work in the Pacific. I remember the literature of that time, the grant proposals, and also reviews that I got about proposals that I wrote. They were divided right down the middle. There were the East Coast, Caribbean ones, which would not accept anything that was not basic equilibrium, competition-type ecology, and then the West Coast ones, which accepted that there was disturbance, and you have to take this into account before you start talking about competition. Then came Hurricane Allen, and there was that multiauthored paper in *Science* and the reaction of the people in the Pacific was: what on earth are they making such a fuss about this? We've been saying this for 20 years.

Since then, of course, there has been a coming together of the two trains of thought, but previously they were totally isolated. It was very interesting sociologically. I must admit, I got some very nasty reviews from people who were clearly Caribbean workers. The reviewers did not just differ. There were very destructive kinds of comments: Reefs do not work this way! He doesn't know what he is talking about! I tried at that time to write proposals to do comparative work, Caribbean and Pacific. The reviews that came back were of the form: he has never worked in the Caribbean, he knows nothing about Caribbean reefs, therefore there is no point doing this.

I think all big advances in disciplines tend to occur when someone with a very different perspective gets involved. I think that is what Connell was doing by bringing in his experimental way of thinking, which revolutionized the way ecology was done. The experimental concepts and designs were not new, especially among botanists, but by doing them at the right place at the right time—his barnacle work of the 1950s, for example—he just triggered a chord, because it dealt directly with ideas that people were interested in. It also challenged ideas and provided a way to look at things. Another example is [Josiah] Harper, in plant ecology, who took demography into botany. Zoologists were demographers at that time, not plant people; this way of thinking revolutionized plant ecology. Harper made some statements—I can remember them from many years ago—basically to the effect that the good botanists for the rest of the century are going to be people who were trained as zoologists. I think there is a lot of truth in that in general. When you go into another discipline, you take a very different way of thinking and you look at things in a different way. You are not imbued with the conventional wisdom.

It is no accident that Andrewartha's and Birch's view of ecology, formed in Australia, was so different from that of the North American ecologists. These systems have some very major differences. Similarly, when you go from one discipline to another, you get very different perspectives. I feel that having a mix of biological and geological training has helped me both ways. I think of things in different ways from people who have only been trained in one or the other. The fact that I did a lot of botany, at a time when most people either did botany or zoology, but not both, has also made me a lot better biologist because I have a sense of at least two major groups of organisms. They have some very big differences.

One of the real disasters in teaching in the past 20 years is the way that universities have gone more and more toward separating practical work from the lectures. At Santa Cruz, 20 years ago when we arrived, virtually every biology course had a lab or field component associated with it. Now many of our courses are pure lecture courses with a smaller number of separate, pure lab courses, but only a small number of the students will take the lab course. You can't build on people's experience because they haven't done it in the past. They'll never do it formally, so that is a huge area of their education that they have lost. As an undergraduate, every course I took had lab or field automatically as part of it. With the added constraints of faculty having to do more and more and with financial constraints most of our support staff are gone. Labs are expensive, lectures are cheap. We are teaching more students, but we have fewer TAs. The reality is that effective field or lab training for most students doesn't happen until graduate school.

What surprises me sometimes is how the pipe dreams become reality.

ROBERT R. WARNER (b. 1946)

Professor of Marine Biology
University of California, Santa Barbara
Interviewed in St. Croix 6 August 1996

Bob Warner outlines his transformation from using the field as a place to make collections to using it as the site of experiments. He became an experimental field evolutionary biologist and basically defined his own approach, whose validity he had to prove to other people, especially to the field ecologists at UC Santa Barbara, when he arrived on the faculty. Defending your position is a significant element in how science is conducted, and Warner clearly articulates the motivation provided by needing to prove himself to skeptics. In the mid-1980s he began using St. Croix as a major field site for his research, and he continues to be a close colleague of mine.

The turning point at Berkeley was when I took an ichthyology course taught by George Barlow. It was the first ichthyology course he ever taught. The TA was Armand Kuris.

Really!

Yes, my closest colleague now at Santa Barbara was my TA then. Of the eight people who took the course, six ended up with Ph.D.s working in ichthyology: Gene Helfman, Hugh Ellis, myself, Dave Soltz, Lynn Montgomery, Dave Noakes. It was just an amazing class; we had a great time. We charged all over northern California collecting fishes. That was when I really was convinced I wanted to do something having to do with being out in the field, but I wasn't an experimental field biologist yet. I liked collecting, actually, so I decided I was going to be a systematic ichthyologist because then you could go out and collect fishes.

I went to Scripps [Institution of Oceanography] and tried looking at fish in bottles for a few weeks. It probably took about eight weeks, and I decided, I am not cut out to be a systematic ichthyologist.

How did you get interested in sex change?

Oh, we had a seminar in advanced ichthyology, [Dick] Rosenblatt's class, and the topic just came along. I thought I'd review it. I realized that there was lots more to be done—and so I did! But I still wasn't the kind of field biologist I ended up, because I still just went out into the field, collected the animals, brought them back, did the lab work, and did a bunch of theory. I enjoyed being out in the field because the field meant access to the animals. I didn't spend any time looking at the animals.

One of the nice things about Scripps was that everybody arrived equally uninformed; none of us were oceanographers. Scripps insisted that we become oceanographers. There were two years of coursework in marine biology, biological oceanography, chemical oceanography, and physical oceanography. You had to take this stuff, and so we did. I still surprise some people when we are interviewing somebody, and I can ask a semi-intelligent question about ocean circulation. In that sense, Scripps was a fine place to go because it gave me a broad base about different aspects of the marine environment, things I wouldn't have learned elsewhere.

Once the Ph.D. was done, I got a two-year fellowship to STRI [Smithsonian Tropical Research Institute] in Panama. I began working there at the same time as Ross Robertson. Ross, to me, is still the quintessential field biologist. He's crusty, he has a very strange personality, it was worse then than it is now, but he is good in the field. He likes being in the field, and he is especially good at making observations. Ross happened to end up at STRI at the same time I did, also interested in sex change in fishes. I'd done a bunch of theory and a bunch of collecting, but Ross had done field experiments. He'd taken the dominant male of *Labroides,* a cleaner fish, out of the harem and watched sex change occur. This got me really excited. I started off working closely with Ross. Again, we did collecting per se, but we began to watch animals as well. I realized all of the potential for observations and experiments in the field rather than just as sources of animals, and so Ross was a major influence on me. We did a series of experiments together. We also did a huge number of collections together with Isabel's help, actually. My first experiments were done with Isabel [Downs; Warner's wife, whom I first knew from the spring course at HMS] on the little parrot fish.

Right about the time I was finishing up at STRI, I met Steve Hoffman. He was a student of Bill Hamner's at one time and was looking for an advisor. He was working on sex change in *Bodianus,* a wrasse that lives off the coast of Baja California. We began to talk more and more about it, and it is odd to say I was influenced heavily by my graduate student, but in this case I was. This is what Steve was good at. He was part of a group that would go down each summer to Baja and essentially camp out [see MADIN] for three months on an island and do field research under very primitive conditions. He loved it. The more I learned about how they were doing it and what they were doing, the more fun it seemed. You are off in the middle of nowhere and are still doing good work. I came to Santa Barbara, and Steve came with me as a graduate student. By then I was spending a lot of time writing grants to do field experiments.

The next thing that happened was the real transformation as far as I'm concerned in terms of my approach to fieldwork. By now I was committed to work in the field, lots of observations, some manipulations. But I was still an evolutionary biologist—evolution of behavioral traits, evolution of life history traits, things like that. I came to UC Santa Barbara, the "center for field experimental ecology."

That's what it was called?

No, but it was just the center for that because Joe Connell had a tremendous influence on everyone. But, I encountered from Joe, and the rest of the ecologists, an enormous amount of cynicism about evolution. Here's why. The idea is, in field ecology, if you want to show competition, you do the experiment. You have the proper controls and if you set it up right—let everything else vary but do your manipulative field experiment—you can show whether there is competition or not. But here I am working on evolved traits, like sex change. Their claim is that everything important in evolution happened in the past. Field biology and manipulative field experiments are the way to prove things, and you can't do that in evolution, you can't prove it. I took this both as a personal challenge and a major affront! So for the next two or three years, I proved to myself—and I think convinced them—that there's a way to do this.

Actually, it is investigating evolution by analogy. The basic idea is really quite simple. Hypotheses in evolution propose that certain environmental conditions have posed certain problems in terms of having babies—and organisms evolved adaptations, solutions to those problems, over evolutionary time. Nobody has any problems with that. The reason organisms are diverse in different traits that they show is that they've been exposed over history to a diversity of problems, a diversity of conditions. But, still, that's not testable, because I don't know what the conditions were. But what happens if the organism is faced with a diversity of conditions of those things that we claim are important in the environment? What if, within their lifetime from generation to generation, things change? Then what are you going to do? Well, evolution can get around this. It builds in phenotypic plasticity; it builds in conditional responses. Imagine then that I have an array of competing hypotheses in evolution, let's say for group size: animals run around in large groups to avoid predation, to conserve warmth, or to conserve moisture. You can have a whole series of alternatives. Which one is likely to be true? The basic lesson that I tell myself and my graduate students is: if you are interested in group size, go and find an animal that is variable in group size. Sometimes it's alone and sometimes it runs around in groups; in other words, find an animal that is variable in the evolutionary trait of interest. Assemble your hypotheses about the various environmental conditions that are candidates for the important factors affecting the trait and then manipulate those conditions. The organism is likely to respond; we already know it is variable in the trait. Which condition did it respond to? Which ones didn't seem to have an effect? So you can start throwing out hypotheses and work just like an experimental field ecologist, except you're an experimental field evolutionary biologist.

I wrote an outline of this approach, and it's in Peter Sale's book *Ecology of Coral Reef Fishes*. It gives the protocol and shows how you'd use it. I just used fishes as an example. The reason I've continued to work with, coral reef fishes is that they're incredibly easy to work with, and they're incredibly variable, even within a species. So they're the perfect animals to use this kind of approach. And, you know, the water is warm!

So, it was actually incredibly satisfying because (1) it gave me an intellectual basis to continue to do field experiments, and (2) it established a good amount of credibility of this approach at Santa Barbara. It was also very satisfying when we finally decided to call ourselves the Department of Ecology, Evolution and Marine Biology. It's a long and cumbersome title, but evolutionary biology came into acceptance at this very heavily field-oriented department. So those sorts of influences, such as people teaching me approaches to the field, approaches to living in the field, approaches to doing field experiments, and the intellectual underpinnings, the rationalization for doing things—these have been a tremendous help to me. The whole evolution of behavioral ecology as a field is interesting. I found out that I was a behavioral ecologist only after some time. The field was evolving as I was evolving, so by 1978, when the [John R.] Krebs and [N. B.] Davies book [*Behavioural Ecology: An Evolutionary Approach*] came out, I realized that here's a field that I fit into very well. I wish there were more behavioral ecologists in marine systems, but there aren't.

In terms of influencing others, I think the thing that I'm good at is experimental design. We run whole seminars where people don't come with finished products, they come with problems, and then we talk about potential pitfalls. If you do it this way, there might be practical difficulties; there might be a control missing. What are those damn reviewers going to say when you try to publish the paper and there's this hole? We try and fill in the holes and make it logically complete. The only thing that they have to do at the beginning of this is to make the question extremely clear, and that is very hard for some people to do. Often, I'll say to graduate students, "You have to be able to establish the question, the importance of the question, and how good you are in 100 words or less, because that's how many words you'll have in a meeting when you go talk to Shmoo Brains, the world's expert on the subject." You've just got 100 words! And they can do it well, after a while.

The thing that I think I really push, and hope that other people continue, is the idea that field experiments are really a marvelous way of sorting through a whole series of alternative hypotheses. Just like [Bob] Paine and [Joe] Connell, and all those people who really got community ecology going through manipulative field experiments, you can do the same thing with some clear thinking and some good design in evolutionary ecology. Just be careful of your experiments.

The fact is, as the tools become available, you're a fool if you don't use them. Molecular biology is a tremendous tool for answering some of the questions that have been outstanding for a long time. I observed a group spawn in bluehead wrasse. I didn't know how many males there were, maybe ten in there, and I knew there was a female in there, too. OK. We assumed that each of those males contributed a tenth of the sperm, and that's the assumption we've worked on since 1975. As theory progressed, it got more and more complex, to the point that we could not make that assumption. I really needed to know what the distribution of paternity was. At this point, a graduate student came along and said, "I want to do some DNA work, and it's time that we got into it," so it was a perfect mesh. There's a lot of luck. What surprises me sometimes is how the pipe dreams become reality. Like with Lisa Wooninck and her DNA work. I said, "If we can just get paternity, that would be great." And maybe down the line—I had this suspicion—we could look at the sperm itself, independent of the paternity, and see how much sperm each of these guys contributed, by running the whole sperm cloud at once. We searched the literature, and it just hadn't been done. Well, it turns out that it can be done; this is going to be a major part of her thesis; and it's going to be a real revolution. You catch all of the sperm; the DNA microsatellites are so specific and so variable that we know each of the individual genotypes that are in there. And so, not only can we say how many males participated, but we can also specify the relative contributions coming from each male. This is really going to be great for people working on a variety of animals. You can take the membrane off the egg of a bird, for example, and in that membrane are embedded all the sperm that didn't make it to fertilize. Run that, and you can specify how many males mated with the female and how much sperm each contributed. It's fabulous! It's really a great tool. That was a pipe dream, but it looks like it's going to work.

The major question in the recruitment of larval fish to this island is: where do these guys come from? Are they from St. Croix or have they traveled in the plankton in their larval stage from some of the other islands upstream, the eastern Caribbean islands? It would be wonderful if we had a tool where we could get that information. We are working on developing that tool, a laser attachment to the mass spectrometer. We slice the otolith, go down into the core, and we can get, perhaps, a three-to-five-day resolution on the trace metals. And if Guadeloupe has a different coastal metal environment than does Martinique than does Saba, we might be able to find it. As fishes leave their natal island, they go through blue water [oceanic water between the island shelves], where there are few metals. Then they pick up the new regime, wherever they settle. That's the new pipe dream. That one's maybe three or four years off. We want to reconstruct the biological connections among the islands. That to me is a great unsolved problem in marine ecology, the connectedness of local populations. How connected are they? How dependent are they on the production of other populations? The managers need to know this, the ecologists need to know this, and as an evolutionary biologist, I need to know this, too. That's what we're working on. It's very exciting.

If you know that it is a really important question, then you'll go there, you'll pursue it.

Pamela Reid (b. 1949)

Research Associate Professor of Marine Geology and Geophysics
Rosenstiel School of Marine and Atmospheric Sciences
University of Miami
Presently: Associate Professor of Marine Geology and Geophysics, RSMAS
University of Miami
Interviewed in her office 26 February 1998

Like several others interviewed, Pam Reid gave up a good position in the oil business to go back to graduate school. She loved all aspects of graduate research, including the fieldwork on "her mountain" in the Yukon and the lab work back in Miami, however, the writing phase was perhaps the most difficult part of her life, and it almost discouraged her from continuing in science. Fortunately, she began a long and productive research collaboration with IAN MACINTYRE, leading her into the emerging discipline of geobiology. She has now gained a full faculty position at Miami, after being in a "soft money" position for many years.

Why didn't you become a banker on Wall Street? Or a diplomat in Paris?

OK, so how did I get into geology? My mother used to take us fossil collecting and, although I was not enthusiastic about this as a kid, I guess these early experiences are all a part of who you become.

Where did you grow up?

In Toronto. My mother, however, grew up in the Middle East, and from an early age I was exposed to people who had interesting lives.

The first-year geology professor at Queen's, Al Gorman, was a charismatic professor, whom everybody loved. He was the reason a lot of people, including me, decided to take their first geology course. This seemed a more reasonable path to the adventurous lifestyle I was seeking than working for *National Geographic* as a traveling chemist. I remember somebody giving a lecture about beach sands. I thought that it was amazing that people went around the world and collected beach sand as a job!

One summer, I worked in the civil engineering department. There was a great group of engineering graduate students, mostly Europeans. Three of these grad students, an Englishman, a Frenchman, and a New Zealander, were going on a climbing expedition to Baffin Island. At the time, I didn't even know what rock climbing was. Suddenly I was hanging out with guys with helmets, ropes, boots, and all the other rock-climbing gear. I was awestruck! These grad students were also doing other interesting things. For example, a master's student in the group was going to a conference in Russia. So, I realized that I too could travel as a student. A few years later, as a master's student, I went to a meeting in Israel. To this day, I am still very good friends with some of the original group from the civil engineering department, and they will always be a very special part of my life. These people opened up a new world for me.

After finishing a bachelor's degree, I worked for Shell Canada in the oil industry for three years in Calgary. I quickly found out that I did not like the rules and regulations associated with working for a large corporation and spent much of my time planning my weekends. The highlight of living in Calgary was the mountains.

Ian Macintyre also worked for an oil company.

For Shell. In fact, a lot of important carbonate geologists have worked for Shell. As an undergraduate, I took organic chemistry, which most geologists avoid. In fact, that's why I got the job. Organic geochemistry was an emerging field in oil companies at the time. I am, however, not a good oil company person. I hate being told what to do and, even more so, when to be at work. Working as a geologist for an oil company involves a lot of arm waving, which is not my style. I like details. There are usually so many details that we don't understand that I can't get beyond them. In oil companies you generally do not have the luxury of investigating the details. It is a good thing I didn't stay with Shell. I wasn't happy with them, and they weren't happy with me.

Ellen Prager called me when she was working for Shell and said, "I don't think this is right for me. I think I am going to give up this high-paying job and go back to graduate school."

I know. It is a hard thing to do. I worked for three years and paid back my student loan. If I had stayed with Shell, I would have been making lots of money by now.

While living in Calgary, I had a wonderful time in the mountains, and this mountain experience was invaluable later when I did my Ph.D. in the Yukon. But before going back to school, I wanted to travel for a year. I decided that I would take a freighter to Lebanon to visit relatives. However, war broke out in Lebanon, and I ended up getting off the freighter in Greece and spending a totally unplanned year there. At that time, *Calypso* was working in Greece. I thought they did science, but they don't. They make movies. I ended up making friends with some of the *Calypso* crew, however, and spent a lot of time with them during the year, which was great!

The year I spent in Greece was a very important experience for me. Until that time, I had a rather narrowly defined image of myself: "I'm Pam Reid; I'm a geologist; I work at Shell." That puts you in a certain category. But, when you are traveling and you are not doing anything in particular, you have no definition. It is a really interesting experience. "I'm Pam Reid, and I'm just hanging out." People can't categorize you. Until then, my life tended to be very structured. I could tell you six weeks ahead exactly what I would be doing because I had a whole program laid out. But in Greece, I was traveling with no plans. It was a strange and wonderful year!

While I was a master's student, an International Association of Sedimentology meeting was held in Israel. I decided to try to go to this meeting and then go on to

Faculty colleagues at the Rosenstiel School of Marine and Atmospheric Sciences of the University of Miami in the year 2000, Pam Reid shares a story with her major professor, Bob Ginsburg. Photo by Christophe Dupraz.

Lebanon, the original destination of my freighter trip to Greece a couple of years earlier. As part of the IAS meeting, I also signed up to go on a field trip to see reefs at the tip of the Sinai, mainly because it sounded like an exotic destination. That trip changed my life. It was a totally magical, wonderful experience. Bedouin, camels, desert, mangroves, and tropical fish! I had never seen brightly colored fish before. I had never seen a coral reef. And the people on the field trip were fabulous. These reef geologists really seemed to love what they did. I came back from that meeting and applied to do a Ph.D. studying reefs with Bob Ginsburg at the University of Miami. I came to visit RSMAS [Rosenstiel School of Marine and Atmospheric Sciences] with my own money. It was not like today when graduate students are hot items and are courted by schools all over the country.

I flew from Vancouver to visit RSMAS. Bob Ginsburg arranged for me to stay with a graduate student. I remember driving through the streets of Miami in a rented car, practically in tears, because I realized the school was great, and I knew that I was going to come here, but I was very sad at the thought of leaving Vancouver. I did not identify with Miami. I was a mountain person, and Miami seemed a very strange place to live.

When I finished my master's thesis, I wanted to get a summer job before going to Miami. A friend advised me to talk to Dirk Templemen Kluit at the Geological Survey of Canada in Vancouver. This led to another significant development in my life. It turned out that Dirk was looking for someone to study Triassic reefs in the Yukon, where he was doing a major mapping project. I had been accepted as a grad student in Miami and was planning to work on reefs. I would not have applied to work with Bob Ginsburg, who is a modern reef expert, if I had known I would have a dissertation project working on ancient reefs in northern Canada. But it turned out to be fabulous. Everyone in reefs passes through Miami, and Bob Ginsburg is a very special person.

I remember a story Bob told once. He said that when he was first in Miami, he thought he must be very popular because all sorts of people stopped by to see him.

But when he moved to Johns Hopkins for a few years, no one came by, and he realized it was not him but Miami that drew the crowd.

Bob tells that story, yes. But, it is true that everybody does comes through Miami, and Bob knows everybody so they stop in to visit. It is wonderful for the students here.

The field site for my thesis was Lime Peak, a mountain that is about 15 minutes by helicopter from Whitehorse. It is a fabulous mountain, a cross section through the most gorgeous Triassic reef in North America, really similar to some of the Tethyan reefs in Europe. I got a lot of intellectual help from the European Triassic mafia. I had helicopter support and a field assistant for three summers at Lime Peak. It was a deluxe operation. My fellow grad students had to carry their supplies in to their field sites on their backs and carry their rocks out. We carried rocks back to our camp each day in small day packs. Every two to three weeks, the helicopter would come and take us and our rocks to town. We'd get clean clothes and groceries and fly back out again. It was great.

The summer before I started my Ph.D., I worked for Dirk doing general mapping in the Whitehorse area. This was not a happy experience. There were ten or twelve of us in a field camp, and there were a lot of personality conflicts. One day we were all sent out. Two fellow students, Sheila and Tim, were sent to Lime Peak, which I resented as it was to be *my* future dissertation area! As it turned out, these two had an unfortunate mishap. They were walking along Thomas Lake at the base of Lime Peak when they were attacked by a grizzly bear, which was defending a moose kill. A bear will defend a kill even if a helicopter comes right down on it. The bear pushed Tim into the lake and started mauling Sheila. He had her pinned to the ground; she lost some fingers and got very scratched up. Tim flew into a rage and attacked the bear and amazingly enough the bear took off. They got back to their tent and radioed a helicopter, which took Sheila to a hospital. The following winter in Miami, I thought a lot about this episode and worried about Lime Peak and grizzly bears.

I was very lucky to find a wonderful field assistant to work with me at Lime Peak. Jennifer O'Brien was a first year undergrad at UBC. I was very nervous about bears when we arrived in Whitehorse for our first summer. Dirk was annoyed with me and thought I was going to terrify Jennifer about working in the field. We learned to use a rifle. A friend who had worked in bear country in the Arctic suggested we set up an electric fence so we learned how to do that as well. Off we went into the field with our rifle and electric fence. We did not camp down by the main lake but instead chose a site higher up the mountain, in a little valley. We used the helicopter to bring jugs of water up from the big lake. We used very little water. We had a good summer, and things just got better and better in the next two years. We both loved working in the field.

The first time I looked at carbonate rocks in the Whitehorse area, I could barely distinguish the fossils from the in-fillings of cracks. By the end of three summers at Lime Peak, we had covered just about every inch of the mountain and had documented the exact location of almost every fossil. Jennifer worked with me for the entire project. When I first met her, she was thinking about dropping out of school, but she soon decided to study geology. She eventually went to grad school and married her thesis advisor. So working in the field changed her life, too. Lime Peak quickly became "my mountain." I loved this mountain and the rocks. During these first three years of my Ph.D., I was really happy. This may have been the first time in my adult life when I was completely happy without regard to any romantic relationships.

But I had a hard time writing my dissertation, the last two years of my Ph.D., after the fieldwork. In this time period, I had a very strained relationship with Bob Ginsburg. He thought that I should be working harder, but I was working as hard as I could. I never did understand all of the pieces of the depositional history of Lime Peak. My thesis was very long. When finally it was finished, Bob wanted the text totally rearranged. This was very difficult to handle. Finishing my dissertation took me to the limit of my ability to cope. Somehow Ph.D.s do that to people.

It took me two years to recover from writing my Ph.D. dissertation. I didn't think I wanted to do geology any more. I got interested in technical presentations and considered developing educational programs for scientists. Scientists generally do not receive formal training in technical presentations and need to learn basic rules such as "tell them what you are going to tell them, tell them, then tell them where you've been." In addition, a good presentation must answer the question Bob Ginsburg teaches all of his students to ask: "So what?" In other words, "What is the relevance of this work?"

I finished in 1983, so we were the same age when we finished.

You must have taken a break, like I did when I worked for Shell and went to Greece. That break was really good for me. In many ways I think it is good for kids not to go straight through school because when you go back, you are very happy to be there. I love the academic lifestyle. I feel that I'm still in graduate school. I love to do my own program. I hate to be told what to do. No amount of money makes up for that. I would never be happy doing a job in which somebody told me what to do and, even more important, when to do it. I'm really happy in academics. It is a luxury for me to be able to afford it, though. The reason I can is because I'm married, and my husband, Jack Fell, is a tenured professor. I'm research faculty at RSMAS and have no guaranteed salary. I'm 100% soft money. Other research faculty have been forced to leave RSMAS because they can't depend on raising enough grant money to pay bills, support kids, etc.

To go back, in the summer of 1985, I went to Antigua with Walter Adey's group. I was interested in the Lesser Antilles because it is a modern volcanic arc. I was scouting out possible research projects. When we were in Puerto Rico on the return, Walter was showing me an outcrop. I started hammering, and a rock wall fell down on top of me. I went to the navy hospital. I was still really tired from finishing my thesis. I was getting seasick and was not enjoying the boat trip. So, I flew home from Puerto Rico.

I decided that Walter's project was not for me and got in touch with Ian Macintyre, who is also at the Smithsonian. I knew that Ian had worked in the Caribbean and had some samples from the Lesser Antilles that I might be able to work on. This led to a study of algal nodules. I talked to Ian a lot about cements and precipitation. Ian was one of the first geologists to recognize submarine cements in the 1960s, and at that time he had just described peloidal precipitation in reefs. These discussions led to one of the memorable moments in my geological career. These are the high points of science: when a light clicks on and you have a major breakthrough in thinking. I have had a few moments like this.

The University of Miami used to own the former quarantine station on Fisher Island, and for many years it was the domain of Bob Ginsburg and his students. I had my office in an old house with a fireplace. Most students and faculty went home at 4:30 every day. I could never understand that. I often work long hours, and I used to stay overnight on the island a lot. While looking at thin sections one night, I realized that a lot of the micrite [small crystals] in my Triassic reefs was probably precipitated, just like Ian Macintyre's peloidal precipitates in modern reefs. Mapping of petrographic thin sections showed that these precipitates comprised as much as half of the rocks in some reefs. This was amazing. At that time, massive precipitation of micritic carbonate had not been described in ancient reefs. At about this same time, Henry Chafetz was also showing that precipitation of peloidal carbonate could be caused by bacterial activity. I became very intrigued with the possibility that microbes played a major role in the construction of Triassic reefs.

By this time, I was regaining my enthusiasm for geology and applied for a Smithsonian postdoc to continue working with Ian on carbonate precipitation. We became interested in the origin of lime mud in Belize. I had been talking to Lisa Robbins, who was a grad student at RSMAS working on proteins in forams. I had the idea to use proteins to investigate the origin of lime mud. Lisa decided to look at proteins in aragonite lime mud in the Bahamas, while I would work with Ian on Mg-calcite mud in Belize. The protein aspect of this work did not produce definitive results, but the project evolved into a fascinating study, which Ian and I have been working on for the past ten years. Lime mud in Belize lagoon is composed of anhydral crystals of Mg-calcite. We started looking at skeletal grains in the area as possible sources of this sediment. Forams are supposed to be composed of rods of Mg-calcite, but when we looked at forams in Belize lagoon using scanning electron microscopy, we did not find many rods. Gradually we realized that the rods were recrystallizing to anhedal equant crystals. This was a new idea because, at this time, recrystallization was not recognized as a process of shallow sea-floor alteration. At first we thought that the process may be peculiar to Belize lagoon.

But, then came another one of the great moments in my science career! I love working at the Smithsonian because when I am there, I can work long hours with no distractions. I stay with a family friend, Mary-Averett Seelyee, who has been incredibly generous over many years, letting me stay with her for weeks at a time. I generally work all weekend long when I am in D.C. and enjoy having the entire paleobiology lab to myself. While working in Ian's office one weekend, I suddenly realized that the recrystallization we were seeing in Belize could be an important source of cryptocrystalline grains that are a major sediment type in the Bahamas. We decided to write an NSF proposal to investigate whether recrystallization is important, not just in Belize, but also in Florida and the Bahamas.

In order to recognize diagenetic changes in dead skeletons on the sea floor, you first have to understand what's happening in live organisms. We decided to look at the green alga *Halimeda* and peneroplid forams. We chose these skeletal types because both are common and have open crystal structures. It is hard to observe changes in the size and shape of crystal in mollusks, corals, etc., which are very tightly packed. Much to our amazement, we found that skeletons begin recrystallizing while the organisms are still alive!

Ian gave me your paper on Halimeda.

The live foram paper just came out in January. And I finally just submitted a major paper about what is going on in the sediments. We have been working on this recrystallization project for a total of ten years. The whole project turned out to be a long, long process and a lot of hard work.

Do you think bacteria play a role in recrystallization?

I think organics play a role, but whether it is bacteria or the slime that they produce, I don't know. Organics likely play an important role in the formation of many minerals.

While I was doing the recrystallization project with Ian, stromatolites were discovered at Lee Stocking Island in the Bahamas. Shortly thereafter, my husband and I were sailing in the Exuma Cays and asked a park ranger if he could tell us exactly where to find the Lee Stocking Island stromatolites. He told us he had seen similar stromatolite-like structures off Stocking Island, 50 kilometers south of Lee Stocking Island. So we sailed down to Stocking Island to investigate. Lo and behold, there were carbonate build-ups along the shoreline that could be stromatolites.

The Stocking Island reef presented an ideal opportunity to look at relationships among stromatolites, coralline algae, and other reef-building biota. This setting was quite different than the recently discovered stromatolites at Lee Stocking Island, where stromatolites grew in tidal channels and were not part of a reef complex. I contacted Bob Steneck, a well-known coralline algae expert, and we decided to write an NSF proposal for a collaborative project in which Ian and I would drill the Stocking Island reef and Bob would study the interactions of stromatolites and coralline algae. This got funded, and so began my long-term involvement with Bahamian stromatolites.

In this project, we are asking fundamental geological questions that have biological answers. No previous work on stromatolites has been done at this level of intensity. A geologist could not attempt to do all of these analyses alone. We do not have microelectrodes and all of the other toys the microbiologists routinely use.

It is not just stromatolites. I don't think this is done at many places at all.

Definitely. Few geologists know much about microbiology. It is really a struggle for us. I would love to take time off and take a microbiology course. I am the leader of a team project with microbiologists but often don't know what they are talking about.

On the other hand, they don't know what we, the geologists, are talking about either! It is very interesting to work together.

It is dawning on me that this is the next phase. Biology and geology independently are integrative sciences. They have very parallel courses in terms of the kinds of skills you need and the tools used. Geology usually looks at a bigger picture than biology, but in fact right now we are getting to the point where these two are fusing at all different levels.

Yes. I think that is wonderful, a very exciting area, but I don't see it reflected in current reports of future directions at NSF. I've always tended to go in my own direction scientifically. I do not do science based on what is currently in vogue. If you know that it is a really important question, then you'll go there, you'll pursue it. When writing a proposal, you have to believe, without a doubt, that what you are proposing to do is the most relevant thing imaginable. That leads back to Bob Ginsburg: a proposal must not leave the reviewer asking, "So what?"

Postscript (January 2000): I have switched to a tenure track position at RSMAS, and my research continues to expand. The ONR optics project is great. We are getting involved in hyperspectral remote sensing with a fabulous field program at Lee Stocking Island. Our stromatolite project has been highly productive. We have come up with a new model for lamina formation in modern marine stromatolites and are in the process of writing a major renewal grant. Geobiology is finally emerging as a hot topic of research at NSF!

If you study a part of the world and haven't seen it, you are guaranteed to form a new image if you go see it.

PETER MOLNAR (b. 1943)

Senior Research Associate, Department of Earth, Atmospheric, and Planetary Sciences
Massachusetts Institute of Technology
Interviewed in his office at WHOI 12 December 1999

In the 1960s graduate schools in geology began recruiting students with backgrounds in physics (see NAGLE) to address the new questions emerging with the development of the sea-floor-spreading hypothesis and culminating in the theory of plate tectonics. Peter Molnar, an undergraduate physics major, credits the excellence of his liberal arts education from Oberlin College with the ease with which he started grad school. It also undoubtedly contributed to the broad views he holds today, not only of science but of education. He leads us through the excitement and scientific openness characteristic of the late 1960s and early 1970s when plate tectonics changed the way we study the Earth. He contrasts that to the atmosphere of today, when we have a "whole bunch of little clubs and they don't talk to one another." He concludes with the prediction that "climate is in the same position as continental drift and plate tectonics were 45 years ago."

When I was nine years old my father took an extra week of vacation without pay to drive from New Jersey to the West because he wanted to show my mother and me the mountains. My mother had had an aneurysm rupture when I was two. She was lucky to be alive, but no more kids. So it was easy. There were three of us and we drove west. I saw the mountains and I was hooked. We drove first to Rocky Mountain National Park. I was a nervous wreck on the drive to the top of Pike's Peak. From there we headed west. We went to the Grand Canyon and to Zion and Bryce. Then we went north up to Yellowstone and the Tetons, and then we came home. When I was 11, my father had a business trip to Europe and we went along. I saw the Alps. I was really hooked.

I didn't really appreciate how important these earlier experiences were until I was about 40, going through a midlife crisis and wondering, what the hell am I doing? I just looked back over my childhood. Can you imagine the image of looking across the horizon, and what do you see? Towering in the background were the Rockies of 1952. In front of them were the Alps of 1954. The rest was just a blur and didn't matter. Those were the experiences that really shaped me, and I realized that I had done the right thing. They were the epiphanies. I would add one more to them and that was seeing Gothic architecture in Europe in 1954. My father really hooked me into that. My hobby has become architecture from the classic period up to about 1500.

I went to Oberlin College. I was a physics major because I couldn't read or write. I've never been a good reader, I'm a slow reader, but I've learned how to write. My friends always joked that I was a mere technician and an illiterate. I had no hope of doing anything except math and physics, but somehow I knew that I wasn't good enough at any of these things. I had a friend who said, "You ought to take geology. If you take geology, you'll look at the whole world in a different way. You'll just see it differently." Senior year, I thought, I'm not going to go on in physics. I don't have a chance. I'd just be a technician in that. I thought geophysics sounded like fun. I went and asked the physics teachers to give me some advice, and they said, "What is geophysics?" I said, "Gee, I don't know, but I think it is just applying physics to the Earth."

I went to Columbia to study geophysics, not having a clue as to what geophysics was. I was so ignorant. I had better training than almost all the other students, because I had gone to Oberlin, where teachers were paid to teach, and they did a good job. I didn't go to the "big schools," where teachers were paid to do research and hence you weren't taught that much. I went to a school where students were bright, so you learned from students; of course, that's where you learn the most, from other students. I know I had better teaching because I saw what the teaching was like at Columbia. I was better prepared and that meant there was no pressure on me. In fact, I was tired. I had worked hard as an undergraduate. My first semester at Columbia, I saw a lot of theater, I heard a lot of concerts. I didn't have to study very much. I sat in on a course for a whole semester in modern painting because it was fun. I took advantage of being in New York. But the second semester, I got serious.

Then all of the developments in plate tectonics (it wasn't yet plate tectonics, but sea-floor spreading and transform faulting) were becoming clear. I remember a friend saying, "Hey, you ought to stay for the seminar tonight because they're going to talk about the evidence for sea-floor spreading." I didn't know what sea-floor spreading was. I was out of it completely. This was 1966. The seminar was by [Jim] Heirtzler; this was Lamont [Geophysical Institute] data. He presented it very gently. I'm not sure he was convinced himself. There was argument from Joe Worzel at the back of the room. It was clear that there was something here, but it really didn't influence me that much. Then, several weeks later, Lynn Sykes presented the seismological evidence for transform faulting, and I suddenly realized, my God, continental drift is real and something is happening. I got excited about it. I went looking for something that I could do, but I wasn't smart enough to choose a problem on my own, so I went and got advice from Jack Oliver and Lynn Sykes, and then I was fine. I had my own research problems, which were armchair based. I just looked at data in the laboratory.

These were data generated by the instruments around the globe?

Worldwide standardized seismic network.

Someone else I interviewed said that that had originally been put in for military reasons.

Oh yes. It was the Berkner Panel, which recommended in the early 1960s a worldwide standardized seismographic network, because how were you going to be able to discover, detect, and discriminate nuclear explosions? My father [J. R. Molnar] was on the panel. So were Jack Oliver and Frank Press. This was put in for that purpose, but the data were made available. Seismology was a primitive science. I don't think I'm the only one who thinks that the military realized that the best thing it could do for seismology was to pour a whole lot of money into it, just let the field grow, and not hinder the progress of it. Let people do what they were doing and not try to guide them.

Collect a lot of data. Support people to do something interesting with those data and see where it led.

Exactly. I was one of those people. I was doing this, but I had to clear my head, so I went back to Alaska for the summer. When there were field programs using seismographs, I would usually go, but I have to say that they never were a big part of my research. They were my way of connecting to the real world, seeing a region. If you study a part of the world and haven't seen it, you are guaranteed to form a new image if you go see it. Whatever you do, you automatically form images from whatever data you have. I am a visual person. It was important for me to go. So, I did my fieldwork. It was vital to me, although probably not vital to science.

I got hooked on plate tectonics. I did a couple of studies that gained me enough notoriety that I had no problem getting a job when I was done. In fact, I remember CARL BOWIN urging me to come to Woods Hole. I wasn't ready to do that. I needed more training, but I remember telling him that I thought plate tectonics was dead [as a research area], and I wanted to move on. I had a two-year postdoc at Scripps.

You say it was dead, because the mechanism had been described?

I came into the area in 1966. We didn't know it was plate tectonics then, but people had been developing the story of sea-floor spreading. There is a good book by Naomi Oreskes [*The Rejection of Continental Drift: Theory and Method in American Earth Science* (1999)] on the story of plate tectonics, which argues that it is not the absence of a mechanism that provided resistance for so long. She argues that it was geophysicists being stubborn and geologists listening to them. There is truth to that, but there is more to it. Geologists have a long history of being angry with geophysicists for being snobs and rightly so. So why do they listen to them? They shouldn't listen to them. I have another opinion. I think the reason continental drift didn't get anywhere was because it didn't matter. Most people, "Joe Six-Pack" looking at his rocks in the field, couldn't care less whether the continents drifted. It wasn't going to change his interpretation of the rocks in the field. The petrologists didn't care. What was continental drift going to do? It wasn't going to change any petrologist's rocks. What about sedimentologists? It wasn't going to explain anything to them.

Scientists were narrow in scope. They weren't interested in anyone else's questions. They were only interested in their own questions, which was a fault in the way science was done. I think we've gotten beyond that. That is what plate tectonics has done for us. It has made us realize that we can bring our techniques and solve someone else's problem. That's good. I think continental drift failed because it didn't matter; plate tectonics mattered because sediments could be explained. You had subsidence of sea floor. You had rifting. What happened when you rifted a continental margin? It subsided. You could make what some people would call models, but I would say, you would have theories that could explain what you observe. You could make predictions, and you could be quantitative. You predict that in this case you would have so much sediment. You go out and measure it, and you have twice as much. Go back and revise your theory. You could do the same thing with petrology, with the rocks coming up in the midocean ridge. You could suddenly account for why they were composed they way they were, where they came from, what their mineralogy was. None of that came from continental drift. It was just a rather qualitative idea floating around out there, while plate tectonics was very precise about a lot of these things. It provided tests that have pervaded the field. I think it made scientists realize that each had to look at the data that other people had. I have a letter from a paleontologist who was quite angry with me because I don't know any paleontology, and he said, "Why do I have to learn about fault plane solutions of earthquakes and triple junctions, and you don't have to know any paleontology?" In a sense, he had a point. But, the real point was that he had to learn that stuff because if he was going to be on top of his field, he had to know what we could say about his field with our data. Of course, since then, I've become somewhat interested in his field because he has answers to questions that I want to get at; thus, I now read his papers.

Plate tectonics was called a revolution. In the sense of the French Revolution, who really benefited from the French Revolution? Not the peasants, of course. The people who really benefited from the French Revolution were the aristocracy. You could argue, "who benefited from plate tectonics?"

Ha. The geophysicists.

Well, yes. I wouldn't necessarily liken them to the aristocrats, but how did Joe Six-Pack working in the field benefit from plate tectonics? Not very much, because it didn't explain his rocks. When he stood up and said, "Plate tectonics doesn't explain what I see in my field area," he got ridiculed, but he was right.

I wouldn't say that. Every geologist I interviewed from the 1950s, the generation before you, said that their dissertation fieldwork all fell into place when the plate tectonics theory was developed: STEVE PORTER working in the Brooks Range, DAN STANLEY working in the Alps, FRED NAGLE working in the mountains of the Dominican Republic.

That's because they were looking at a different scale from most geologists. They were working with whole mountain ranges. CLARK BURCHFIEL told me he was out in Bulgaria when plate tectonics came along; he suddenly saw that the whole Mediterranean could make sense, but he wasn't talking about his mapping of his quadrangle west of Death Valley. Plate tectonics didn't help him with that.

Plate tectonics forced people to look on a large scale. The joke among us is that a student maps an area the size of a rupture of a magnitude 5.5 earthquake. That is a small earthquake. If you are looking at a scale of a mountain range, that is something different. Not that that is the only scale you can work at. My point is that plate tectonics wasn't a revolution in the sense that it did anything more than make people believe in continental drift. It didn't instantaneously change the way a lot of things were done. It changed people's belief of one large-scale image, which was important.

It gave people a context in which to put some of the questions they couldn't solve before.

Yes. Plate tectonics slowly forced people to look at processes. Not that people weren't looking at processes before. Plate tectonics just speeded up the process of people looking at processes.

It is the question of pattern and process. You can't really understand process until you've detected pattern.

Right. But there have been a lot of processes that have been ignored.

One of the exciting fields right now is geomorphology, which is process-based among the leaders of the field. I used to think of geomorphologists as a bunch of dead wood.

Stamp collectors.

Exactly. I didn't realize there were some very good people out there trying to understand what is going on. These are problems that I consider to be exciting now. Other people would say, "They've been exciting problems for 50 years," correctly.

Do you think the theory of plate tectonics fostered a more integrative approach to some of these problems?

Temporarily. There is no question that in the late 1960s everybody talked to everybody. You'd go to meetings and there would be people there who couldn't solve a differential equation but were eager to know what other people were doing. You'd have people who were doing all different kinds of science having to know what was going on because plate tectonics brought us all together. I think it is sad that we have all dispersed. Now, of course, you have these subgroups. I see our field as having a whole bunch of clubs just like a little town with its Elks and Masons and what-all that don't talk with one another. The people doing the GPS work have their annual meeting, and they pat one another on the back. You ask someone not in the club, "What is going on over there?" and they won't know what the work is. It has all gone isolated.

Part of the problem is that the problems have gotten harder and more narrow in scope. Maybe that is a clue that the field is winding down. I've talked with people and they say, "It is too hard to know everything."

There is too much information.

Right. That means they are doing the wrong thing. It is time to move on, to change to a new field, something where there isn't so much information. When I went to Scripps back in 1970 or so, Bill Menard told me how he decided one day that he wanted to study the floor of the Pacific Ocean. He went downstairs at Scripps, where they had all the echo soundings, and after a few hours during which he looked at all the data, he was the expert, because he was the only one who had looked at all the data! So he started writing papers. He said that it was really pretty easy. There wasn't any literature to read because he wrote the whole literature. He didn't have to read anybody else's papers. He didn't have to be bogged down by this enormous amount of literature. He wrote papers. He got out of that 15 years later when the literature had increased by at least an order of magnitude. He used to define an exciting field as one which doubled its literature in five years. So, in 15 years it had grown at least eightfold.

I do think that one of the problems we have is that people are bogged down, they are getting too narrow in scope because they are afraid to stretch out beyond what they know. The whole system is limiting in a sense. The tenure system is a deadly system. You have senior people telling younger scientists what they have to do, and they wind up getting into ruts and following narrow advice, rewriting their Ph.D. theses, narrowing their work in order to get enough publications that are interesting enough so that they can get tenure. By the time they get tenure, they are boring. They have no scope, no perspective. They are in a rut. They can't see out.

It is also hard with the funding the way it is, because the funding agencies say, "This person knows nothing about this field."

Funding is really hard on it. So we are winding up with a bunch of narrow, uninteresting people, and all they can do is ask for big projects with lots of money to do something big.

I find that, the older I get, to learn something new is really hard. You can store some facts in there, you can pull them out, but it is difficult to learn new ways of thinking, for instance, the new rules that guide all science. I try to teach myself climate, and I have to relearn, by going back to the principles, every single time whether the oxygen isotopic ratio should go up or down when it gets warm or cold. I can't store that well enough to just recall it and plug it into an understanding of something else. Whereas, if I had learned that when I was 23, it would be embedded in there right next to Mickey Mantle hitting .356 in 1956 and winning baseball's triple crown. I've got that stuff stored in there so deep. What I learned in graduate school, I remember: the shapes of seismograms, all kinds of useless knowledge, but all kinds of basics, too. So if an undergraduate education has failed to do that, it has failed. You could argue that the two important things of undergraduate training are to get the basics and to become inspired.

In the mid-1970s at MIT, we knew we needed a geologist who had a bigger view of the Earth than most geologists did. We called the person a "regional geologist." We needed somebody who went into the field but didn't just look at a single quadrangle. He looked at a quadrangle as being crucial to solving a problem on a larger scale. Clark Burchfiel was recommended. We hired him, and he was great. We've been lucky ever since. He's trained a whole string of people since who have this same view, this same approach to looking at problems, that is, mapping something on one scale to answer a question on a larger scale. This isn't unprecedented, of course. [James] Maxwell added one term to one of four equations, the Maxwell's Equations, which explain light as a wave phenomenon. Maxwell only added one term, but they are the "Maxwell's Equations." This idea of finding the key and, in a sense, turning the key to solve a larger problem is common in all scientific fields. Clark's approach to geology has always been that way. You find that crucial corner, map that, and all the rest will fall into place.

My father used to tell me that you could hire a theoretical physicist at the age of 24, and he would be ready to go; a good experimental physicist wasn't really any good until 30 or 35; but an engineer took until 40 years old, because he needed to get all this experience. I think a geologist is much the same, because you have to see so much. There is a lot you just don't learn in a classroom. In contrast, a student graduating now in seismology knows most of seismology. Unfortunately, that is not enough either, as you have to know enough to know what problem to solve. If all you know is seismology, chances are you won't know the "right" problem.

Do you believe that there are bright young students who can put it all together without having ever been in the field?

Probably not. If you take plate tectonics, of course, Harry Hess spent plenty of time in the field and plenty of time on ship, so he knew what the bottom of the ocean looked like as well. Fred Vine spent plenty of time on ships; he was sea oriented. Jason Morgan did not, although he takes students into the field at Princeton as part of their training, and he has certainly done fieldwork with GPS. Dan McKenzie didn't have much formal field training, but he did spend a summer mapping in Spitzbergen as an undergraduate, as an assistant. A couple of years ago, Dan McKenzie was here, and I got together with him and one of our other people. The other fellow asked Dan, "What do you prefer? Do you prefer teaching geology students geophysics, or geophysical students geology?" When I started, I was told, "You are better off starting with physics, and you can learn the geology later." That is what all the geophysicists said. In fact, McKenzie's answer was just the opposite. He said that, in his experience, it has been easier to teach the geologists the geophysics than the other way around. Of course, he gets the cream of the students. He gets students who have no fear of mathematics.

There is this enormous arrogance in the physics community that makes it hard to deal with them sometimes. There is no question that physicists have a track record of trying to dabble in the earth sciences and failing. Carl Wunsch at MIT once gave me a long list of these physicists who would come to Woods Hole in the summer and think they would dabble in oceanography, but they never did anything useful because they really didn't understand how the Earth worked. They were accustomed to designing experiments in rooms or on tabletops. They wouldn't know the problem with a lot of the data we deal with, which is sparse and inadequate in a lot of ways. The meteorology data, well, it isn't any worse than geology, but a lot of it isn't data at all; it is filtered in a lot of ways. I think the bigger problem is: you don't do controlled experiments. You make observations. That is a different approach. There is more to it that that. I think it is an attitude of mind. Some people can pose questions that may start with equations and motions, start with Newton's Laws and work their way through, or start with Maxwell's Equations and work their way through, if they are theoretical or have some underpinnings that are quite simple like that. However, the problem arises if you choose to study fluids, which are quite messy. Fluids are a nightmare. Of course, 20 years ago, physicists didn't work with fluids. They've just started to do that. It wasn't part of their domain. They were not interested. It was 20 years ago or so when they realized that fluids was the best way to see chaos in action. Then fluids became a big thing—not to understand fluids, but to study chaos.

When I look at what is known about climate, I see climate as being where plate tectonics was a few years before it was recognized. It is wide open. It is politicized, which makes it hard. There is so much bad science done (but which is politically correct) that it is hard to do good science, which appears to be too narrow in scope because its implications can't be seen immediately. However, it has to be done to get some foundation. What is understood about climate is really quite small. I see that as really exciting. What does climate do to the landscape? How does climate affect the surface of the Earth? Geomorphology is another field that is really taking off. Harvard is looking for a geomorphologist; Princeton is looking for one. At MIT, we happened to hire one a few years ago, and there is no question that we definitely look smart for having done so. There are other places, like Berkeley and the University of Washington, which are way ahead of us because they've been doing this for a long time and even training people.

My interest in climate is total. My research interest is the paleoclimate because I don't have a hope of making a contribution to the twenty-first-century climate or next year's climate, you might call it. I don't see that I have the time to learn all of the new techniques that I would need to do that, plus it is damn hard at age 56. I am interested in the dynamics of atmospheric and ocean circulation and how erosion and weathering affect the chemistry on geologic time scales, where I can work, and I think I can bring something. My view is that climate is in the same position as continental drift and plate tectonics were 45 years ago. There is something there. There is a hint of connections, but they are not understood. A lot of people just turn their noses up at it. The really good physical fluid dynamicists, who should be working on this problem, are to a large extent so turned off by climate as a slovenly field that they won't even touch it.

Almost 20 years ago, in the early 1980s, when we at MIT had an earth and planetary sciences department and a meteorology and physical oceanography department, the dean wanted to merge the departments, mostly for economic reasons. The administration paid for department heads, and this meant they would have to pay for only one instead of two, so they'd save some money. Almost everyone was opposed. I changed my mind completely from being "against" to "for" when Carl Wunsch said that geophysical fluid dynamics was in jeopardy at MIT if we didn't merge. Geophysical fluid dynamics clearly was the wave of the future. Everything—not just the Earth's magnetic field, climate, or weather but also the lower crust—is a fluid, and the mantle is a fluid over any time scales that are interesting to most of us. Pollution is a fluid. It creeps into the groundwater. All that jet fuel coming out of Otis [Air Force Base] up north of us is fluid moving down toward us in Falmouth. Fluids are fundamental. A decent geology program in a very short time will have a course in geophysical fluid dynamics. It has to; it is just so fundamental. It is part of the field. So we couldn't afford to lose geophysical fluid dynamics. That was clear.

Yet, the geophysical fluid dynamicists turned up their noses at it. When we went to merge, there were two eminent meteorologists at MIT. One of them you'll know, Ed Lorenz from chaos, and the other, Jules Charney, arguably more eminent but less well known because he didn't discover chaos, but he did fundamental work all through meteorology, gorgeous work, mathematical work. One of our people, John Edmond, was with him traveling on an airplane and said, "Yes, we are both against this merging of the departments, but we could take advantage of this. We could go to the administration and say, 'OK, we'll merge if you give us something.' Maybe we can go into a new field like climate." Charney looked at my friend John and said, "Climate? Climate is beneath the dignity of this department." That was the attitude. I tell that story time and again, and I always have to apologize for Charney, because he sounds like a jerk whereas he was a smart guy, and at that time he was right. Climate was a bunch of lousy data. It was not science. Surely, there were a few people that were doing good science, but by and large, it wasn't good science or didn't seem to be. It certainly didn't have the foundation that meteorology on a weather time scale had, and we didn't have the potential for understanding. So climate was left to others, kind of the way continental drift was left to the stratigraphers and the paleontologists, who could not answer the question. It is not their fault that they couldn't answer the question, they just didn't have the data. It took something new.

What climate needs is people trained in fluid mechanics but who know how to look at the right kind of data. I don't know if this means you've got to teach meteorologists geology or you've got to teach geologists fluid mechanics and the proper meteorology. Somehow, you've got to get the combination that some of us got in the 1960s when plate tectonics came along. We'll go back to your question, "Who do you want to hire?" and my reply is "What kind of people do we have to train?" We have to train people who have the ability to understand what the meteorologists are doing and what the contemporary meteorological data look like but who also know what the past record looks like and know what to believe in it. Know what an oxygen isotope tells you or a series of them tell you. Know what a pile of loess tells you and understand more than just that it is a bunch of dirt which blew in from the West, when it was cold and windy. It seems obvious, but yet it is kind of like insisting that our lawyers have a full four-year course in ethics. You just don't have time to do all of this. Yet, I don't have any doubt that there will be a bunch of smart young people who are going to do all of this, learn both fields, and they are going to come in and clean up.

Some people think that science is about certainty, some think that it's a conspiracy, and some think it's incomprehensible. After all, science is nothing more, nor less, than organized curiosity.

Paul F. Hoffman (b. 1941)

Sturgis Hooper Professor of Geology
Harvard University
Interviewed in his office 26 January 1998

Paul Hoffman accepted a chaired professorship at Harvard because he wanted the opportunity to teach nonmajors about the process of science. In recounting his career, he gives us a glimpse of how he does scientific research. He explains how field experience accumulates with time (he himself has spent most of the past 40 summers in the field), but new students with the proper guidance can develop the confidence to recognize the odd phenomenon and produce good work. He also discusses how he thinks science ought to be conducted and administrated for producing the best and most effective work.

I spent four and a half months in northern Ontario, in one of the last paddle canoe reconnaissance field parties, in the summer of my freshman year. That was absolutely the clincher. I mean, to combine my interest in rocks and minerals with full-time regional scale mapping outdoors, that was glorious. I worked in the field every year as an undergraduate. I have, from my parents, this interest in the northern Canadian landscape. We did a lot of camping trips to northern Ontario, the north shore of Lake Superior, and also in the Blue Ridge and the Appalachians. My parents had a summer cottage. My mother still actually lives in this cottage the year 'round now. She is 81 and lives in a house with no electricity, running water, or telephone and no direct road access. She hauls everything in there through surreptitious trails so no one knows she is actually living in there. She is laughing at this ice storm now [which paralyzed parts of the Northeast] because it doesn't matter to her. I spent the summers there; both there and at home, we were supposed to be out-of-doors. We were not allowed in the house until after dark.

The Geological Survey of Canada offered field parties in the Arctic, the Rockies, or wherever, so I joined up with them the second year. I started, at that point, spending a minimum of three months a year in the field. Of course, it just made the class work so much more real. The next really big influence was a prof at McMaster University in Canada, where I took my undergraduate degree. Most people working in the Canadian Shield were interested in metamorphic or igneous petrology because a lot of the rocks are high-grade rock. I hadn't imagined that I would become interested in sediments and stratigraphy until I took third- and fourth-year courses in sedimentary geology from Vinton E. Gwinn. He was an American who came from the oil industry and was a sedimentary and structural geologist with a tremendous enthusiasm for regional tectonics and Earth history. Tragically, he was killed a few years later when a point bar he was trenching collapsed. He explained how, if you were to begin the investigation of the tectonics of a large area, say the Appalachians or the Alps, the place to start would be the sedimentary rocks adjacent to the mountain range. First of all, they record the history of the mountain belt in terms of the sediment that is derived from it, and also they provide a stratigraphic basis for working out the structure. That really opened my eyes to an approach to the tectonics of the Canadian Shield that hadn't been used very much, because the sedimentary rocks had been rather neglected. I knew from my experience with the Canadian survey that there were lots of very well preserved and exposed sedimentary rocks. I imagined myself working out the tectonic architecture and evolution of the Canadian Shield from scratch. I said, "Well, if it were 300 years ago, and I wanted to work out the history of the Alps, how would I proceed?" I operated on that basis.

My interest right from the start was in tectonics, despite the fact that at the time I was a student, in the early 1960s, tectonics was very unfashionable. The seeds of plate tectonics were already being sown, of course, but most geologists were unaware of it. This was also true, actually more true, of the graduate school I went to, Johns Hopkins. Tectonics was considered disreputable. They had something of an aversion, as much as they prided themselves on being field oriented, to regional tectonics. Geophysics and geology were in different departments, as they still are in Canada, and the geologists were not very aware, nor would they have been sympathetic to, the new findings in geophysical oceanography.

That was just the culture of that geology department at the time?

Yes, and of this one also [Harvard]. It was an odd situation because, as you know, the theory of continental drift had first been proposed by [Alfred] Wegener, [Alexander Logie] Du Toit, and [Pieter] Joubert, mainly on geological grounds. It was rejected primarily by geophysicists. Its revival began with the paleomagnetic results of [Stanley Keith] Runcorn's group in the mid-1950s and came quickly to a boil in the 1960s with all of the unexpected results from the systematic geophysical exploration of the ocean basin, culminating conceptually in the sea-floor-spreading hypothesis. Most geologists were taken by surprise, and there was not a very positive response.

My ambition, expressed in an undergraduate term paper, was to understand the tectonic evolution of the Precambrian Shield that is primarily exposed in Canada. I saw the study of Precambrian sediments as a new way to approach the tectonic evolution of the shield. So Vint Gwinn had a very powerful influence on me, and his approach shaped my career. One day Gwinn arrived in class with a newly published book in hand, which he left on the table after class. It was *Paleocurrents and Basin Analysis* by Francis Pettijohn and Paul Potter. Pettijohn was at Johns Hopkins and had resurrected paleocurrent analysis, whereby the orientations of primary sedimentary structures are used to infer patterns of sediment transport. This technique is well suited for Precambrian sediments, which are often well preserved but of limited areal extent because of later erosion or burial. The problem was to work out the geometry of the original basins of deposition given only small areas of preservation. Paleocurrent analysis offered a way around this problem because sediment transport in a given formation is often uniform over large regions and can thus be inferred even from limited areas of preservation. From the sediment transport direction, the regional paleoslope can usually be inferred.

I applied to Johns Hopkins for graduate work because of Pettijohn, their geologically well rounded department, and their tradition of supporting Canadians doing theses sponsored by the Canadian Geological Survey. Their first Ph.D. in geology (the first in North America, I believe) was Andrew C. Lawson, who later gained fame at Berkeley for his synthesis of California geology in the aftermath of the catastrophic 1906 San Francisco earthquake. He was a Canadian, and his thesis work in the Precambrian Shield had been sponsored by the Canadian Geological Survey. I intended to do likewise.

Those years, 1964–1968, were thrilling. Not only had my dream of doing independent research begun, but I was able to indulge my interests in politics, African-American culture, and marathon running. But first of all I had to choose a doctoral thesis. I went to graduate school wanting to do research. I didn't want to take any more classes. I didn't like taking classes. I hadn't really liked school very much. I was just sort of waiting, not very patiently, to start doing independent research. For my thesis, I had to pick out a particular area to make my start. I'm not sure I accepted the fact that I was not going to be able to map the entire Canadian Shield (nearly two million square miles) by myself!

But you had to start somewhere.

I had to start somewhere, and I had three areas that I was particularly attracted to because they had thick and well-preserved sedimentary successions. I chose the East Arm of Great Slave Lake in the Northwest Territories because it was the one project that I felt I could get funding to operate on my own. I knew that I didn't want to be in the position of being an assistant on someone else's field party. As a student assistant with the survey, I had witnessed cases where a student assistant doing a thesis project was caught in conflicts with party chiefs who were permanent survey geologists.

My thesis involved a test, using paleocurrents, of the proposition that the sedimentary basin in the East Arm of Great Slave Lake was tectonically analogous to the basin associated with the Appalachians. As the Great Slave Lake basin is five times older than the Appalachians, the comparison would probe the evolution of tectonic regimes through half of geological time. In the Appalachians, there is a paleocurrent pattern that had been revealed by students of Pettijohn. In the oldest parts of the basin, paleocurrents were directed from the continental interior outward toward the margin. After the mountains rose up, the paleocurrents were reversed and flowed back toward the continental interior. At the time, this pattern was not understood in terms of continental break-up and collision but only in terms of the so-called geosynclinal cycle. My reasoning was, if the two-billion-year-old succession in Great Slave Lake showed the same paleocurrent reversal with time, it would indicate that a tectonic regime similar to the Paleozoic had operated much earlier in Earth history. Moreover, if the early sediments flowed to the south, as I expected, and the later ones flowed back to the north, then it would imply that a very ancient continental margin formerly lay to the south of Great Slave Lake. Thus, I had a specific research strategy, including a testable prediction and an appropriate methodology.

In fact, within the first couple of weeks in the field, the test had been made. Yes, there was a reversal of the paleocurrents, and the overall lithological succession was very much like the Appalachians. But all of the paleocurrents were oriented 90° to what my model predicted! The early sediments flowed to the west, and the younger ones flowed back to the east. The observations implied that there was indeed a collided continental margin but one that lay to the west, not to the south.

Somehow, right from the beginning, I sensed that it was important to have a preconceived hypothesis to test and that the test meant potential falsification. I'd like to think it was because I was very interested in philosophy, particularly logical positivism, in high school and as an undergraduate. I'd like to think it paid dividends, but I don't actually know if that's true. In any case, my thesis led directly to a series of field projects and further testable hypotheses that the survey supported over the next 20 years. Pettijohn encouraged his students to be independent and that was perfect as far as I was concerned. On the other hand, the many field trips we had at Hopkins, both in the Appalachians and in areas of recent sediments, provided an essential basis for comparison with the very ancient rocks I was working on in Canada.

In those days, an undergraduate field assistant carried the boss's lunch, trimmed samples, and did other routine things. We did not do much independent mapping. Later, when I became a research scientist with the survey, I engaged my undergraduate field assistants much more actively in the field, including independent mapping. The first two or three weeks, they would accompany me on traverse, where I would teach them as much as I could about the local geology, about what we were doing, and why. Thereafter, it's a matter of having them learn by struggling to map on their own. In fact, it doesn't matter how much field experience you've had. When you go into a new area with new rocks, you flounder to begin with. It's extremely frustrating, and you just have to work through it.

And you would say that is true for you today?

Absolutely. If I go to a new field area, it's like I don't know anything. Part of the challenge as a party chief is training new field assistants. What you can expect is that they will do a good job with rocks and structures they are familiar with. You can't expect them to do very well with something that's new. So you have to plan their traverses very carefully to begin with. You can get very good work from them, and you'll be repaid because they will learn quickly, if you are careful and don't put them in situations that are simply too frustrating, where they are going to see something they've never seen before and won't know what to make of it.

It's very important that they learn well, because the crucial field observations are always the ones you don't expect, the ones that go against your current interpretation. It's an awful lot easier to cover them up with moss, so to speak, or to turn and walk away. You really have to be careful, so that when they find something that they don't expect, or don't understand, they will have the confidence to realize that it's potentially important and will stop and try to work it out while they're there. That is a very good discipline that we have from working in the Arctic. Access is difficult, so when you are on an outcrop, you have to assume that you'll never be back there again. You must try to work things out and to see everything important there is to see the first time.

I like to make the point that fieldwork is one of the most intensely scientific things you can do because you have to figure out what every outcrop means in terms of your understanding of the entire region. There are millions of outcrops, and you can't see them all, so you have to see the right ones. The next outcrop should be the one that would be most devastating if it isn't what one's current understanding predicts it should be.

You just went into how you can introduce field assistants to this approach, but the more experience you have, the better able you are to make these choices.

Sure, and that's another way in which field geology is somewhat different from many other types of science in that it is very dependent on experience. So you should expect to become better and better as your experience accumulates. So it's important to start early and to work in the best field areas you can find. And it's also important to continue as long as possible.

Another thing in my life that has been extremely valuable is my involvement in athletics, which also came from the encouragement of my parents. I've been a competitive distance runner for 45 years, a marathoner for 35, and it paid tremendous dividends in my field research. First of all, I approached fieldwork with a physical intensity that was not common at the time. If people walked 10 miles a day, I would walk 30, because I knew I could do it, as a national-class marathon runner. I stressed to field assistants the importance of being fit. I knew that students could do much more than they thought they were capable of, but I also knew they had to be trained properly. I am not the kind of person who wants to run people into the ground. I mean, you learn as a competitive athlete that there are proper and effective ways to get fit and there are ways that just lead to injury or people giving up. You have to treat people properly, but their capability is far greater than they would ever imagine. Also, I noticed with people with the survey, most of them tended to give up active fieldwork in their late 40s or early 50s because the discomfort or the effort just became too onerous. But I saw John Kelley was running a three-hour marathon at age 60. I figured you should be able to do fieldwork until you are 60. So my plan was to do 40 consecutive field seasons.

How old are you?

I'll be 57 this year.

And are you going to have 40 consecutive seasons?

No, I had a five-year gap when my son was young, and I undertook a synthesis of the Precambrian tectonic assembly of North America. I've always had plans but I've never been afraid to change them.

But you still go in the field now?

Oh yes, I do two or three months of fieldwork every year in Namibia, in southwestern Africa. Fieldwork offers an ideal balance among physical, aesthetic, and intellectual rewards. When I'm not in the field, running helps me maintain that balance. I need to get out-of-doors every day and hammer the roads for an hour or two.

It is true, what you said a while back. At the survey, some people saw field geology as strictly descriptive. They saw their role as providing information to the mineral or petroleum exploration industries. I never saw any conflict with providing geological maps and synthesis for industry and also tackling scientific problems, like the verification of Precambrian plate tectonics. In those days, we were well paid, we had tremendous logistical and technical support, an outstanding library, and inspiring geology. I felt we should be second to none in terms of the quality and significance of our geological work. If you hire motivated people and support them properly, they will have the freedom to pursue research objectives that may not pay off immediately.

There is a very different process in providing research grants in Canada than there is in the United States. There is something to be said for the Canadian system. In Canada, university researchers are evaluated every year on the basis of their productivity for the previous six years. On the basis of that evaluation, the level of support is adjusted incrementally up or down from year to year. It is much easier to plan ahead, particularly insofar as graduate student support is concerned. In addition, the funding received is not tied to any particular research activity, so there is more flexibility, and there is more freedom to pursue innovative research that carries higher risk of failure. Now, most NSF grants are very specific, and they are for a relatively short period of time, one to three years. NSF grants are highly competitive, so you spend an inordinate amount of time writing proposals. You are expected to show results on the time scale of the grant cycle, which can lead to premature publication or overlapping publications. That is not a very efficient way of reporting research results because there is a great deal of repetition. I think that the key is to hire the right people, who have the right motivation, and then to provide them with the proper support and recognition for their work. Then the incentive is their own success and the recognition of their peers.

I picked up the New York Times *about a month ago, and there was this article where the "new Proxmire" is complaining, why didn't Clinton save money by line-item vetoing these various research projects? I'm reading the topics and I'm thinking, hmm, this sounds interesting, and hmm, so does this, and then I'm thinking, who do you think you are to be qualified to judge these topics? He is probably a lawyer.*

Oh yes, almost certainly. One reason I came to Harvard was that I wanted to teach courses for nonscience students. In fact, I spend more time teaching core classes for nonscience majors than departmental courses. I think that science is widely misunderstood, including by most of the people making public policy. Some people think that science is about certainty, some think that it's a conspiracy, and some think it's incomprehensible. But all of these misconceptions can be overcome, I believe, by appealing to people's latent curiosity about the world they live in. After all, science is nothing more, nor less, than organized curiosity. All the scientists I know say, "We love this work so much, it's amazing anyone pays us for it!" This may not be the best thing to admit in a puritan culture. My parents told me, "You'll spend most of your life working, so you might as well enjoy your work." Don't work for the money; work for the pleasure and satisfaction of the work.

Sometimes the most interesting and important things aren't the quantitative, they are the qualitative things that lead to something new.

LAURENCE P. MADIN (b. 1947)

Senior Scientist
Chairman, Department of Biology
Woods Hole Oceanographic Institution
Interviewed in his office 18 February 1998

Larry Madin describes a new approach, blue water diving, to studying living plankton in the open sea. It was developed and perfected by his advisor and his cohort of grad students in the early 1970s. As he elaborates on how this was done, he illustrates some of the problems encountered, particularly those imposed by bureaucratic regulations. He then describes the newer tools, including submersibles and ROVs [Remotely Operated Vehicle], which are adding a new dimension to the study of live plankton at sea but not supplanting the old methods. I first met Larry Madin when he visited St. Croix to use a submersible to study planktonic organisms occurring at 1,000 meters' depth. I particularly appreciated the fact that he sent back video footage of the animals they studied in this environment for us to show our students at West Indies Lab.

When I was in college and took invertebrate zoology, I first discovered that there were a whole lot of very unusual kinds of organisms that I didn't know anything about. Even better, I discovered that there were some kinds of creatures that hardly anybody knew anything about. So that was quite interesting. I'm thinking in particular of pelagic tunicates, which is the thing I've pretty much worked on ever since. When I first saw some, my professor, Michael Ghieslin, didn't know anything about them either. This seemed to me an excellent thing to keep in mind, that there were a lot of unknown mysteries out there.

When I went to graduate school at Davis, one of the first people I met was Bill Hamner. That was just when he was beginning to start this blue water diving idea for studying plankton in the field. Our first experience was in 1970, when several of us went with Hamner and his family to Baja California. We spent the summer doing the first organized effort at doing blue water diving and looking at zooplankton in situ. It was field biology in the true sense because there was no laboratory there. It was camping on the beach with only the most rudimentary of facilities for collecting the specimens and handling anything that you brought back. It was almost entirely field observations, simple field experiments, and some collecting. A lot of it was "Well, what can we do with these things?" because no one had ever tried it. So we were experimenting with everything, including what kind of boats we should use. We had a gasoline compressor. We were scuba diving; we were filling our own tanks.

It was very observational, but it was that sort of preliminary thing people need to do to understand: What kind of work can we do? What are the animals like? Where do we find them? What are they doing? Can I learn to recognize them and identify them?

But it must have been exciting, because it had never been done.

Oh, it was! It was a group of new graduate students and undergraduates working together on this, so it was new for everybody. People started sorting out things they were interested in working on. We saw different kinds of gelatinous animals; we saw marine snow; we saw fishes. That experience, I think, convinced all of us of the value and also the practicality of doing that kind of fieldwork: open water scuba diving and looking at plankton in an undisturbed way.

The following summer, that same group of us and a few others went to Bimini for a year and a half, 1971–1972, to do the same thing but in an organized way. We had a laboratory, Lerner Marine Lab, to work from. We had boats with engines, instead of canoes and zodiacs. We had better diving gear. Basically, we were set up to do a longer term, more comprehensive study. But it still began with people going out and looking around to see what was there and then making decisions about what they were going to work on. In a sense, we kind of divided up the fauna. Bill Hamner was interested in jellyfish, and he was also interested in everything else, tuna, turtles, and whatever came along. Alice Alldridge and I flipped a coin and decided she would work on larvaceans, and I would work on salps. Ron Gilmer decided he would continue with his work on pteropods and heteropods, which culminated several years later in a very nice book. Neil Swanberg, an undergraduate, wanted to work on ctenophores. My wife, Kate [Madin], was working on larval stages of things, and she was also doing some writing and editing for another project of Bill's. There was a guy who was working on nurse shark feeding behavior and on damsel fish.

We'd make one or maybe two dive trips a day out in the Gulf Stream—it was only about 20 minutes away from the lab by boat—where we could do our diving, make observations, and then just bring things back in. We had some cultures going in the lab, and we would make laboratory observations and do experiments. It was still mostly field based, but we were able to follow up on things a little bit better. We could

Blue water diving (divers, tethered for safety, making observations of organisms in clear and deep oceanic waters) has been practiced by Larry Madin and colleagues since the 1970s. Photo by Larry Madin.

work out ways to do photography. We worked out a lot of the techniques that we've used later. First of all, for diving itself, so that it was safe and convenient and people didn't get lost. It was a tethered system; you'd have three or four divers working at the end of long tethers and one just watching over them all, sort of the buddy for everybody. We had a big inflated rubber raft that they were tethered to, and we had someone else in the boat, the Boston Whaler, who would keep station around it.

The raft was floating with the current?

Yes. It would essentially drift along with the current, except for the wind resistance on the raft, so there was some sort of a relative current between the divers and the water. This was good because that way you'd move past more water and more animals and see things that you might not otherwise see.

Was it frustrating sometimes, though, if you were watching something and it would drift off?

Oh, yes. Frequently, you'd be at the end of the tether and reach something and you just couldn't get it. Or you'd drop something and swim to try to catch it, and you'd find you couldn't, you were too late. Usually that was just things like bottles, but a couple of cameras got dropped. The temptation is very great to take off the tether and go after it, but that was a strict no-no. I don't think anybody ever did that.

Or at least confessed?

No, people would have seen it. We were always there in groups of four or five, very clear and warm water; it was great diving.

We also did net sampling. We thought we should have some sort of baseline plankton sampling. We'd go out maybe twice a month and do these 24-hour cycles; maybe every four or six hours you'd take a tow to 100 meters. That was sometimes pretty exciting, in the middle of the night out there in this little 20-foot Boston Whaler. The weather would come up, and these giant ships would loom up out of the darkness. There were a couple of times we saw pretty big ships pretty close, closer than we would have liked, since we were quite small and inconspicuous. Following that, I was at UC Davis another year or so finishing up.

The whole idea of coming here to WHOI was to try to do the same kind of fieldwork, except from big ships at sea, instead of coastal like we had had to do it. Frank Cary, a scientist here, had come down to visit us when we were in Bimini. He was interested in tuna. When he came back up here, Rich Harbison had just arrived to be his postdoc. The two of them went out on the SEA [Sea Education Association] ship, the *Westward*, for a short leg, and they decided they'd try this. They fastened ropes around their waists and jumped off the side of the ship. Despite the fact it was a little technically awkward, I guess it was amazing. They had the same reaction Bill Hamner had had and we had had when we first did it. You see all these amazing things that nobody had noticed were there. When you catch them in a net, they are just a little blob. I arrived, and we started doing it pretty regularly.

So you were doing the diving from the ship?

From a boat from the ship. They'd put us off in a zodiac, and we'd go off and do our dives much the way we'd done them in Bimini.

When IAN MACINTYRE *was a postdoc at Duke working on the* Eastward *off the Caribbean islands, he said they couldn't do any diving.*

It was a little bit of a struggle to get the WHOI ships to accept it. At that time it was an unusual thing to do from an oceanographic ship. Most oceanographers do not like to get their feet wet. They like to stay safely up on deck and put things over the side. The idea of putting people over the side and having them go off somewhere . . . The first ship we did that on was the *Knorr*. It was pretty much a biological cruise, and there were all kinds of new things being tried, and we were one of those new things. The senior people who were involved, like John Teal, went diving with us.

Did he think it was great?

Oh yes, definitely, so that helped. That was in the Gulf Stream in the Sargasso Sea, a Woods Hole–to–Woods Hole cruise. But after that, each time we were on a different ship, we'd have to try to convince the captain: this is OK, we've done it before, it's safe. Once we had gotten them convinced that it was OK generally, there were always arguments about, well, the weather's too rough, it's too windy. We had a different idea about what was too rough than they did very often. Then they wanted to do things like keep a line attached to the ship. We'd say, "That won't work. There is so much windage on the ship that we'd get towed through the water." Then they wanted to stand by, 50 feet away. We'd say that just won't work, "We'll drift down on you and get under the propellers. You've got to be a half mile at least away."

Our next hurdle was to convince them we wanted to do it at night. The first attempts at doing this were really ludicrously complicated. We actually built shark cages with these dutch doors, so we could open them, reach out, and grab things. The cages were built with their own flotation and lights. It was just ridiculous. And the most ridiculous thing was that basically these cages were built out of chicken wire, garden fencing.

So a great white shark could crunch right through them.

Right. It was purely psychological and not even for our benefit but for the captain's. It was such a cumbersome thing to try to get these things in and out of the water. We only did it twice, I think. Then we realized there wasn't any necessity. Eventually we convinced the ships we could do this at night, just the same as in the day, except we had flashlights. All the WHOI ships got used to it. That became a standard thing, and it was very productive because, at night, for an organism in the upper ocean, it is completely different than in the day.

It's that way on a coral reef.

Well, it is the same thing with plankton. It was clearly essential that we have some idea of what was happening between day and night. People in other places did it to some extent, but there still isn't a lot of it being done other than here. There are some groups that do it in Bermuda and in California but not so much from big ships out at sea.

The next logical step as far as we were concerned in that kind of field biology was to use submersibles, so we could go down a little deeper. That's what brought us down to St. Croix using the submersible *Johnson Sealink*.

You sent us back a film of those dives. It was a completely different world (as compared to diving depths). I remember weird octopuses and fishes.

If we could get down another 1,000 or 2,000 meters deeper than that, it would be yet another different world. But that's a little difficult to do for a while. It gives you a really different sense of what the world is like. The intellectual thing is realizing that this is most of the planet. This is 70%, 10 billion cubic kilometers, 12,000 feet deep on average. You realize, here I am, and this is what the Earth is like mostly, and the other part is sort of a minor part of it. I guess my motivation is partly selfish. I want to be down there, seeing it, because I think it is really cool to be able to see it. I think it is also useful to have people who have some familiarity with these things to be down there. Seeing it in person, even if it is through the window of a submersible, you still can have a better perspective, a three-dimensional view or better peripheral vision. You just can see things better with your eyes than you can with video cameras. You are better able to maneuver and manipulate around some organisms.

It is routine now for undergraduates to be very well versed in all kinds of molecular techniques. Those tend to be laboratory centered. We've seen here in our graduate students a focus toward mostly laboratory-based work or nearshore coastal kinds of work. There haven't been as many students lately who have a real heavily field-oriented offshore program in biology. Partly that's for practical reasons, not being able to secure the necessary ship time. So in a sense, it may not be wise for students to pin their hopes on something like that. When I first came here, in the 1970s and early 1980s, we had a lot of cruises where we would have the whole ship for a couple of weeks a year, anywhere we wanted, and we didn't have a particularly fixed plan. You pick up ideas that you may then pursue later on. They were very useful. The submersible cruises were like that. They were even more inherently exploratory, because you really didn't know what you were going to see or what you were going to do with it. We would have plans of things that we were going to try to do, but the plan would say, I am going to go down to this depth, steam back and forth, and count all the chaetognaths. Well, that plan goes out the window as soon as you've seen something that you've never seen before. These days people tend to want to be very quantitative in everything they do, which is fine, but sometimes the most interesting and important things aren't the quantitative, they are the qualitative things that lead to something new.

Why do I go to sea? I have to be there—have to live in the place, have to experience it—to develop the kind of intuitive, artistic response to the science that needs to be done.

PETER H. WIEBE (b. 1940)

Senior Scientist, Department of Biology
Adams Chair for Excellence in Oceanography
Woods Hole Oceanographic Institution
Interviewed in his office 28 February 1998

Combining a childhood love of the sea with a passion for tinkering and a joint honors degree in math and zoology provided Peter Wiebe with the skills to contemplate a thesis topic at Scripps. Choosing the open ocean over the nearshore, he addressed a problem posed by a lecturer: How can one adequately sample plankton living in a patchy environment? His account of developing the technology, employing it in the field from oceanographic ships, and learning computer modeling is almost a textbook case of how a research project—and indeed a research career—can develop. In light of both the unknown in the ocean ("only 50% of the ocean is available to study by oceanographers") and our eventual need to know, Wiebe discusses the current national trend to work nearshore. He says, "As a country, we have largely turned our backs on the open ocean."

How did I get here? I grew up on the California coastline in Salinas. From the earliest times I've been involved with the ocean; I started swimming when I was about two. I read [Jacques] Cousteau's book *The Silent World* and that just took me into outer space.

I applied to Scripps for grad school in marine biology. At the start, my principal contact was E. W. Fager, who said, "We've looked over your records, and we think you should be in the oceanography part of the house, not the marine biology part." I said, "Sure, it's OK with me." However, I had no idea what oceanography was, and in fact, I had only an inkling of what marine biology was. I just was terribly naive. In fact, I was so naive at that time that if I were that naive now coming in as a student, I probably wouldn't get accepted. The students today are so much more tuned in to what they are trying to do and why, but I don't think they necessarily should be. I'd much rather have somebody who thinks they have an interest but doesn't know quite what they want to do. That's not the way they structure things around here in the WHOI/MIT Joint Program, though. You have to have an advisor already lined up who has the money to pay for your education.

I went to graduate school, and I started out diving like crazy. I was everybody's buddy for their thesis projects, diving nearshore. I went on my first cruise in the first semester I was there. That brings me back to childhood. My father loved to go to the ocean: clamming at the beach in Seaside, where the Pismo clams are; going out on charter fishing boats into Monterey Bay. Before high school, I would go out with him on these deep-sea fishing boats, and I was classically seasick. I'd get out in the boat and within ten minutes I'd be retching over the side. My dad would feel so bad, because he didn't get seasick. After this had repeated itself a number of times, he would come in on a Friday night and say, "I'm going deep-sea fishing tomorrow, do you want to come?" I'd say, "Oh, yeah!" and he'd say, "Are you sure you want to come?" Yes, I wanted to go. Every time, I'd be just absolutely miserable, but I could never say no. I always kept doing it. Well, the same thing happened when I went to graduate school. I got miserably seasick. But yet, given the chance to go, I went. I don't know why. I just loved to be out there.

I got to the place where I had to begin to really think seriously about a thesis problem. I had been working on the settlement of a species of barnacles on *Dendraster excentricus*, the sand dollar, in the shallow water sands. But, I woke up one day and said, "I'm at the world's largest oceanographic institution, where the access to the sea is so tremendous, what am I doing trying to work in the inshore area when the whole of the open ocean is available as an ecosystem to try to understand?" I had been in [John] McGowen's lectures for several years, but the first semester he gave one lecture about small-scale patchiness in the ocean, how we didn't know anything about it, and how important it probably was. That sat at the back of my mind. In fact, I still have the lecture notes from that particular lecture, which triggered off in me this argument: Do I do a nearshore project or do I go to the open ocean and start to develop my career in that area?

The open ocean won. I figured I could create a career niche that would have much less competition and much more career opportunity, ultimately. That was a conscious decision I made. It was about two and a half years into my graduate education before I finally put this all together and came to this conclusion. What I did for my thesis work was to go out and try to map, on a scale of tens of meters, the distribution of planktonic animals in the field. I designed a project that allowed us to actually do the mapping with a brand-new piece of instrumentation, the Longhurst-Hardy continuous plankton recorder. Alan Longhurst, six months before I actually started to do this thesis work, had built the first one. I saw it; I went to his talks; I knew what he wanted to do. I thought, I want to get that piece of gear, too. When I told him that I'd like to build one for my thesis work, he replied, "I'm going on a six months' sabbatical. Why don't you work with my tech, because I want to build another one, too." I said, "I think I see some places where we might improve it a lot, so we can actually make it far more quantitative." This is where my background, my engineering skills, and my tinkering came in. Longhurst's tech, Bob Bower, and I did it. Finally we got to the place where Bob said, "Look, leave me alone for a whole week, and I'll have both of these things running." It gave me an instrumentation package that allowed me to go into the field and do the experiments I had in mind.

I started this fieldwork, but at the very start of my thinking about it, I started creating mathematical models to try to simulate what I thought might be going on before I ever actually went out into the field. I started doing that with pieces of paper, trying to sketch out what I thought might be going on. Then I started making 3-D models out of cardboard, and I tried to sample through the thing, to see what I had to do. Ed Fager watched me struggle with this in 1963 or 1964. He talked to Bob Arthur, a professor of physical oceanography, and said, "Don't you have a graduate student who is doing 3-D mathematical modeling with computer models?" It turned out to be Bill Holland, now one of the world's premier ocean modelers. Fager said to me, "We're going to help you learn. You can take a Fortran class, but in addition, Bill Holland can spend a little time helping you think about it." I had actually, earlier, taken a Fortran course, a three-day-per-week class, but I got nothing out of it. I didn't have a clue, when I came out of it the first time, what was going on. The minute the connection was made to the problem I was trying to address, and the use of the computers to do this with, I knew how to proceed.

When I sat with Bill and started looking at what he was doing, I immediately knew what it was. By the time we got done, I basically took the code away and started modifying it on my own and running these models, always going back to him, showing him the results, and making sure I had done everything right. He got out of the loop pretty quickly, but my first paper is published with him. I still use that paper because it is a good summary of all of the errors associated with taking replicate net tows in a given region of the ocean, to look at how patchiness affects your ability to sample the world. That was my thesis. I built a piece of gear; I used computer models; I did the fieldwork. I incorporated the field results on plankton patchiness in the model to reconfirm the first model results. So that was a satisfying career as a graduate student.

After I passed my oral exam and was actually given permission to go and do my thesis work, John McGowen, who was my principal advisor, said to me, "The next couple of years are going to be the best years of your life. Think about that and relish them, because when you get your Ph.D. you are going to go into a world where you are never going to have the freedom to do your research the way you do right now." He was absolutely right. Yet, I've been doing this for a long time, and the enthusiasm for doing it has not gone away. I am still in my graduate student mode of thinking and working. I often thought that working 12, 14, 16 hours a day was going to stop after I got my Ph.D. It never has. There is a drive; it is really a joy to be able to discover things and to show people what has been learned.

Almost from the start, my career has developed working with teams of people as opposed to working by myself. When I first got to WHOI, I was brought into a group that was block funded. The first two years I was here, I did not have to write a proposal. I did not have to justify my existence. All I had to do was work hard on problem sets that I was interested in, but were more or less defined by the leader, who at that time was JOHN RYTHER with assistance from David Menzel. This was a project funded by the Department of Energy, the old Atomic Energy Commission, and was called "Cycles of Organic Matter in the Sea." If it sounds a little bit like JGOFS [Joint Global Ocean Flux Study], it is. The idea was to try to figure out how carbon was fixed and how carbon moved within the water column down to the sea floor.

No new questions, just new ways to look at things?

Yes and no. I think there are some new areas. I mean the discovery of the hydrothermal vents and those communities came out of the blue. Nobody ever suspected their presence. The cyanobacteria, the very small phytoplankton producers, the microbial loop, actually, that's new. LARRY MADIN's work, the role of gelatinous zooplankton in the water column and the approaches that have been taken to study it, that's new. We just didn't have the comprehension of their importance in the ecosystem, and we still don't have a really good understanding. It is a very difficult problem. So, I think there are a lot of new things that have happened in the last 20 or 30 years. People say that we haven't really made very much progress, but yes, we have.

I think we've made progress, but I think a lot of the questions aren't new.

You can go back to the late 1800s for a lot of the questions. There was a British scientist, [William] Herdman, who worked in the North Sea, the Irish Sea, and the English Channel, who was worried about plankton patchiness. This was 100 years ago. He was saying that the idea of the uniform distribution of plankton in the sea was still a vexed question. Plankton biologists were struggling with the idea of how to sample the ocean given the patchiness structure. Well, you can follow the evolution of that idea. We are still trying to deal with the patchiness structure and still trying to understand how animals are distributed on a small scale. The difference is that the technology is getting better and better, so we can actually address the question. You see things like the development and application of instruments like the video plankton recorders and high frequency acoustics systems. They allow us to make visual images of the distributions of the organisms in a way that no one has been able to do before. It allows us to address that problem set much more comprehensively.

Why do you think it's important for you to be in the field?

I have been to sea nearly every year since I came into this business. In fact, I'm going to sea more now than I think I have at any other time in my life. This year [1998], I am going on six cruises, next year probably seven. I think it's really important that I be there. For one thing, I don't have enough money to send anybody else. For another, the instrumentation we are using now is complicated enough that, because I don't have enough money to hire technicians, I'm the only one who can actually go out and do what I want to do. I just can't put the instrument on the boat and say to somebody else, "OK, go and make the measurements." I wish I could in some cases, but I can't.

Do you still get seasick?

Oh sure, but not for very long. My seasick remedy is the computer.

Why do I go to sea? It's not just to collect the information. It's actually to experience the place. Every time I go to sea, I watch the acoustic records come in, and I see something different. I'm a phenomenologist; I have to be there to see what's going on. Every time I go, something new crops up. If I hadn't been there to actually witness it, I wouldn't be able to say something new came in. So, I have to be there—have to live in the place, have to experience it—to develop the kind of intuitive, artistic response to the science that needs to be done.

The idea of ecologists working in a terrestrial environment usually involves people who go out into the environment they are studying. They can sense it, they can hear it, they can see it, they can smell it, they feel it. They have lots of sensory inputs, awareness of the science they are doing. You walk into a meadow from a forest, you immediately can see the transformation that is taking place. You can design your experiments to try to elucidate what that transformation is quantitatively and what the processes are that give rise to the boundaries between the forest and the meadow. We've got none of that in the ocean. Yet, when I'm out there, standing on the deck of a ship, watching the waves, watching the birds, and looking at the sea surface, I'm actually looking into the sea, visualizing what I think is going on with my remote-sensing instrumentation helping me to do that. We see transformations, for example, on Georges Bank every bit as dramatic as the ones that you would sense going from the forest to the meadow. You go from the top of the bank, which is well mixed, where the water column is isothermal and isohaline all year round, across into an area that is stratified, and the distance traveled is about a 100 meters. That's about the boundary line from a forest to a meadow, maybe not quite as sharp and distinct, but sometimes it is. The difference between the biology on one side and the other is dramatic. It is not something we appreciated when we started this GLOBEC [Global Ocean Ecosystem Dynamics] project. That comprehension has come out of this very intense set of surveys.

Science is a very creative process. You don't just look at all the data and make some kind of wonderful pronouncement about what you have now seen. You have to have some idea of what's going on before you ever got there. You have to have some inkling of what it is you are going to trip over and what you are trying to explore. I don't know where that creative process comes from. It comes partly from training, but it also comes from curiosity, and it also comes from some kind of imagination because, especially in ocean science, you have to be imaginative. We can't see a lot of the phenomena. We have to imagine and then figure out a way to actually show what we imagine is taking place. We have to make that step.

As a country, we have largely turned our back on the open ocean. We are putting most of our resources into the coastal regions. The navy is doing this and NSF is doing this. But, my bet is that we will soon again find our attention focused on the open ocean, where it really belongs. We are going to need to live off the ocean in ways that we cannot live off it today. We are at the end of hunting and gathering. We have basically effectively utilized all of the available fisheries resources. There really isn't anything left to take. We are going to have to become much wiser in our use of these resources. We are going to have to rebuild them, which means we are going to get less in the future for a while. So what are we going to do? We have to go out there. I don't see that there is any question of whether it is going to happen; it's just a question of when. How soon before we can actually get the funding agencies to buy into this perspective and support it? Because they don't want to now. They are so under the gun to keep the current fleet of vessels going—in fact, they are laying them up—that they don't want to even be faced with another initiative that would cost them more money. So there is a lot of education to be done. I don't see a problem in getting the young people out to sea so much as I see a big problem in getting the funding agencies into the right framework so that they actually provide the facilities that would justify getting those people out to sea.

What we know is—and this comes out of a study that was done some years ago—that only 50% of the ocean is now available to study by oceanographers. Now, you say, what are you talking about? We went through and looked at the sea state worldwide. It turns out there is hardly a single boat that can operate beyond about sea state 4 or maybe 5. But you get sea states up to sea state 9! We don't really understand the physics or the biology above the sea states that we can work in. About 50% of the ocean is at sea state 5 or above.

At any given time?

Worldwide. We need to be able to study these sea states. They are tremendously important.

You have got to know the functional levels of biology from the molecular to the global level to be a real biologist.

JAMES T. CALLAHAN (1945–1999)

Program Director of Field Stations and Marine Laboratories
and of Biological Research Collections and Ecosystem Studies
National Science Foundation

Interviewed in his office 3 November 1997

Spending his entire career at NSF, Tom Callahan played an important role in ensuring support of both field biology and field facilities. He discusses the fact that biology (and every science) has a complex of content-context relationships at different levels of resolution. In biology, if one doesn't know something about the functional levels of biology from the molecular to the global levels, one is not a "real" biologist. Callahan concludes with his vision that biology will be the ascendant science in the twenty-first century. Unfortunately, due to his premature death, he will not witness the 50th anniversary of NSF.

My mother used to like to tell the story about the summer after I had turned three years old in December, I came into the kitchen one day with a snake gripped firmly behind the head, nearly as long as I was tall, asking her, "Is it a copperhead or a rattlesnake?" I've always been really interested in herps, reptiles, and amphibians, fascinated with them, and with insects. In fact, those are the kinds of things I'm working on even now, nearly 50 years after I started.

In the spring of 1972, I ended up being called to an interview for a job as assistant program director at NSF in a program that was then called Ecosystem Analysis. I went for the interview. Two weeks later they called and offered me the job. Here I've been ever since. The NSF Act, the Organic Act, was passed by Congress and signed by President [Harry] Truman in 1950. The first research grants actually went out the door in 1952, so NSF was functionally 20 years old when I started. I've been at NSF, well, more than half of its existence because I just passed my 25th anniversary, and I guess I'll be around until NSF turns 50. I should point out—at a level of significance, numerically and dollarwise—the first year of grants made from NSF contained grants in environmental biology and ecology, systematic biology and ecology. I've done the research back through all of those old records and annual reports. They were there from the very beginning.

I have some concerns. First, many schools are turning away from undergraduate biology preparation like we had in the 1960s, for instance, even though organism-type courses had been lopped off the curriculum at Stanford, we still needed to take a broad spectrum of courses like botany, zoology, cell physiology, etc.

That's kind of a shame, and we have noticed it. It has become especially prominent and seditious over the last decade. What I call the traditional "-ology" courses have really undergone a dangerous level of deletion from a lot of biology departments' repertoires: mammology, ornithology, herpetology, ichthyology, etc. I think that's an awful thing and to the degree that I can I certainly talk and act against that trend.

Another concern of mine is that some of the people who got their Ph.D.s in the late 1940s or early 1950s played a vital role on faculties in the 1960s and 1970s without doing a great deal of research. They were tremendous teachers and often provided vital background information, particularly their expertise in certain groups of organisms, to other, younger faculty.

I think that was probably more the general case rather than what would have been the academic opposite, that is to say, the highly productive in terms of papers and large grant research–driven. Now we've gone the other way, probably too much so. The traditional metaphor of academic credentials in departments was a three-legged stool: teaching, service, and research. The research leg of that stool got a whole lot longer than either the service or teaching. I would never denigrate the value of research, but what I want to see is the integration of research with the education and the service components.

All research does not have to be big-money research.

No, it doesn't. As a matter of fact, the stuff I'm doing now is done on nickels and dimes, begging and borrowing, stick-and-string ecology. I think there is a lot that can still be done in ecology, well publishable with stick-and-string approaches. The questions, the only constraints on questions, are how you think them. You've got to get out into the ecological systems and have the questions present themselves to you. You can't do it behind your desk or in a lab. You've got to get out and get dirty. Field stations and marine labs are the places to go out and get dirty.

If you don't go out and get dirty, lab results . . .

Are devoid of context. If you don't see, feel, and understand the critters working in their natural environments, how can you possibly make real meaning out of what you may be constrained to having to measure or observe in a lab?

The way I see these things in terms of disciplines, as disciplines. I only see one discipline. Biology. You've got to be a real biologist. You have got to know the functional levels of biology from the molecular to the global level to be a real biologist. Otherwise, you are a one-, or at most, a few-trick pony, and so far as I'm concerned, your utility to academic biology, to real biology, is very limited. If you are a one-trick pony to the degree of really knowing how to do PCR [polymerase chain reaction] and gene sequencing, you are less a scientist than you are a technologist. There is nothing wrong with being a technologist, so long as you admit that that's what you are.

To me, you are not really a biologist unless you understand how an organism functions. That's the whole. Levels above and levels below are important, but you need to relate them to the organism.

There are levels of resolution in biology, just as there are levels of resolution in virtually any other science. Each level of resolution has one immediately above it and one immediately below it and you have this complex of content-context relationships that runs up and down the levels of resolution. So any one level of resolution is devoid of context if it doesn't have an appreciation of knowledge of the levels above it. Those provide the context in which that level operates, and it is devoid of its best meaning if it doesn't have an appreciation of and knowledge of the levels below it because those are what provide the content that it operates on.

Biology, I think, is the most synthesis driven, or should be, of all scientific disciplines. Biological scales can be dealt with easily. Biology doesn't discern to the level of particle physics, with quarks, anti-quarks, and such, but biology is what gives context to particle physics, because otherwise why bother? It is a biological system that is perceiving particle physics.

I think there is a very good likelihood, as we approach the end of the decade and the century, that biology may be to the twenty-first century what physics has been to the twentieth century, in terms of the ascendancy among the sciences.

You really are a Pollyanna.

I really believe that this is a genuinely distinct possibility. It won't be purely organismal biology, nor should it be. But consider the prospects of biological materials, neuronal cells, for instance, becoming part of central processing units in some future generation of computers. It is almost bizarre, really, but certainly doable. Humans might be sent into extended missions in space (and I'm not talking about four months on a space station) to Mars, that kind of thing. Given current propulsion systems, it is probably a 12–14-year round trip to Mars. I really think there is a real possibility that biology will be the ascendant science in the twenty-first century, and we better get ready for it. I believe another aspect to its ascendancy in the twenty-first century is the way it has to do with ecosystems. I think we're seeing the front end of it right now. I believe that, at the corporate level, let alone the person level, it is very likely that vast fortunes are going to be made in work derivative of fundamental understanding of ecosystem processes, and that has to do with reclamation, restoration, and conservation. I think ecological restoration is going to be a major industry of the twenty-first century and that huge fortunes will be made.

But it won't take place unless we understand the processes.

That's right, and you can't do it without real biologists. Maybe I have my head in the clouds, but I hope I'm around in order to find out.

SUMMARY: CHANGING TIDES

Although it is recognized that students are better prepared today quantitatively than ever before, there is concern that students have been forced to specialize too early. Breadth in the undergraduate years has been sacrificed. For graduate students, the time and opportunity to make and thus learn from one's own mistakes is lacking. Today, many students go through graduate school as technicians, doing a specialized aspect of a professor's research and never having the opportunity to develop their own creativity. Exploratory opportunities, where a student can be exposed to new ideas, new techniques, and new sites are fewer today, both on shipboard and on land. Each scientist interviewed is still actively engaged in investigating some aspect of how the world works. The science they continue to produce is exciting today.

Women of this generation had the opportunity to become field scientists, yet it is clear from Koehl's experience that not everyone thought that "the field" was a woman's place. Fautin also mentions a revealing comment made to her by her major professor while she was in graduate school, "You are married to a rich doctor, so you don't really need a job." Interestingly, many of us from this generation did end up working our careers around a direction set by our husbands (some of whom also were and are scientists). Certainly many opportunities to do field research were made available to women; some are the senior faculty of today, making new discoveries and inspiring the next generation of scientists.

By combining new technologies and new approaches, by being creative and innovative, by having the opportunity to explore remote areas of the planet, the scientists of the *Sputnik* generation have built upon the sound observational and conceptual science produced by the generations before them. They have obviously loved the way they have been able to live their lives, combining the physical and intellectual challenges of their field research with the chance to teach others. Of great significance to their science and their lives, however, is that they admit they are humbled by nature. The awe, respect, and love they feel for the natural world is honestly revealed. *Sputnik* keyed a short-lived national commitment to science on which this generation was able to capitalize. However, what has sustained them throughout their careers (despite frustrations encountered) is their innate love of and respect for nature.

5

The Last of the Golden Years: The 1970s

It's true. You can look at the same organism, the same interaction for 20 years, and you see something new. It is really remarkable!

Drew Harvell

In their unfettered enthusiasm for unveiling the intrinsic order underlying natural processes, the scientists of the 1970s are most similar to the World War II generation. Both generations came of age during a time of great social upheaval and change. The World War II generation had to overcome the harsh realities of the Great Depression and World War II; in contrast, the scientists who tell their stories in this chapter grew up in the 1960s, when no rule was made that could not be broken. Tradition in academia, as in much of the rest of society, was torn asunder during their college years. At many colleges, including the nation's most prestigous, grades became meaningless. Distribution requirements, for breadth as well as for a major, were reduced or almost eliminated. Instead of conforming to societal expectations, undergrads of the 1970s did their own thing; they "found themselves." Although they attended a broad range of public and private institutions as undergrads, each found a small, academically focused group, which provided a nurturing atmosphere. It may have been a small department in a large state university, special field programs or field camps, or a job as a field assistant. As undergraduates they developed a sense of what they wanted to do and in many cases were directed to the experts in the fields of their choice. Choosing a graduate school was a fairly deliberate decision for this generation as contrasted to the naiveté of previous generations; each had extensive field experiences before deciding what to do, where to go, and with whom to work. It is interesting that modeling was often the first step in graduate training, and ideas and concepts about nature, rather than nature itself, often was the career-motivating force The first Earth Day was celebrated in 1970, making people cognizant of environmental problems looming on the horizon but also cultivating in the general public an interest in and respect for nature. Support for field education was significant; opportunities for field education and research available to students at the high school, college, and graduate levels are vividly portrayed. If a person in the 1970s generation developed a sense of academic discipline on his own, this could lead to innovative thinking. In the lives of these scientists, it translated into creating and carving out their own special niches, avoiding direct competition.

There is a lack of understanding of the use of the experiments that evolution has done to teach us about basic biological principles.

Margaret McFall-Ngai (b. 1951)

Associate Researcher, Kewalo Marine Laboratory
University of Hawaii
Interviewed at UC Berkeley 17 November 1999

Maggie McFall-Ngai and I were fellow grad students at UCLA at a time when students and faculty regularly discussed and argued science, but we also shared good times away from academics, playing intramural sports and becoming connoisseurs of the best Friday afternoon "happy hours" held in Westwood Village. McFall-Ngai admits that it was the order of nature, not love of the outdoors, which drew her to biology. Her philosophy that there are basic ways that animals can solve problems has led her to investigate fundamental issues, which have important medical implications, including vision and symbiosis. However, she illustrates that the specialist viewpoint prevailing in medical research and in biological funding, emphasizing cellular and subcellular techniques using laboratory animals or cell cultures, has blinded this community from taking advantage of the experiments that evolution has done to teach us about basic biological principles.

I have always thought that biologists are of two sorts. One are the naturalists. They love nature; when they take vacations, they go to where nature is. Then there are people who are interested in the order of biology and how it works. Not that those two things are mutually exclusive, of course. When I was a kid, I loved Latin because it had an intrinsic order to it that was very pleasing to me. Although I was always a very athletic person, I was never a person who wanted to go hiking in the outdoors. When I go on vacation, I want to go to an art museum. Another thing, as a kid, I was a bleeding-heart liberal. In 1969 I started off at a small liberal arts college as a sociology major, but early on I couldn't understand how people distinguished themselves in sociology. Everybody got As. I went over to the head of biology and said, "The other thing I really like is biology." He said, "You're coming over from sociology? What makes you think you could make it in this field?" I said, "At least there seem to be ways to distinguish yourself in the sciences." So, I switched majors fairly early on and became really interested in it, especially in invertebrates. I loved invertebrates.

I graduated and didn't really know what I was going to do. I cast about with hundreds of other people in the [San Francisco] Bay area looking for jobs in laboratories. I landed a job as a chemist for Pepsi Cola Bottling Company. For a year, there was no vacation. Half of my job was R&D [research and development] and the other half was as a technician. For an entire year, five days a week, I measured brix and pressure per square inch, incidence of carbonation on the inside of bottles. It was pretty boring. I said, "I can't do this for the rest of my life. This is going to drive me crazy."

Several of the other women I have interviewed have said the same thing: "I can't be a nine-to-fiver."

It is really interesting. In academia you work so hard, 60, 70, 80 hours a week, but the time is your own, except for specific teaching or committee requirements. It is much more flexible; the nine-to-five is just awful.

For graduate school, I got into Berkeley and UCLA. Berkeley had a very interesting attitude, from my viewpoint: "Prove to us that you deserve to be here." When I interviewed and talked to people at UCLA, they said, "Come and join us. We have a really exciting graduate program." I liked the idea of coming into a place that felt as though there was room to get better, there was excitement and a frontier attitude, rather than something that was completely formulated.

At UCLA, there was a guy who looked at invertebrate endocrinology. He principally worked on insects. In my interview, I told him I wanted to work on marine invertebrates. He said that would be fine, there was all kinds of interesting stuff. When I got down there, however, it was insects or nothing. I had a very hard time. I don't like entomology very much. I took a course in Cnidarian biology out on Catalina [Island]. There were only five students in the course, and it was taught by seven well-known people, including my future advisor. I loved it.

My advisor, JIM MORIN, said to the people in his lab, "You work on luminescence or nothing." Once again, I thought, I really don't like luminescence; it wasn't very interesting to me. But then I came across this group of fishes that nobody had ever studied, which had a circumesophagael light organ. It utilized a whole bunch of its other tissues to use the light in its behavior.

You said you "came across" it. Through the literature? through seminars?

Through literature. I started to cast about, looking for something I would be interested in that had luminescence.

In *The Naturalist*, E. O. Wilson said that you should search for areas of science that are not well populated. Get in there where you can stretch. That really resonates with me now. That book just came out a couple of years ago, but years ago I remember thinking that there are two possible ways (I seem to put everything into dichotomies!) one can be successful in science. You can be a really good bandwagon person, be extremely competitive, be a *Drosophila* biologist with big elbows, go into *Drosophila* vision research and the progression of the morphogenetic furrow, get on top of that heap, and just be king of the mountain. Or, you can be a pioneer and do something that nobody has done before; it is a risk.

I think women are particularly well suited for that because often in the other situation they are not allowed. For one reason or another, they have a hard time. I have a very good friend on the faculty at UC San Francisco. She was the graduate student of a person at UC Berkeley. She discovered something extremely exciting in

Drosophila biology. She started to work on it, and she presented it at a major meeting. Her former advisor and another one of his peers went up to her and said, "We are going to do this. Do you want to be second or third author? We'll offer that to you." The thing was that his lab was so big that he could take her idea, run circles around her, and publish it first. She was his "child"; he had been her major professor. I found that so impressive that that would happen in these competitive fields. It is really interesting, this idea of a pioneer, the desire to do something in a field that is sparsely populated.

I felt that way in Len's lab. I didn't want to be a Hydra *person; I wanted to carve a niche for myself. But I look at what I did, and I think, anybody could have done that.*

But they hadn't. It is like Rothko's art. People look and think, I could have done that. It is amazing; those simple things can be incredibly inspired.

This group of fishes I became interested in had only been described anatomically. The people who were interested in luminescence had looked at it, but not people who had any training in fish biology. So, I went down to Scripps Institution of Oceanography and took Dick Rosenblatt's ichthyology course, which was fabulous. He is a great teacher.

You know that Rosenblatt is a descendant of Louis Agassiz, through David Starr Jordan, Carl Hubbs, and Boyd Walker?

I think that is such cool stuff. I was Jim Morin's student, Jim was Woody Hastings's student, Woody was E. Newton Harvey's student, and E. Newton Harvey was Thomas Hunt Morgan's student. It is really interesting how it traces back.

One of the things that is hard about field biology is that things can go so wrong. I was interested in working on this group of fishes. Jim Morin said, "I have a whole bunch of friends in the Orient. Just write to them, and they'll send you some fish." Now, is that naive or what? My feeling about field biology is that you never rely on anybody to send you anything. You always go there. It is really important to look at its habitat and think about its biology as you are working along. For two years, I keep writing to these people. Nobody writes back. A few say they'll do something, but they never do. There was a Filipino fish market in town [Los Angeles], on Culver and Sawtelle. I went down there and bought bags of the fish I wanted to work on, and I described their morphology. I decided on the questions I wanted to ask when I got a chance to go to the Philippines.

The center of the distribution of the family that this fish was in, the leiognathids, was the Philippines. Dumaguete University was in the central Philippines. Jim Morin had been there the year before and found out that these fish could be collected from Bais Bay. Jim had identified a fisherman who would collect them for us, and we could go out with him. This was about 45 kilometers from the university. On my first field trip there, I only had three weeks, and I wanted to scope things out. I had been writing to people at the university, and I had told them that I needed to get down to Bais Bay. To get from Manila to Dumaguete was complicated. I had to take a boat. It was two days. I finally got to Dumaguete, but then I had to get this additional 45 kilometers. I was told before I got there that we could take the marine station's jeep down. When I got there . . .

The jeep didn't work.

No, the jeep worked, but it wasn't available to me. When I had written to them, I had been very specific that I needed to go down on a certain day. When I arrived, they said, "It won't be available for a week and a half." I thought, This is really frustrating. Why am I here? For the next couple of days, I would walk down to the station, and the jeep would be sitting there. Why wouldn't they allow me to take the jeep? I said, "I don't care what the problem is, I need to get this done." So, I walked into town, walked into a bar, and sat down. I said to the bartender, "Forty dollars to anyone who will drive me to Bais." That, to them, is $400. Just like that, I had a car. I told them that I was going to go back and get my gear, and I'd be back within the hour. This bar was located three blocks from the administration building at the university. As I walked back onto the campus, it was all over the university that I had done this! The vice president of the university, who was the head of the marine lab, Dr. Angel Alcala, met me on the road. He said, "What happened? Of course we have a jeep for you. It is right here. Anytime you want to go." It was such a weird situation. I never quite figured that out, but I was able to get to the field.

In Bais, I met the fisherman, Gaspar Rodriguez. It was fascinating. He lived out on the bay in this little town, on this little spit of land. It was actually an island off the mainland portion. There was no fresh water, and there were a huge number of families. You'd park the jeep and be really quiet, but within a few minutes you'd have 100 children around you. Gaspar Rodriguez had this beautiful nipa hut overlooking Bais Bay. Of course, he didn't have shoes; his kids didn't have shoes. They were very poor; but he and I would go out on bonkas, outrigger canoes, at night, fishing for the animals. I'd gather them up in trash cans with aerators and then I'd drive back to the station, where I worked on them in aquaria. I was really impressed with what a great education the common person in the Philippines had. That year, I got home and wrote a paper on the work I had done, which I published in *Marine Biology*. In the paper I acknowledged the help of Gaspar Rodriguez. The next year I brought him a reprint. On my next visit, he said, "I loved the paper!" He had read it. I was really surprised. Every spring quarter for my last three years in graduate school at UCLA, I went down there to work in the field.

I worked on the functional morphology of these fishes and determined how the gas bladder became a reflector. I became really interested in the fact that the ventral musculature is almost transparent. Transparency is one of the principal camouflaging mechanisms in the marine environment.

STEVE WAINWRIGHT *worked on transparency in chaetognaths during his postdoc at Woods Hole, but when I asked him for a reprint, he said, "Unfortunately, I never published on it."*

That's right. I gave a talk in a symposium at the ASZ [American Society of Zoologists] meeting on crypsis in the marine environment and talked about the phenomenon of transparency. I talked about how in a three-dimensional homogeneous environment, you have few options. You can be transparent, you can be reflective, or you can have counterillumination. It depends on the intensity, quality, and predictability of light and so forth. After this talk, Steve Wainwright came up to me and said, "Wow. I wanted to do that as a postdoc. It is such a neat thing."

I became really fascinated by the fact that a very large number of organisms in the marine environment can become transparent. There are little surgeon fishes, the acanthurid larvae, which are totally transparent. They come in to settle on a reef, and within 24 hours they have lost their transparency and taken on the color of a little surgeon fish. Little flounders, leptocephalis larvae, lots of invertebrates. You can be thin, you can be watery—there are all these biophysical ways—you can have very few pigments, so I explored this in a theoretical paper I wrote on transparency, reflectivity, and luminescence, when they are used and how they are used. I became interested in transparency, and I thought, there is a niche there. Bill Hamner worked on krill in the Antarctic. When they become parasitized, they lose their transparency in their tails and then they get picked off by a predator. I knew that when animals become physiologically compromised, they lose their transparency, so it is likely to be an active process that is involved in the maintenance of transparency. So, there would be processes that had evolved in the development of the organism, which would allow it to maintain transparency through development to a certain point, and then there had to be physiological mechanisms to maintain it.

The best understood biophysical system was the eye lens. So, I took a postdoc at Jules Stein Eye Institute at UCLA to work with a card-carrying protein biophysicist, Joe Horwitz. Joe had never taken a biology course in his life. I started in his lab in November. By February, I had him out on the research vessel *Velero* with Jim Childress on a five-day cruise collecting deep-sea fish because I had started reading the literature and found that if you take a lens from a rat and put it in the refrigerator, it goes opaque. Now, if you take a lens from an Antarctic fish, it doesn't do that. There had to be adaptations.

The proteins of the eye lens never turn over through the life history of an organism, so that the oldest proteins in your body are in the center of your eye lenses. So, they had to have special properties that allowed them to maintain the function of the lens in a variety of habitats, particularly temperature regimes. I did a study in Joe's lab, comparing in-groups and the out-groups of the major vertebrate classes, looking at the transparency of the eye lens and how it is maintained in animals that live at 0° C all the way to the lizard *Dipsosaurus dorsalis*, which lives at 43° C. It turns out that the stability of the lens is directly related to the preferred body temperature of the animal. It is really neat. To do those studies, I had to go up to the High Sierras and collect the Sierra yellow-leg frog.

Didn't someone in Bart's [George Bartholomew] lab [at UCLA] work on the ecology of that?

Yes, but now it is almost extinct. I went to the Red Sea and collected a little rockfish.

Because it is so hot?

Yes. You had to choose your organisms in such a way that you divorced phylogeny from it. You had to look at two very closely related organisms, which had very different temperature regimes, and a distantly related one that had the same preferred body temperature as one of them. I had to go all over to collect these things, which was so much fun. What was most fun was that Joe Horwitz, this protein biophysicist, whose Ph.D. had been on the effect of heat on the circular dichroic properties of tyrosine, was so open to field biology.

He came up through a physics or chemistry background and then got into biophysics?

Exactly. It was amazing.

That whole idea is being lost now, that if you look across animal evolution, there are basic ways that animals solve problems. You can learn from the ways animals solve problems to do practical applications to human biology. Remember when pre-meds used to have to take comparative anatomy?

Yes. It is the most amazing thing that the whole medical community doesn't catch on to this. I'm in the middle of Jules Stein [Eye Institute], and I walk in to look at the most stable proteins in the human body, and no comparative biologist has ever touched the system. I got up at the Association for Research in Vision and Ophthalmology meetings, half basic scientists and half clinicians. The basic scientists are almost entirely cell biologists or molecular biologists figuring out how the retina works or how the lens works. I presented this work on the in-groups and the out-groups of the major vertebrate classes. People came up to me afterward and asked, how did I know these animals? How did I know how to construct this question? They were absolutely fascinated.

While I was at Jules Stein, I became aware of how valuable the comparative approach can be. When I first came out of Jim Morin's lab at UCLA, before I had done any postdoc, I went for two job interviews, one in Georgia and one in New Hampshire. I was offered both jobs, but when I talked with people on my committee, they said, "Don't take those jobs. You can do better than that." That was a very tough time. Turning down jobs was scary. I had a friend who was so smart and such a good biologist, and I'd think, she doesn't have a job. This was really scary. I thought that, by working with Joe Horwitz, I would increase my chances of getting a job because I would be integrating biochemistry into comparative approaches. Wrong! Joe Horwitz was a biomedical person, so he was perceived as an outsider. People were afraid that I would come into a biology department and do biomedical research, and this would not be acceptable. I couldn't understand why, after having been in his lab a couple of years, I wasn't getting any interviews. I went back to my Ph.D. committee and said, "What's happening?" They said, "You've got to get back into the fold." So, I went and did a postdoc with George Somero at Scripps on the biochemical basis of thermocompensatory changes in the lactate dehydrogenase molecule in barracuda species.

When I got my job at the University of Southern California, I decided that I was going to wing walk; I was going to do two things. First, while I was at Jules Stein, I had realized that no comparative biochemist had ever touched rhodopsin, the visual pigment. BILL MCFARLAND and Ellis Loew were doing physiology, which provided a great background for what I was envisioning. Second, while I was a graduate student, I was frustrated by the fact that I couldn't raise the animals and ask, "How do the host fish and symbiont [bacteria] get together?" I realized that a certain species of squid that I had heard about in Hawaii was not only the *only* animal–luminous bacterial association with which you could do that, it was also the *only* animal-bacterial association that you could do that with. So, I decided to wing walk and do both: rhodopsin and symbiosis.

The rhodopsin was a field project. Jeremy Nathans, a very famous biologist out of [David] Hogness's lab at Stanford, was now at Johns Hopkins University and asking the basic question about spectral tuning: How is it that rhodopsin absorbs at the wavelength of 505 nanometers? Their approach was to take the 350 amino acids of the rhodopsin molecule.

And start substituting?

Basically, yes. Now, being a student of Bill McFarland's work and Boyd Walker and Ellis Loew, they have shown that the marine habitat is the only place where rhodopsin has responded over evolutionary time to the difference in wavelengths of ambient light with depth. If you want to take a comparative approach, and you want to use these fishes, what you want to do is find a closely related group of fishes that spectrally tune. In Hawaii, the squirrel fishes spectrally tune their rhodopsin. The surface fishes' rhodopsin absorbs at 505 nanometers, the next group deeper in the water column is at 500 nanometers, the next deeper group at 495 nanometers, the next at 490 nanometers, and so forth. Bill McFarland had shown that this spectral tuning correlated with the habitat. The fossil record on these fishes and various other indications had shown that they had diverged in the last 10–60 million years. We calculated that the theoretical number of substitutions in the rhodopsin molecule due to neutral processes would be somewhere between 3 and 12, depending upon the species. So you would be able to identify candidate amino acids that are likely to be involved in spectral tuning because you only have three changes, and those three changes must be the ones involved in spectral tuning. Then you can do your site-specific mutagenesis. You know the molecule is functional and you know which amino acids are responsible.

I began to work on that and found out a lot of interesting things. This was an instance where the comparative approach—and knowing and understanding the field biology of organisms—was a great complement to what people were doing in basic cell biology. When the physical chemist who had developed the point charge model I mentioned earlier heard me give a seminar on this, he had the same reaction as the cell biologists had had at the vision meeting: "How did you know about these fishes?" I said, "I read Bill McFarland's work." There is this disconnect between people that I find fascinating.

But a little frustrating from the point that there is all this comparative organism–level biology, which helps contribute to these answers, but if people don't take it seriously for itself, the work isn't going to get done.

Absolutely. In that regard, I put in a grant to the NIH to do this work. Steve Fisher, who is a comparative retinal biologist at UC Santa Barbara, was the head of the study section. I was given six percentage points above the fundable range. I got a call from the administrative officer, who said, "We have a big problem with program relevance. This is not human research." I said, "This is your only hope in hell of finding out how rhodopsin spectrally tunes as far as I can see. You've got to give the cell biologists some candidate amino acids, or you are going to spend buckets of money, and you'll never know the answer." But, I wasn't funded the first go-round. I was visiting Joe Horwitz at Jules Stein, and one of the people in the study section came up and congratulated me on getting the grant. I said, "I didn't get the grant because this comparative approach was not appreciated."

The study section was blown away that I wasn't funded. The next go-round, I made the disease arguments also [there had been some recent research results on disease that involved amino acid substitution]. The study section was so appalled that I hadn't been funded on the first go-round that they gave me a priority one for funding. I got a call from the program officer saying, "We have a problem with program relevance." I said, "The study section is telling you that I have a priority one. What more do you want?" She said, "I'm sorry. It is just that you are working on fish." I was frustrated, so I said, "I'm not doing a good job at explaining the rationale to you, obviously. Otherwise it would be abundantly clear why this approach is a valid one."

I was at a meeting and I went up to Matt LaVail, a professor at UC San Francisco, and I said, "I need help from somebody who knows how to approach the NIH." I told him my story, and he said, "I will hold your hand and make sure you get a fair shake. What you have to do is write a letter saying that you respectfully request to go before counsel." In other words, you want this to be aired in front of the scientific community. The program officer called me back and was horrified that I had requested to go before counsel. She didn't know that I knew I could do that; in fact, I hadn't known before I had talked with Matt; it is like an appeals process. I got the grant, but they cut my budget by two-thirds. They did what they could. You are absolutely right. There is a lack of sensitivity as to the value of the comparative approach. So, I dropped the rhodopsin stuff because the squid work started to get so exciting, and the squid work was getting supported.

You try to tell people that all animals have essential associations with microorganisms, but we just haven't been able to study it. The reason is because it is often a consortium, so it is extremely difficult to figure out what one microorganism is doing against a background of a whole bunch.

There would be all these competitive interactions.

Yes. It's a community. So, you want to get something really simple: one host, one microbe, which is what this squid-bacterium system that we work on is. Unlike the cases of the insect symbionts or the hydrothermal vent symbionts, in which the bacterium can't be raised and so no molecular genetics is available, this was a neat system with which you can do microbial genetics, and it is the only one I know of. I have had situations where people have been sitting there listening to me explain to them the power of this particular system. One guy said to me, "This is great. Maybe you can get it in a mouse model." I stood there dumbfounded. I didn't know what to say. I'd just explained why this couldn't be done in a mouse, because the mouse naturally has a community of microbes. There is a disconnect.

There is a lack of understanding of the use of the experiments that evolution has done to teach us about basic biological principles. I'm hoping that people can become more open to field biology. The evolution of developmental mechanisms is one area where evolution and developmental biology have come together. My viewpoint is that developmental biologists have focused on looking at how the cells of an individual organism talk to each other. If you go back to the origin of the metazoa and the Cambrian explosion after that or the diversification of body plans, you see that all of the body plans arose in a microbe-rich environment. These microbes are the most metabolically clever organisms that there are. They just know how to make everything. Are you going to fashion your developmental mechanisms in the absence of the incorporation of some subset of these clever microorganisms in the environment? There is every indication that we did incorporate them. For my money, I think that the interface between the microbes, historically, and the animals is going to be a central theme. Try to get the developmental biologists to listen to that. Forget it. They are not there. Looking at the environment and the motive forces behind the selection for these various body plans is something they just don't think about. It is an overload.

Ned [Ruby, Maggie's husband and the head of the microbial genetics lab studying the *Vibrio* bacterium, symbiotic with the squid, whose biology Maggie's lab works on] and I were talking about the future. I said, "We've got 15–20 more years to do this." Where do I want to have gotten at that point? It is a scary thing, because I always probably put it higher than I will have gotten by that point. One of the most fun things is to be able to work on a symbiosis with someone who is an expert on one side of the symbiosis, so that you can just focus on being an expert on the other side. To be a good microbial geneticist, boy, is that tough. They look at the morphological and biochemical work we do and are puzzled. We never have lab meetings together, because the research is too different; we don't understand one another. Once a year we have a powwow, we call it, we get together and talk at a general level about the significance of our findings and the broad outlines of our work. That is one of the great parts of this adventure, to be able to work closely with another lab and to try to make each aspect of it as rigorous as possible.

Nature does not respect scientific disciplines, yet we need to pigeonhole people to fund their research.

John Dacey (b. 1952)

Associate Scientist, Department of Biology
Woods Hole Oceanographic Institution
Interviewed in his office 19 December 1997

John Dacey grew up steeped in an academic philosophy from early childhood as son of a physical chemist who directed the academic program at Canada's Royal Military College. This scholarly tradition translates today into his concerns about the way science is currently practiced. Bureaucracies place growing demands on the time and creative energy of the individual scientist. This selective force changes the process for academic advancement and rewards skills in empire building over individual scientific initiative and creativity. A firm believer that science advances through the recognition of the rare event, Dacey worries that scientists and research institutions are becoming defined by the funding sources.

I'm very keen on the notion of the rare event. In oceanography, lots of people are aware of this. You can't measure things when things are happening, because you are getting seasick or you are not even there. A lot of things only happen in moments that you don't see necessarily by your sampling. If you are averaging, you don't see them. So the question of taking the ocean apart in a way and trying to understand the underlying processes from real patchy measurements of things, like concentrations of things, is frustrating.

It's interesting to me to come to the ocean from very much an experimental point of view. I was, more than I have probably been lately, much more of a field experimental scientist. For a lot of that research, I was standing wet in a pond, sticking things in, and watching things come out of plants. That is real field experimentation, I think you would agree. You have treatments, and in the ocean often you can't do that. Some people actually use tracers in the ocean. There are some big programs, SF_6 [sulfur hexafluoride], for example. Recently, in a lunch seminar, Jim Ledwell described a big SF_6–based experiment to look at how deep water moves in the ocean. That is field experimentation in the ocean. Normally it's hard to do. I am taking a narrower approach, thinking about gas exchange on shorter time scales.

My father was a scientist and very much my model. I always wanted to be a scientist. I can't remember wanting to be a fireman or a hockey player or anything else. I saw myself as a professor but here at WHOI, I am not what I thought I'd be. I had some kind of Oxford model in mind, I think. It really came to the surface when I visited Oberlin one time to give a talk. It struck me, as I walked across the campus kicking leaves, that this is the type of place where I always thought I would be, the professor in the tweed jacket, smoking a pipe.

I grew up on the grounds of the Royal Military College in Canada. When I was a kid, my dad was a professor there, then chairman of the chemistry department, and ultimately, while I was in high school, he became principal. The commandant is in uniform and is responsible for all the military affairs and overall administration. The principal is a civilian, responsible for the academic affairs. I started at Queen's University in Kingston, my home town. It's a good chemistry department but this was the late 1960s and early 1970s, and chemistry wasn't as "relevant" environmentally —whatever that meant—and biology was more so. When I was in high school, my brother-in-law Gary Sprules, the husband of my eldest sister, who is eight years older than I, was a Ph.D. student at Princeton with Robert MacArthur. Gary had a big influence on me, because here was somebody who was younger than my father, and he was interested in animals and ponds in Colorado. I ended up with a combined honors degree in biology and chemistry, getting a B.Sc. from King's College, Dalhousie University.

I remember as a student, I had a very idealistic view that I was going to be a renaissance man. I find now I'm jammed more into a hole than anyone. My whole purpose, all the way through, and this started in college, has been to not compete. I don't mind winning, but I want to win in a game that nobody has ever heard of. Working outside the pigeonholes, you set your level; you don't have to worry about some other bastard being a jerk, because you've defined your own rules, and you work within your own designs. I want that. It's seeking creativity in new directions, I think. I'm very worried about science not allowing that any more. We don't give people the time to define themselves. The constraints of funding now force us to define ourselves increasingly within the bureaucracy of government, within the paradigm of some panel of people who may know you or not.

I think the fact that I have never had a full-time technician and that I've had only one graduate student in my career at ostensibly one of the best research institutions in the world is a bit weird. There's something wrong with that. The reason that's true is because I'm not willing to spend the time to raise the money to do it. That's a larger, cultural issue.

You can do your work without it?

Ah, but I don't do enough of it. However, I used to say, and I think there is some truth in this, you can't hand the problem off. I'm thinking about technicians now. It is very difficult to find the person who's going to recognize the twist in the data that is important.

To me, the measurements that are important are the ones that don't fit, because they are either telling you that you're completely wrong or that there is something else here that is important.

Now it's hard for me to see what's down the road. What is a conventional education any more? I think universities are important places in a sense, but what are they going to do? It has to do with philosophy. I have the image of the universities being

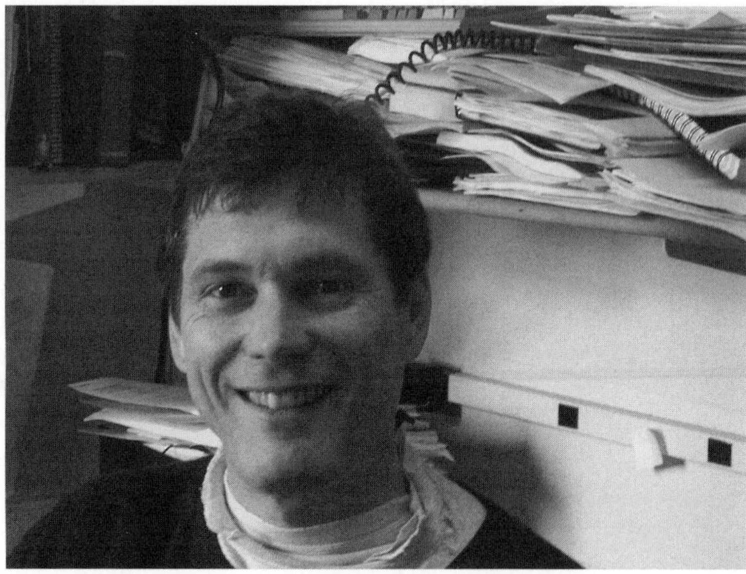

In 2002, John Dacey discusses his frustration with the current tendency to pigeonhole people into narrow scientific subdisciplines. Photo by E. Gladfelter.

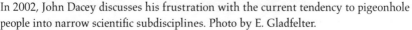

the guardians of the library in a sense: the gatekeepers to knowledge. The monks with hoods, professors in gowns, became the university. Everything was built around the books. But now, in the electronic world, books and information are completely distributed and, perhaps more important, instantly distributed, by the internet. What is the future of universities?

It's hard to say. Another thing I find disturbing about universities today is the avoidance of teaching by professors because research is considered more important.

Is that because the laurels go to the publications and not to the teaching? That's driven by a different gene in a way.

I remember my father frequently saying how rewarding it was for him to have people come back and tell him what he'd done for them. In fact, I recently received email from a woman who must have found my name on the internet. She was a military kid at the RMC. She wrote me and said, "Pardon me, but are you the John Dacey whose father was John? It was because of your dad that I took chemistry." Now she works as a chemist for the pulp and paper industry in Quebec. I hadn't really thought about it that much, but definitely, where we lived, my father was the senior intellectual. I heard stories about how many different people he would encourage to get a degree. He makes me think we pay too much attention nowadays to the single dimension of an academic's success—the publication record. Where is the value on the whole man as humanist?

Richard Hill, a long-time colleague of mine, is an excellent example of dedication to teaching, and he puts an awful lot of energy into it. I met him in 1973 or 1974 while in graduate school taking his animal physiological ecology course. Dick was not on my committee, but it was because of the strength of the academic connection as a professor and that sort of relationship that our collaboration together has lasted through the years. He worked on animal energetics, and he now has difficulty with animal rights activists. Now it is my good fortune that he enjoys working with me on marine problems. Hill, though, takes his teaching extremely seriously. He applied for some jobs and had an offer from Bowdoin, a premier college in Maine. He said it was overwhelming how much enthusiasm they expressed for him at Michigan State when he went back with that job offer. It is horrible that you have to go with some sort of threat before anyone takes you seriously. State colleges, probably universities everywhere, are disaster areas for the wheeling and dealing in power and politics. Dick said, "If I were 35 now, I'd go crazy, but I'm 55 and I can survive." It seems to be a common expression. Accomplished people are happy to leave the academic world.

But another thing, of course, is money. Universities want people to go out and bring in money. It is the whole business of how research is done. The whole country suffers. It's good for somebody who teaches to have some kind of understanding of research.

And maybe get a paper out now and again.

Exactly, a good paper, though, not just to count. But this is all people can do: count. Nobody can read; there's no time to read, if you are out trying to write these things that nobody is going to read.

Although there is plenty of money for research, there is a problem with the distribution. The people who can do the research are not necessarily given enough money, and it is being spread around for political reasons. The balance and the whole academic structure in science today is based on getting graduate students. At most colleges, the pressure is to generate the Ph.D. students.

Of course, what's driving it is grant overhead; your problem here at WHOI is that you don't have technicians. If you had technicians, you'd be bringing in overhead on their salaries plus on your salary. That is why people are probably disappointed that you don't have postdocs in your lab.

Of course. The WHOI administration would like a lab filled with people, and I have a lab filled with equipment!

One of my enjoyments now of not being in a university is that I can sit back and snipe at them a little bit. I think that most of our woes have been brought on by university faculty not having convictions and the courage to say, no. Basically, to stand up and say, "This is right and this is wrong."

It is attractive to get more money, more people, and more power. So you change the nature of the institution.

Well, you design the institution. I sat last year on the Strategic Planning Committee for Science and Engineering, and we produced a document, which has vanished. It was to have been a five-year plan that the WHOI director, Bob Gagosian, could present to the trustees. It was a good exercise, but two of the members got on the whole thing that strategic planning for science is an oxymoron. You can't plan, because if you plan you might tie yourself to something that might sink, and you need flexibility. Of course, this is true. Nobody disagreed that you want some flexibility, but you've also got to make some investments.

Research proceeds because of the creative nature of individuals.

And risk taking.

Yes, risk taking of individuals, but individuals cannot do science without some support. And the institution's role is . . .

Right, exactly, you've touched it. You are allowing me to integrate this in a way that I hadn't.

The money should be raised—I won't argue that we don't need money—but the money should go into pools to support internal competitive proposals. If we are the "best" oceanographic institution in the world, we should be funding our own research. We shouldn't be paying people to sit here and generate funding [by writing endless grant proposals], we should be funding them to actually do research. We should be allowing people to say, "I've got a really brilliant idea. I want to go and do it." Rather than saying, sit there, think up this idea, develop it, and then write a proposal to Washington. That bureaucracy is extremely conservative and very slow.

In terms of this academic business as a vacuum, when I first came here, I saw the last few of the Monday night journal clubs. These were evening seminars on topics of broad scientific interest. People came back to Woods Hole at 8 P.M. for the seminar and stayed afterward to talk and drink beer. There was a kind of open, optimistic exchange of ideas. That would have been in 1979. I'll bet you by 1980 they weren't doing those any more. Now people stay at home. They are exhausted by the day's events.

My move from the freshwater wetlands to oceanography occurred by passing through the salt marsh. I worked on several questions in the biogeochemistry of salt marshes, including the role of plants in water turnover and carbon cycling, and I ultimately became interested in the emission of volatile sulfur species. Dimethylsulfide [DMS] is the major sulfur gas emitted from marine systems, and it is known to have potentially important consequences in the atmosphere. Its oxidation products tend to facilitate the development of aerosols and cloud nucleation particles in the atmosphere and as a result tend to represent a cooling mechanism by reflecting sunlight before it can reach the Earth's surface. My experience with the emission of DMS by marsh grasses indicated that cell disruption increased the formation of DMS, and I translated that concept into my first paper on truly marine issues, demonstrating that zooplankton grazing on phytoplankton is important in the dynamics of this gas in ocean water.

Since that time, there has been an explosion of research on the biological and chemical mechanisms of turnover of DMS and its precursor, DMSP, in sea water. I have kept an interest in those problems, but I have also moved back toward my early interest in bringing the biological and physical worlds together. In this case, I am using the flux of a biogenic gas, DMS, from the ocean as a way to understand the physical processes of gas exchange. There is great interest in gas exchange driven largely by the CO_2 budget of the atmosphere. The concentration gradient for CO_2 between the atmosphere and the ocean surface can be measured with great precision, but the *rate* at which CO_2 exchanges between the phases is not known within a factor of about two. I have been working with others at WHOI on a micrometeorological method for estimating DMS flux. The rate of DMS flux itself is important, but that flux can also be used to estimate the flux of other gases (such as CO_2) by estimating a more basic physical parameter of the sea surface, the gas transfer velocity. I like to say that by studying the behavior of DMS in the lower 10 meters or so of the atmosphere, we can learn about the physics of the upper 100 micrometers of the ocean.

I got into oceanography because it is inherently interdisciplinary, and people seem to understand that. Here I am in a biology department. I'm interested in gases, and I'm going to do gas in the atmosphere and sea surface exchange.

It sounds more like chemistry, physical oceanography, or physics.

Nature does not respect scientific disciplines, yet we need to pigeonhole people to fund their research. Program managers need domains to administer. Physical oceanography is not 100 microns at the sea surface, it is this big circulation. Chemistry is not these exchange processes, it is a process that affects what is in the water. Similar reasoning in atmospheric chemistry. So work at the interface (physically and metaphorically) has become too interdisciplinary for an interdisciplinary field. For me, it's become the source of a little bit of anxiety because you still go to panels. Program managers have a horrible job. It's not to criticize any of them; it's the bureaucracy that is the problem. They have to pick people to sit on the peer review panels, and they pick people who don't understand this stuff and they can't. Who understands it? If we are the experts, who the hell understands it?

To me the worst thing we could do at the Oceanographic [WHOI] is make ourselves look like a university. We are competing with universities. We shouldn't try to make ourselves look like them and be better than them. We should try and make ourselves look different. In the age of electronic education (not something I look forward to, by the way) what the hell are universities going to become; why look like them?

What are our strengths? How can we build on them? How can we market them as an institution?

We are marketing information and knowledge.

And innovative ways to do things.

Yes. I remember as a child hearing my father say that universities were making a big mistake by depending on the government too much. The universities in Canada, like they are here, were always in desperate need of money. They came to depend on the government. Once you depend on the government, you are ruled by it, because you fall subject to its whims. The notion of academic excellence is way the hell down on the list of other responsibilities.

I do want it to be on tape about this rhetorical business we always hear about scientists "standing on the shoulders of giants," because we certainly are, but many of us are also busy kicking those giants in the sides of the head. That's my modern metaphor for people who stand up and ignore what went before or are critical of it without placing the work in the context of the field at the time. I mean, it's mad not to acknowledge how important the stuff was that went before. It's terrible. I hate it.

You refer back to the work of the 1800s.

I try to do the best I can. I struggled with Old German, which is awful. It's important to understand how any academic field evolved, but of course there is only so much time you can spend on it. But it is the responsibility of the individual scientist to be honest about the relevant things. In fact, I thought what I was doing in my thesis research was novel until I found a lot of it done in a primitive sort of way 150 years earlier. I found the references either by reading *Hydrobotanik*, which is a 1959 two-volume set by [Fritz] Gessner and awfully difficult reading, or by reading a book about aquatic botany by Agnes Arber published in 1920. She said something like "this is an interesting topic and it needs to be studied some more." That was the last reference to anything related to my study in the English language until my dissertation. The time that people could take to write back then was more generous. Agnes Arber's book is quite readable. Here is a book by Darcy Thompson [*On Growth and Form*] that was given to me by my brother-in-law Gary Sprules when I was a high school senior. People don't read things and appreciate what they are reading about science for the writing itself. There is this overevaluation of data and not enough [appreciation] of synthesis in some way, but also I think of the expressiveness. What we have now is the visual world. If you can present your data visually, then you've got it made.

In geology field camp, students have to face the absolute terror of knowing there is no answer in the back of the book, and they have to participate.

RICHARD ALLEY (b. 1957)

Professor of Geosciences
Pennsylvania State University
Interviewed at WHOI 10 March 2000

It was clear from Richard Alley's seminar, given in a spirited, humorous, and thought-provoking style, that here was a man who was passionate about his work. He and others have used ice core data to determine climate change over the past 150,000 years. He explains that he is but one small part of the ice core community, a collaborative "big science" effort that has proved to be a success in research, administrative, and sociological terms. Before his time, the U.S. ice core program had collapsed; he describes how it was resurrected and redefined. Alley emphatically states that field camp is *the* course for a geology student, yet lack of administrative support threatens their existence, and they have already become much too short in duration.

I've been interested in rocks since further back than I can tell. I eventually became involved in a rock and mineral society called the Rolling Stones. It was basically run by one woman, Ann Kramer. She'd take her family and anybody she could get, and we did all the places around Ohio. We'd take big swings around the western states and come back with big petrified logs. We did it on the cheap, camping. My parents' house is still full of my rocks, along with old rock and mineral polishing pieces and the tumbler. Ultimately, I loved writing, I loved journalism, and I loved geology, but the geology won.

After my freshman year at Ohio State, Ian Whillans, the glaciologist, gave me a wonderful summer job, which was very nice of him, doing computer and laboratory work studying ice cores. I was there every day. There was a really good Antarctic geologist, David Elliot, who was director of the Polar Studies Group at Ohio State. He would walk by each day and see me working. He ended up needing a field assistant for an Antarctic trip. He was collaborating with Doyle Watts. Doyle was doing paleomagnetism in the polar rock cores. We were working in the Antarctic peninsula, which is warm in the summer, so there is no water. The drill for taking rock cores was a converted chainsaw motor. It had a rotary bit with diamonds, and it was water cooled. It needed someone to take the Sears pump cans for doing sprays in your garden and to pump it so the drill would be cool.

That was your job?

Yes. So, I got to go to Antarctica in my sophomore year to pump water.

Have you been to the peninsula? It is the most glorious place. The mountains rise right out of the sea many thousand feet with glaciers dripping off them. There are whales and penguins and seals. We just went and did geology. A couple of the islands we'd go in on, and David would say, "I don't have a record of people visiting this island before." We were studying the tectonic history of the area from the sediments. We went ashore at Hope Bay, Esperanza at the tip of the peninsula, and we worked on Mt. Flora there. We worked off the research vessel *Hero*. She was an old boat. She leaked a little; she rolled a lot, but she could get in close. Then you'd zodiac in. The ship had a ladder hanging over the side. Even if you got cold, you still had to climb the ladder to get back in the ship. It was a wonderful experience.

I got away from geology eventually. I'm doing ice now, not rocks. But David was extremely good. He has seen everything and done everything, and he always gets it right. In fact, I had him in field camp later. Geology at Ohio State always required a good field camp and that was mostly where they sandwiched in the field. I did field camp and then I TA'd field camp. The truth is I did better as a TA in terms of learning the stuff than I did as a student. That was a wonderful experience. It was in central Utah and it was long, eight weeks.

Now they keep making field camps shorter and shorter because the students have to get home to get a summer job to pay for tuition. Field camps are not really what they used to be. I've always believed that, for a geology student, the field camp is *the* course. It is almost the only class, because it is the point where you ultimately put students on an outcrop, give them a base map, and say, "Come back in a week with the geology." They have to face the absolute terror of knowing there is no answer in the back of the book, and they have to participate. A lot of them come back and can't do it. They may become a go-for for someone, but they are never going to be a geologist. The ones who can come back and say, "This is what is there, and this is how I interpret it," they're the ones who are going to become geologists.

Do you think there are a lot of undergraduates getting that today?

Hmm. I hope they are. At Penn State, I still think we do a reasonably good job, because we have a couple of professors who are really dedicated to it, but we've made it very hard for them because we have shortened the camps to a few weeks. We don't own our own vehicles any more; we have to get them from fleet services. Those dedicated professors are still doing the right job with the students, but it is by dint of their own motivation. The geology majors have to take the field camp, but the department has other majors, and they don't.

I had been working on ice all the way through with Ian Whillans, but I had decided that there was no way that anyone could get a job studying ice. I started on my master's degree, and I said, "Ian, you've been really nice to me, but there is no future in glaciology. I'm going to do something else." I stayed at Ohio State because Ian had always told me to do the Ph.D. with the best person in your field. But I didn't know what my field was, so I didn't know who the best person was. I went to each of the faculty members and asked, "What are you doing that is so fun that I could spend the rest of my life doing it?" When I had gotten around—this took a year—I went back to Ian and said, "Would you take me back?" He said, "Yes, I'll take you back; that is all right." So I went back to studying ice and did a modeling study.

After that, Ian pointed me in the direction of a couple of "the world's best people," and I ended up with Charlie Bently at Wisconsin. I arrived with a proposal for fieldwork in Antarctica. I wanted to determine how snow turns to ice, shallow cores. I went down and did the fieldwork, got some data, packed it up, and shipped it home. But there was a problem. When you work in the Antarctic, everything goes through the National Science Foundation's contractor, and they melted my ice core on the way home. I was doing physical characterization of the ice: size, shape, and arrangement of the particles, the pore spaces, and so on. When they discovered that the freezer had broken, they just fixed it and didn't tell anybody. Come April, there is a phone call, "The truck is here with your ice." So I ran down and threw it into our freezer, and I opened it up. I should have had these wonderful cylinders of snow turning to ice. Instead I had little puddles of refrozen water at the bottom of the plastic bag. I was not terribly happy. Then I started screaming and jumping up and down, and I got invited on a short trip to Greenland to get an ice core. By the time I got back from Greenland, they had approved another trip to the Antarctic. So rather than getting one complete story from Antarctica, I had the field data from the first year, then I had the Greenland and another Antarctic core from a different site. I ended up with lots more experience and lots more ice.

It ended up more as a comparative study?

Tremendously so. I could do here to here to here. It was focused on the physics of how snow turns to ice.

In doing this, I had done this large number of snow tests and had worked out the physical stratigraphy: how can you see a year in the snow? They'd been doing it since IGY [International Geophysical Year, 1957–1958]. It has been around forever, but some of the IGY people were really good, and they got it right, and some of them weren't. What we know now is that the good ones had the experience, had the eye, and the bad ones weren't as good as observers. But what it did was to cast a shadow over this technique, maybe it isn't reliable. When I was studying this, most of the books said, "Yes, there are annual layers, which you can see, but they are not reliable. You won't count them, you won't make it work." Tony Gow was the only exception; he really believed in this the whole way through.

When we got to Greenland, I started and then Tony finally got funded. He is the grand old man of the whole field. He and I, Deb Meese, and my assistant, Chris Shuman, did most of the visible characterization of the ice core, along with Kurt Cuffey. We established again, in Greenland, that these are annual bands. We established that we know how and why they are forming. Chris came up with this beautiful way that we could map the formation from space. This is so slick. Everything radiates all the time. Mostly it is an infrared that we can't see. And the snow is doing it, too. Chris can take passive microwave records and tell you when the surface is getting fuzzy. He can tell you that it does happen in the summer; it only happens in the summer; it happens all over Greenland. This is a sunshine summer indicator. Once we had that pinned down, I sat through a whole summer in Greenland counting layers, and I hit the volcanoes right. We've got the dated fallout from the volcanic eruption in Iceland; we know when that occurred. It was 206 years old when we started, and I counted 204 years. It works! From that, you can go ahead and say, "I can date the core."

From the dating, you can see how much snow accumulated each year. You can see abrupt changes in that. You can ask how much of the change in the dustiness is caused by the changes in the snow accumulation. Once you have the dating and the snow accumulation, then you know the ice flow, and you can make sense of the bore hole temperatures. Then you can tell how much temperature changed in Greenland. Then you can look at how does temperature relate to snowfall, to dustiness. It all falls together.

Eventually, clearly, the dating relied on much more data than just mine. It relied on the electrical conductivity, Ken Taylor's work, the isotopes from Piet Grootes's work.

What causes electrical conductivity?

It is the acidity of the ice, but it is a combination of the rate of input of acids and of dust, which neutralizes the acid. That, in turn, makes it a complicated signal. Occasionally, the annual cycle becomes harder to see because it also picks up volcanic fallout. If a volcano in Greenland erupts, it dumps a lot of ash rich in sulfuric acid. It picks up everything. It is the road map. Ken Taylor built this very quick, very reproducible digital recorder, and he can make a road map for everybody. The core goes by Ken's station (this is in a lab carved out of the Greenland ice sheet, where the scientists were making their observations); he turns it into a line on the computer. Everyone afterward can go back to Ken's "map" and ask, "Where am I?" It is good for annual dating, but it is not the best because it is seeing these other things, like volcanoes.

Deb Meese eventually chaired the dating committee. The real key to dating the bands, initially, was our ability to stand there, look at the core, and count the layers. There is what is called the brittle ice zone. As you go down, the bubbles are squeezed by the weight of the ice above them. If you bring up ice from 1,100 meters down, the pressure in it is 100 atmospheres. When you bring that up and try to cut it, it breaks. Sometimes it shatters even if you haven't tried to cut it. What happened in the project was that we left that ice for a year, and then we put it back together. But that made it harder. The electrical conductivity measurements can see the cracks. While you still can see the annual signal, it gets noisier. The conductivity is sampling some spot about a centimeter across and when it gets within a centimeter of a crack, it is affected. The human is adaptable. You can look around that crack and see what is happening. So, for the brittle ice work, we really relied on the visual work, my ability to see the layers. Ultimately the fact that we knew how these layers formed—because we went to the field and we watched them form—was key in the dating. Dating was key in the dust record, in the snow accumulation, the temperature—the whole thing uses that.

Let me play devil's advocate. You mentioned this fancy, new ice core analytical facility in Colorado. Are you going to be processing your ice cores in the future there?

Yes. The ice core community is now using that.

Everyone you've worked with on these questions—determining past climate records from ice core data—has experience in the field. They've been there on the ice caps as the core was being taken. They know the problems sometimes encountered when core material is brought up from depth.

Basically, yes. When this generation starts to move on, we will have to get others into the field. What is happening right now, for example, is that three of my students went down to Antarctica last year strictly as core handlers. They weren't doing science. They looked at what was going on, but they basically went down to be on the drill to see what happens. They've worked in the ice core lab as well, so they have seen the processing end. They took time during graduate school to pop out and see the field process.

Those of us with experience in the field can work for a while in the ice lab, but eventually you have to keep getting people out into the field, in some way or another. You should not start to think of the ice core as a chemical sample, which is supplied from the lab. You can call them up and ask them to ship you a piece of the ice core and they will, of course, with various clearances from NSF. They are very good and very efficient. But you need to understand that this came out of an ice sheet, and that that ice sheet is moving. Right now our community has that because of our experience in Greenland. Paul Mayewski, who was the chief scientist in Greenland, got us all up there and we all worked there. That will last us for a decade, then we'll have to see.

The American ice core community is working really hard in Antarctica now. Kendrick Taylor is leading the effort. The goal is to get an Antarctic core that has the same snow accumulation rate, the same time resolution, and the same length as the Greenland core. The question, ultimately, is how quick were the changes in the south, and were the changes in the south before, after, opposite to, or the same as the changes in the north.

STEVE PORTER *brought this issue up in his interview, also. What was the timing of the glaciation?*

A really good question. We have enough evidence to know—and this is the royal "we"—that the largest of the thousand-year to few-thousand-year changes in central Greenland and in part of Antarctica are opposite. The Byrd Station core shows warming when Greenland cools and cooling when Greenland warms. What one cannot tell is whether those are identically synchronous. Does one lead a little bit? We can't see that. We are going to need more ice from the right core to see that. That is the big goal for the U.S. ice core community—to get a southern equivalent of the central Greenland ice core.

You have very carefully and very graciously included your colleagues in all that you mention about ice core work. They obviously play important roles. You have contributed a piece to this puzzle, but they?

They have a very big piece. Let's face it. The planning for this Greenland project started out when I was in graduate school. I was nothing. There were heroes who made this happen.

What are your feelings on "big science," which must be important to you, because that is the only way you can get to do what you want to do, versus smaller, individual creative efforts?

Big science drives me crazy. You are fighting over whether you get a sample here or there, whether you get a cut a millimeter higher or lower. That drives me crazy, but you have to do it. The truth is that by being locked in with other people who are working on the same problem and require answers, we did that. I could not do a slapdash job of dating because Tony Gow wouldn't let me. Tony would never, ever do a slapdash job of dating, but even if he wanted to, he could not have because I wouldn't let him. What we need, in my field of paleoclimatology, are the type sections that the community is entirely comfortable with. Many good people have worked really hard. It is dated well; it is characterized well; the transfer functions are known well. Once you have those, let the individuals use their genius on their own records somewhere. You don't have to lock everyone in this mass which tromps from place to place, but you do need to put down the type sections. In Greenland, the U.S. and European cores considered together, we have a type section. We need that in Antarctica.

You want a few people to go out and follow their genius and occasionally get together to put down the keystone piece. I think it is happening. I think we will manage to do that in our community. We could use more money, but I think we can do it. It is a statement which reflects good people in the community and good people behind the money. The folks at NSF are doing a good job.

You say that this was being planned when you were in graduate school. What kind of a commitment was made by the grand old men of the field?

A lot of work went into this, building up reports. Of course, building up ice coring originally took genius.

In the United States, we lost our capability for a while. The drill got stuck and then the next drill experiment didn't work. The United States dropped out for a while. Something was lacking. Whether it was the funding agencies, the scientists, I don't know. I wasn't here at the time, but we lost critical momentum. We fell off. Getting it back up to speed was the result of a science board for planning in Greenland. A lot of the top people were on it: Ellen Mosley-Thompson, Hans Weertman, and Tony Gow, some of the really top people. They laid out a basic science plan and a basic rationale. There was an effort which involved a lot of the paleo-oceanographers, as well, Wally Broecker, John Imbrie, and others, who basically worked with NSF and the European National Science Foundation to set the ground rules for the two cores and the joint logistic support and made the money appear.

So, the ocean science community played a role in saying that there are records in the ice core, which we need to help us interpret our oceanographic data.

Yes, certainly, and that helped us. It finally made the ice core work get going again. There was some real brilliance at NSF, as well. Herb Zimmerman did a lot.

They laid out a plan that involved trying to get two groups working on each of the major properties of the ice core. You wouldn't have just one group doing stable isotopes of the water, you'd have two groups: Piet Grootes, Minze Stuiver, and Jim White. You wouldn't have one group doing physical properties—Tony Gow—you'd have two groups, and that's where I fit in. I had come out with the training which allowed me to be in the second group. Tony was the obvious choice; it had to be Tony. The question is: Who is in the other group? And I got that one. This showed foresight; these are good people who dreamed this up. It gets in new blood, because then I got to work with Tony.

Let me put you on the hot seat and consider the term global warming. *Any comments you want to make about science and the public?*

I have a popular book coming out which tells the ice core story. There are real problems facing us today, which we should not hide from, but we are ultimately likely to handle them. We don't want alarmists. We don't want people to believe that the world is going to end, because it is not. What is needed is level-headedness and goodwill on all sides. Global warming is not going to end humanity; we are going to come through. We may be somewhat unhappy with what happens, and there are some things we can do about it. There are people who scare other people, and that is probably not right. There are people who basically lie and say nothing could possibly ever happen, so why worry? The fringes on both sides include people who do not seem to have goodwill, but there are also people moderately near the fringes who have goodwill. The problem is teasing these apart. Who is trying to play the game right, and who is throwing rocks to further his own agenda? It is a little bit knowledge based.

If you know enough about science and scientists, you realize they are people. That means that occasionally some of them lie, cheat, and steal just like all other people. But, ultimately, what they do has to be appealed to nature, so they are not just people. If you know that much, it probably all makes sense. You know there is a little exaggeration on this side, a little exaggeration on that side. Yes, global warming will happen, it will affect us. It is not going to end us, and we might be able to do something to change the effects. If we go that far, we are home free. That is the bottom line.

There at Shoals, they stay in the lab until 11 o'clock at night. Excitement happens.

C. Drew Harvell (b. 1954)

Associate Professor of Ecology and Evolutionary Biology
Cornell University
Interviewed in her office 13 November 1997

After being suppressed during high school years, Drew Harvell's love for natural science was reawakened on a two-week-long college natural history field trip. I first met Drew in the field in St. Croix, where she and her students worked. She compares and contrasts teaching and student response to a subject (such as invertebrate zoology) at the main campus versus the field station setting. It is clear from her interview that she is a talented and sensitive teacher, yet the need to do a large volume of published research drives her to select teaching as the aspect of her career that she wouldn't mind forgoing in her busy life.

I was an undergraduate in the honors program at the University of Alberta in Canada in the mid-1970s, and it had a great marine biology program. In my first summer, at Bamfield Marine Station, I trapped mice with Tom Hermans in deep forests of remote islands, and I helped one of Fu-Shiang Chia's students at Friday Harbor with sand dollar population biology. I also did independent research with Fu-Shiang Chia at the end of that summer at Friday Harbor, studying larval biology of a snail. I learned that I loved staying in the lab until all hours of the night and being obsessed with my study of invertebrate biology. At that point, I knew I loved field biology but not exclusively marine biology. I had also some experience doing alpine biology as an undergrad at Rocky Mountain Biology Lab, and I also loved it.

I finished my honors degree in Edmonton and decided I wanted to work for a year before going back to graduate school. I went back to New Hampshire and secured a nice job in a consulting firm working on monitoring studies to evaluate how power plants affected marine communities. It let me do a little bit of fieldwork and write up reports about the distribution and abundance of plankton and marine benthic communities. For somebody with my training, it was a very exciting job, and I had a lot of responsibility. I had to be able to write, read the literature, and get involved a little in the fieldwork. That job further solidified my interest in marine ecology. I saw there was a real world application for it, in addition to just being interesting to me. I also discovered in that year that I wanted to be an academic biologist. Working in a consulting firm was not stimulating enough. I realized how much I missed the pure discovery environment of academia. That was an important realization for me, and so I did go right back to graduate school.

I went back to the University of Alberta and did a master's with Fu-Shiang Chia.

Drew Harvell (*left*) and her students in the late 1990s studying marine organisms in their own environment, seeing them in their habitats and watching how they behave. Photo by Kiho Kim.

Then I decided that I really wanted to do ecological work, and so I changed schools and did my Ph.D. at UW [University of Washington] with Richard Strathmann. He was really the perfect person because of my interests in invertebrate life histories. It's funny but I can still remember the essay I wrote to get into graduate school. At the time it was so agonizing to write and seemed reflective only of my current intellectual state, but it turns out that it was quite an enduring statement of interests of mine that ran pretty deep. I said that what I wanted to do was to study the developmental basis of ecologically important processes or interactions. I wanted to somehow link development and ecology and maybe evolution. I have ended up working with inducible animal defenses against predation and their life history consequences. It is fair to say my research did achieve the goals I set out, and my course has remained true to those goals. Somehow, even though those interests were embryonic, they were deep rooted enough to sustain me through graduate school.

A lot of people realize, in retrospect, that they had interests in their subjects from way back.

Yes, this is one of the miracles of how our intellectual interests unfold.

Strathmann was the perfect person because he was interested in development; he was interested in ecologically relevant biology; he was interested in invertebrate life histories. The longer we worked together, the more I realized what a good advisor he was. There was another important thing about him. At the time I started, he didn't have any interest in coloniality. That was not part of his repertoire. He was a larval ecologist interested in life history theory, and he was starting to look at breeding systems. It was his openness and his intellectual curiosity that made him such a great advisor for me. He was willing to learn with me and to encourage me to do something that was pretty different from him.

At Washington, I was able to put together the project I wanted, but I also began to develop my interests in tropical marine science through visits to STRI [Smithsonian Tropical Research Institute] and West Indies Laboratory. I became intrigued with tropical predator-prey interactions that paralleled work I was doing in temperate systems. I worked with predators on both scleractinian and gorgonian corals. It was a wonderful period of intellectual development and learning tropical natural history in St. Croix. I was involved in two *Hydrolab* missions [saturation diving facility] as a graduate student. So the programs at STRI, the programs at NOAA [National Oceanic and Atmospheric Administration] and at West Indies Lab really allowed me to broaden in a way I never could have if those programs hadn't been there. As you well know, some of those opportunities aren't quite as well developed any more. Certainly the loss of West Indies Lab has been hard for all of us.

Were there experiences in your childhood that preadapted you to field science?

I was a bit of a loner and into freshwater biology. I spent a lot of my time as a very young child catching frogs, salamanders, and toads, mucking around in ponds, and hatching tadpoles. By the time I got to high school, I would not have identified biology as a topic that I was interested in. I don't know what happened. Somewhere along the line, I lost interest in science.

What did you think you were going to be in high school?

English major, probably. I loved to read and was interested in literature. It wasn't until partway through college that I switched. I said, "That's right, I love biology!"

Was it a particular course?

Probably it was field zoology at Colorado College. It was an amazing field course where we just collected and identified everything. We took a two-week-long field trip in Colorado and Texas that I loved. I found that I really also liked developmental biology, although I didn't like being confined to a lab. Field biology was always my inspiration. I made a very definite decision at Alberta in my master's. I was learning electron microscopy and found it interesting. Then I thought, no, this is not right, this is not what I want to spend my life doing. I liked it, but I realized that what I needed to do would be more ecological, so I could spend part of my time in the field. I think I sensed even then that I had good field insight and no particular microscopical insight.

I can't imagine teaching about these organisms without the connections to the environment, without the students seeing them in their habitats, understanding how they react, understanding how they are attached to the rocks, how the waves wash over them. Every aspect of their biology is governed by them living in the ocean. It is a very bizarre idea to study them without looking at them in their natural habitat. Plus, they don't do a lot of their interesting behaviors when you transport them to Ithaca. They don't spawn usually, they may not behaviorally interact, you can't get the numbers and diversity of organisms or different life history stages here. I think it's important for the students to see the setting and to see the possibilities for how you do research with these organisms, to see how one goes out and collects the animals, puts them into sea tables, and how to observe. If you want to see more of a behavior, you can go out and get more of it. If you want, you can go out, snorkel, and watch them in their habitat. They just don't understand; it's just not exciting enough to do here on main campus. There at Shoals, they stay in the lab until 11 o'clock at night. Excitement happens. I take them out for a weekend, and they spend 16–20 hours in the lab in just two days. That's equivalent in time to a whole lab course, and they do it willingly and love it. They can't get enough of it. Somehow, that never happens here on main campus.

You have many academic activities, in addition to children, husband, dog, house, and so forth. What would you trade in among your academic activities, if you could spend less time on it?

It's hard to say. Although I love my teaching, I think I'd say I would cut back on my teaching because in these years with young children and life at a major research university, I'm really on the edge. I would certainly trade in my teaching for a few years to keep my research going. But you can't trade in the research. You cannot skip the research for a few years and then pick it up later. It is too easy to get behind and very difficult to build up lost momentum. I think I've gotten through the worst of it and managed to keep everything going, but it was very hard when my children were babies. They are now three and five. There are still a lot of years of trying to do things with not enough time.

I'm a little bit sheepish about complaining about teaching because my teaching load is so light relative to most places. I teach invertebrate biology and marine ecology, both of which I love to do. They are given in alternate years, so I really only teach one major course a year. I am now teaching invertebrate biology entirely at Shoals Marine Laboratory, to give Cornell students a chance to get out to Shoals, but also because it's the right way to teach this course, with access to live invertebrates. Then I do a graduate workshop course. Nelson Hairston and I teach life histories of freshwater and marine invertebrates. We take students to the Shoals Marine Lab for five days to do projects at the seashore. Then we take them to Shackleton Point on Oneida Lake, a freshwater field station, and have them do projects there. It is fun for both Nelson and me because we get to see the different habitats and different organisms. It's good for the students because they get a sampling of the kinds of thesis projects they could do more locally. It's not a very heavy teaching load, but Cornell is a very intense place to teach. Students are bright and demanding, and it takes a lot of effort to teach well.

Do you get anything out of teaching students in the field?

Oh yes, a lot. It's the excitement, just the sheer excitement of it! The students being interested and wanting to know about these things is just fun. So the educational experience is rewarding and fulfilling in a way that it is not on campus. Also on campus, the students and professor are frazzled by everything else, making it harder to focus. Certainly, there is no question that the questions that the students ask, the connections that they make, enrich the way I think about problems. They show me new problems. It's true. You can look at the same organism, the same interaction for 20 years, and it always seems new when you go back to it and are showing it to students. You see one new thing that you hadn't seen, and it's just so exciting. When you are doing research, you are on a particular schedule, and you see things that you are looking for, but with teaching, you cast a wider net and do see more novel things.

It is neat when you can get people to realize that they can actually make observations that nobody has ever recorded before, and they can do that in their own backyard.

Michael Donoghue (b. 1952)

Professor, Department of Organismic and Evolutionary Biology
Director, Harvard University Herbaria
Harvard University
presently: G. Evelyn Hutchinson Professor and Chair, Department of Ecology and Evolution
Professor of Geology and Geophysics
Curator of Botany, Peabody Museum
Yale University
Interviewed in his office 3 April 1998

Simultaneously discovering Darwin's writings and the diversity of trees in an arboretum during a two-year hiatus between high school and college alerted Michael Donoghue to the vast potential in the exploration of the natural sciences. Huge, impersonal classes at a large state university almost thwarted his interests. However, when he collected and identified plants on a several-week swing through the Appalachian mountains on a class field trip, and when he was a summer field assistant on a collecting trip to Mexico (for which he was assigned his own project as well), he was hooked for a lifetime. Donoghue discusses the forces conspiring to keep students from the field today, as well as the absolute need for a researcher to develop a feel, an eye, a search image for his organism.

My interest in natural history is a post–high school phenomenon. I can actually trace it back pretty clearly. I took a year or two off after high school, because I thought I didn't really want to go to college. My dad was an anthropologist. We lived in Japan until after I was in second grade, so I learned Japanese. Then we moved to Vietnam for a couple of years and lived in Saigon. We moved to Michigan when I was in fifth grade, as my dad got a job at Michigan State, and I went to most of high school there. After high school, I just sort of took off, wandered around, did a lot of traveling.

With your parents' blessing?

Not so much, no, I wouldn't say that, but they were pretty understanding. Of course, they would have liked me to go right on to college, rather than wander. But that's what I did. I ended up living for a time in Madison, Wisconsin. During that time I got interested in a couple of things. I was reading a lot at the time (now I don't have time to read anything!) and I read Darwin. I read *The Origin of Species*. This is a year out of high school. I was thrilled by that, I just thought that was the coolest thing. So I was really turned on to evolution and natural history from that point on.

You discovered it on your own then?

Yes, exactly, I don't think it was anything from high school or any previous experience. I think my dad actively disliked biology. He was a cultural anthropologist and was generally down on the biological explanations for anything. So I can trace my interests pretty much to getting turned on by Darwin. The other thing is that at that time there was an arboretum near where I lived in Madison. It's built around a series of lakes and there are lots of natural areas. Anyway, I had occasion to go out and started looking at plants. I got interested in the trees and started identifying trees. So, both my interest in evolutionary theory and my interest in plants sort of developed during that time period after high school and before college.

You were sort of your own boss. You had time to do things that you thought were important.

Absolutely. I had odd jobs that I did. I worked at an all-night K-Mart–type store. I had the afternoons to wander around the arboretum. It is nice to have periods of life like that. You can move in new directions.

Then I decided to go back to school at Michigan State. I just wanted to go back and take some biology classes and learn about trees. I had no intention at that time of actually getting a degree. I signed up for a dendrology class in the forestry department, which I loved. Unfortunately, I took a biology course that I thought was pretty miserable. I thought, oh, this is really horrible, maybe this is not what I'm really interested in. It was a huge class and I actually had to watch the biology lectures on TV, in another room, sort of remote access.

I can't imagine a college experience like that.

Oh, it was crazy, not the way it should be done. I was kind of turned off by that and dropped out a little while longer, but I still remained really driven by the idea, I liked plants and I liked evolution. I knew that. I kept coming back and taking a course here and there. It took me about six years or so, but I eventually got a degree. Luckily, I got into some kind of honors college, where I could design my own program. It was a pretty flexible program, and I managed eventually to get enough credits to graduate. I'd been in a bunch of different departments. I was in forestry for a while, soil science for a while.

But then there were two crucial steps for my interests. I got involved with a geography professor at Michigan State, Jay Harman, and to this day, I'm in contact with him. He was interested in biogeography, and he took me on several field trips. The first of these was down to the Great Smoky Mountains. The next one was down to the Ozarks and then down to the southern coastal plain to the panhandle of Florida. These were fairly long trips. We looked at plants along the way. He was very broadly trained. A number of students, but not a huge number went on these field trips. I was completely turned on by that. This was fantastic. To be down in the Smokies in the springtime just when everything was flowering, all up and down. I was totally sold. In fact, it was at that time that I got onto the plant that I then did my dissertation on, *Viburnum*. There are a number of species in the Smokies, but there was one at higher elevations, the "hobblebush," that I particularly liked. So that experience with the geography professor was very, very influential.

The second thing that happened, I met up with a botany professor named John Beaman. John Beaman was very good at sucking students into plant systematics and plant evolution. I took his course, and the next semester he said, "Why don't you help teach the course next year?" I was an undergraduate and I thought, wow, that's very nice. I took him up on it, and so for the next couple of years I served as a teaching assistant in the course.

And you learn more that way.

You learn so much more. I was totally sold on being a plant systematist at that time. John Beaman influenced many undergraduates, and I keep up with him, too. In fact, he's in this building today. He spent the last five years in Borneo collecting plants, and is giving a seminar about this tomorrow.

The most influential thing he did as far as I was concerned was that he took me on a field trip to Mexico. At that time, he was working on the alpine flora of Mexico and Guatemala. He had been going down for many years, so he knew the area well. He just said, "Why don't you come along this summer and pick a group of plants that you can work on? You can have your own special group." So I did that. We traveled all over Mexico climbing up to the highest mountains collecting alpine plants. And I did choose a group of plants; it was this group *Viburnum* that I first met in the Smokies. I said, "Let's go look at those," and he was happy to accommodate.

There are a bunch of them in Mexico. I ended up doing my Ph.D. dissertation on the Mexican and Central American species of *Viburnum*, so this experience with John Beaman was really seminal. He was very nice because he said, "We'll go out of our way to try to find your plants." So before we went, I got all this field locality data.

We went and tried to find these things. Sometimes we found them, and sometimes we didn't. We just had our eyes peeled for these things. We got familiar with the environments where they would grow, and we would pick a spot and say, "We can walk across that river and maybe there'll be some over there." It was just the whole experience of trying to find some particular organism.

That's an interesting skill. My husband loves to go birdwatching, so we've gone birdwatching all over the world. And he'll pore through books ahead of time. But we'll be out in the field, a new area, and he'll say, "This is the kind of environment where we should see a such-and-such." And sure enough, nine times out of ten, we'll see it.

That's when you know you're really with somebody who knows what they are doing, when they can say that, and it comes true. That's a level of knowledge that few people really have. It's very fortunate. I'm that way now with my plants. I can literally smell them.

It is getting to the question: Will there ever be a robot that can think? There are all those nuances that you can't teach somebody else.

That's right, you are integrating so many factors about the environment. They are almost imperceptible.

I was a grad student at Harvard for about six years, and during that time, I did a lot of fieldwork. Overall, I spent a year or so in Mexico and also in Guatemala, Honduras, and Costa Rica. I went also to Jamaica. I liked the plants and was turned on by that, and I liked the idea of field collecting, I was turned on by that, but what was really interesting to me—and obviously I just stumbled into it, because I didn't have any grand scheme at the time—was that this was a radiation, this was a center of diversity for *Viburnum*, and it was underappreciated. These plants were known from eastern North America and eastern Asia, and they turn out to be much more diverse in eastern Asia, but there was this secondary center of radiation in southern Mexico, in the mountains of Oaxaca and Chiapas. I wanted to find the plants and document the diversity that was there. I wouldn't call it a major radiation, but a minor one.

Are you able to go back and explain what the mechanism of radiation was?

I think in the case of these guys it was fairly clear. They are dispersed around by birds. They live at fairly high elevations in the mountains and so what was happening is fairly simple geographic isolation. You'd get some isolated population growing on this mountain, and you'd get some divergence. It was a case of a complicated terrain, living at high elevations, being dispersed by birds.

Island biogeography?

Yes, island biogeography, adaptive radiation, and diverging on different habitat islands, so it was kind of fun. That's what I did. It wasn't anything earth shattering. If my dissertation was interesting for anything, it was for the numerical phylogenetic analysis.

Mainly on morphological features?

Entirely on morphological characters. It wasn't until much later that I got into molecular biology.

What happened next was that I got a job at San Diego State University. That was kind of fun because my field experience there was quite nice, too. The minute I got off the plane in San Diego, they said, "You've got to teach California flora." Of course, I didn't know anything about the California flora, and so the next day I'm in class trying to learn about it. So I'd get out my Munz [the taxonomic key to California plants] and start keying things out. That was great fun.

What a wonderful flora to get thrown into.

It's fantastic. The Californian floristic province has incredible diversity, amazing stuff. I spent a lot of time the first year I was there learning the plants. I traveled all around California, and I took a lot of pictures of plants, and I did a lot of hiking. The funnest part for me about living there was the fact that stuff grew all year 'round. You could take students out at any time of year. The second year I was there, it must have been 1983, it was an El Niño situation, and there was an incredible bloom in the desert. It was mind boggling to me.

Then I moved to the University of Arizona and was there for seven or eight years in Tucson. I taught a field course with Paul Martin, who is a vertebrate guy. In Tucson, it was never difficult to be outdoors. Practically every day you were out somewhere. Then I moved here. It's harder to get away and get out into the country, so I don't get out as much as I would like to. However, since I've been here, I've gotten into fieldwork in China. That has been very, very gratifying because the plants that I love the most are over there; that's really where the major radiation is. There are names that have been given to some plants but that's about the extent of our knowledge. It has been fantastic to actually go over and find them in the field. They are so incredibly diverse over there compared to my experience in Mexico, which as I said, in retrospect, seems like a minor radiation.

That area in China is way underexplored. You know it the minute you get out of the car, and you start seeing stuff. A lot of plants have been described, but we know virtually nothing about them. There might be a name for a particular plant and maybe one collection, but we know nothing of its geographic distribution. Is it common? Is it rare? We know nothing of the biology of the plant. So you really have the sense that every observation you make is a tidbit of new knowledge. It's an exciting feeling to know that you are seeing something that people in general have not noticed. It doesn't matter how obvious it is. I think people in general feel, although they have the wrong impression, that the world is pretty well explored, and in fact, it's not.

When I taught high school biology last year, I showed the kids lots of live fungi, ferns, and invertebrates and told them about a lot of things that weren't yet known, that people still wondered about. I think they liked that.

It is neat when you can get people to realize that they can actually make observations that nobody has ever recorded before, and they can do that in their own backyard. There's just tons of stuff we don't know.

I'm on a sort of crusade now, as I was saying to a colleague this morning. We were talking about training in systematics. He said that training in systematics isn't as good as it used to be. People don't get to look at organisms as much as they used to. I was on a roll about how people don't get in the field enough, young students.

Not only do people think things are already known, if they are not known, just stick an instrument out there, and it will tell us The Answer.

Right. I should tell you one thing I'm doing now, speaking of this educational thing. We're going to take a group of grad students down to the Smokies this spring to just go and look at the plants. We're going to get about ten people, get a couple of vans, and go down. It is kind of returning to my roots. I really feel strongly about it.

How many of these students have had that experience yet?

They haven't. Oh, they may have gone out once or twice, but they've never really made a concerted effort, looked together, had a real field experience where they go out and seriously collect.

To me, there is so much more than just the academic knowledge that is transmitted on a trip like that, there is so much more in terms of how people see things, what kinds of questions they ask.

That's right. And some of these things, just as we said earlier, you can tell when there is an environment and something has changed, and now we expect to see a certain kind of bird, for example. To be out in the field with someone like that and to follow him around for a little while and see what that is like.

Bill and I were looking for a cock-of-the-rock in the Grand Sabana of Venezuela; we had been there before a couple of times. This time, we were with a friend who was also a good birder and he saw one, and we had never seen one. So Bill said, "You two can go off and do what you want. I'm going to stay here, all day if I need to, because I want to see one." And so I went off. I went swimming in the river, I took a hike, and I came back when it was almost dark, and he was still sitting, perched above this waterfall. So I came up, sat with him, and said, "Well, what have you seen?" He said, "I haven't seen the cock-of-the-rock, but there is a hummingbird over there making a nest." He had just sat there all day. Just as it got dark—and I felt so guilty—a cock-of-the-rock came!

That's pretty good.

People just don't realize how much patience it takes.

Oh it does; it takes that kind of persistence. I remember very well when I first was looking for *Viburnum,* I just wasn't looking in the right environment.

But in your seminar you told about the plant in China that you spent, what, two weeks looking for?

Oh yes, we spent a long time looking, and we eventually found one little plant. It takes that kind of perseverance. I remember trying to find these things and just not having the right search image and climbing up an entire mountainside and getting lost. Eventually I clued into the kind of environments to find them in. It's a great feeling when you get to that level of knowledge. I think for most people the world isn't organized that way. You have to be really clued in.

Most of the incoming graduate students haven't had much field experience. Maybe they've gone out a couple of times, but it's not substantial, it's not an immersion. For example, these trips that I took with the biogeography professor when I was in college were really that way. We read a lot of stuff beforehand, and we took along E. Lucy Braun's beautiful book, *Deciduous Forests of Eastern North America* [1950; reprinted 2001, Blackburn Press]. It was our Bible. We seriously made plant lists from one site to the next. It was not a trivial exercise. It wasn't, let's go out and smell the flowers. We were really trying to figure out what was going on. It was a real immersion to figure out the distributions and the patterns. We'd talk about it and compare notes. We'd fool around, too, but we'd have serious conversations over dinner and beer. To me, it was a real eye opener that people were really concentrating their attention on these things. When we went to China this last time, we took along one of our grad students, Rick Ree. It was a great eye opener for him, of course, but especially the experience of collecting plants, digging for them, actually handling them was so great for Rick.

STEVE WAINWRIGHT *talked about the dimension of the field being so much more than visual—the sound, the feel of the air, what you feel with your fingers.*

I agree, the whole sensual experience, to actually dig in, to work, maybe dig around the roots with your knife, it's a sensual experience. I think so few students actually have that in their backgrounds. It was a wonderful experience for the student we took to China. We'd come back at the end of the day and be pressing our plants and we'd ask, "What kind of environment was that plant living in? What other plants were living near it? How tall was it? What color was it?" Stuff that you don't really think about observing. Not until you are forced to record it do you realize that you are not being as observant as you might be. It heightens your sensitivity.

Plus, you forget if you don't record it.

Oh absolutely.

I can tell you from teaching high school biology, kids like live organisms.

Oh, they do, they naturally do. I can see that in my daughter, Nina, at age two. Any organism drives her nuts. She loves it. It just seems to be a natural tendency with kids. It's sad when they lose it as it's so naturally there. We are doing something wrong in our educational system. Any little thing my daughter sees fascinates her. And 99.9% of the people in the world lose that fascination. We obviously aren't doing a very good job at keeping alive people's sense of wonder about nature.

The whole academic system is set up in a way to discourage this, especially in the younger faculty. It takes time and it takes energy. It is not easy to stop everything and arrange some field experience. It is not a trivial thing. If you have tenure and you are kind of relaxed, then you can pull it off. But everybody is so busy. It is really hard to find the time. I can tell you what the main problem is in my own personal case: it's just time and energy level, it's not because I'm not going to get credit in the eyes of my colleagues, or anything like that. The main reason why I find it difficult to get out and take students into the field is that it is hard to get outdoors myself, much less organize a trip. I'm more aware of wanting to do it, to take students to the field and provide the opportunities. Now I'm more conscious of what it meant to me, as I'm getting older. Some people are lucky, and they hook up with the right people at the right time. They have a good field experience that can change their life. Those are the lucky ones. Other people will never have that experience.

Ever since that moment, I've never really lost my inquisitiveness about landscape and how it evolved.

Patrick D. Nunn (b. 1955)

Professor of Oceanic Geoscience
Head of Geography Department
University of the South Pacific
Suva, Fiji
Interviewed in his (UW sabbatical) office 12 May 1998

"Reading the landscape" is the challenge that lured Patrick Nunn into his career. He has broad expertise, ranging from determining the history of the river terraces under central London to the geomorphology of oceanic islands. Nunn, a professor at the University of the South Pacific, actively engages undergraduates in research, using oral history as one of the investigative tools to examine past environmental change. He stresses the value of extensive class field trips, despite the frustration of dealing with a university administration which has no academic tradition of field trips. As a resident scientist in a remote locale with a limited natural history record, Nunn feels isolated and welcomes more interaction.

In 1974, at King's College, University of London, when I began as an undergraduate, I had wanted to do a degree in geography and maths, my two principal interests in high school. But they decided that year not to offer the geography and maths joint honors degree. Therefore, if I wanted to do joint honors, which I did, I would have to do geography and geology, a program which Denys Brunsden was sponsoring. I was very fortunate to fall under his influence. Denys was a large extrovert, an exuberant person, a great believer in fieldwork. I was 18, so my mind wasn't wholly set. He had the ability to take young people and inculcate in them the excitement that comes from doing fieldwork.

I remember a field trip around the Gower peninsula in South Wales. Denys pointed out things in the landscape that I had never seen before. He asked the group of us questions about a particular land form. He'd say, "What's this?" We all just stared at our shoes. Eventually I said, "I think it is a raised beach." He picked on me. He said, "Why on earth do you think it's a raised beach?" I muttered something inconsequential. Then he just tore me to pieces, and he pointed out all the reasons why it wasn't a raised beach and, in fact, it was a glacial drift. I won't say that he humiliated me, because he put me down in such a way that it inspired me to start thinking about these questions. Ever since that moment, I've never really lost my inquisitiveness about landscape and how it evolved. That extends, of course, below the surface as well as to the surface itself, although I see myself primarily as a geomorphologist with interests in the Quaternary and late Cenozoic geology.

How did you learn how to determine what caused the type of landscape you were observing?

It was something that Denys Brunsden and his colleagues at the University of London called "reading the landscape." It is just something that comes, I think, through observation, just sitting down and looking at your surroundings. I think normal people tend to take their surroundings for granted. But there is a way of looking at your surroundings in an analytical sense that gives great fulfillment.

I had an overwhelming desire to do research and do a Ph.D., but there were other options I considered. Perhaps if circumstances had been different, I would have gone a nonacademic route. It certainly wasn't a clear-cut decision. In those days, in 1977, university academics applied to research councils for a Ph.D. grant for a student to work on a particular topic. These were advertised in the newspapers. I applied for several postgraduate scholarships in Britain for several different topics with different people at different universities.

I was interviewed at Oxford for a postgraduate scholarship. I was very excited by the subject. I remember going into the interview room. There was me on a chair and about six quite famous academics facing me across the table. I thought the interview was going quite well until one point, when one of them turned to me and asked, "How do you explain the odium that is attached to so much of modern geography?" I looked at that person, and I could not for my life think of what the word *odium* meant. I sat there for about ten seconds and said, "I'm very sorry, I can't think what odium means," and from that point onward the interview went all downhill. I was very disappointed in that. I thought that to ask a 21-year-old something about the "odium surrounding modern geography" was unnecessarily patronizing. I always regret that

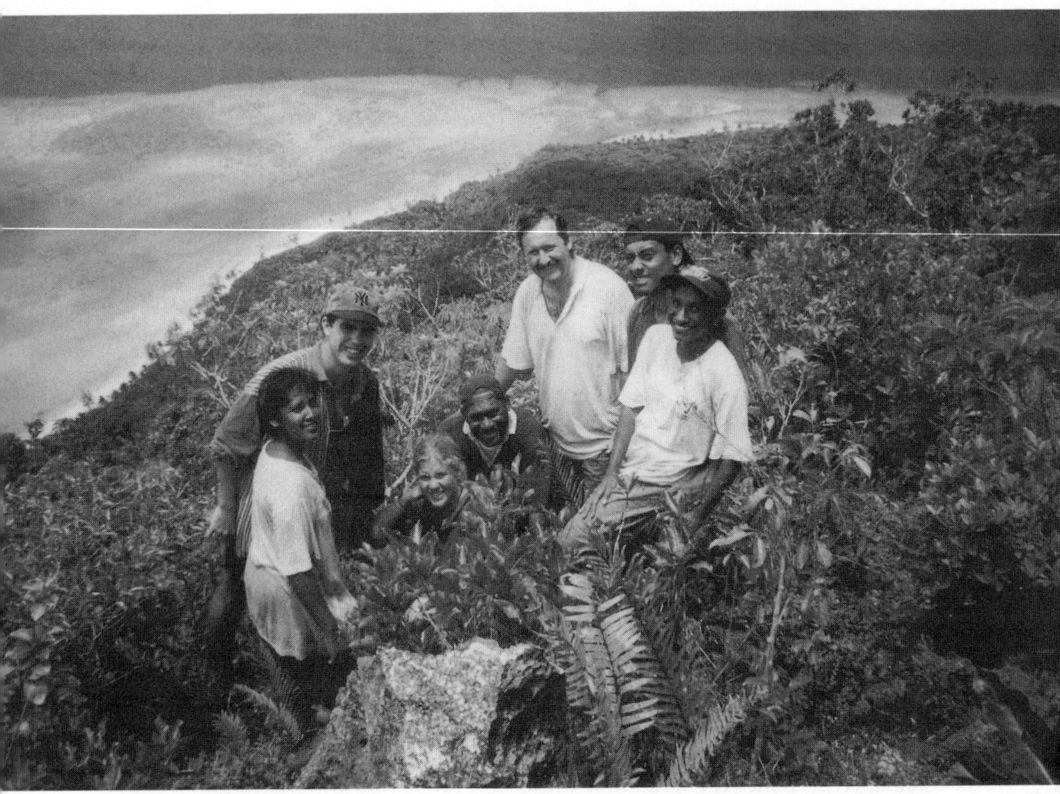

Leading a field trip in physical geography in the 1990s, Patrick Nunn (*center*) and his students from the University of the South Pacific examine outcrops on an island slope 315 meters above the coral reef shoreline of Vatuvara Island. Photo from the files of P. Nunn.

because that Ph.D. scholarship would have been to work on limestone landscapes, which is something I did not begin to reconsider for a decade or so but which I've since come around to, a full circle really.

As it was, I was interviewed also at University College, London, by two people who subsequently had a great influence on me, Eric Brown and Claudio Vita-Finzi. Eric Brown was nearly the last representative of a very influential school of geomorphology, mainly British, called denudation chronology. He had studied directly under the master of denudation chronology, Sidney William Wooldridge. Wooldridge's ideas basically involved going out, reading the landscape, understanding its history simply from looking at it, and correlating surfaces on the basis of their height in different parts of the world. Without going into it in great detail, the whole concept of denudation chronology became discredited largely because people had taken these correlations on the basis of height to what in hindsight appears to be a ludicrous extent. From one side of the world to the other, you have a 60-meter surface and therefore you say it is contemporaneous. When you take tectonics into account, such long-distance correlations clearly become improbable, to say the least, but in the 1950s and 1960s, the global extent of tectonics was far from clear, at least to geographers, and, ideas about Earth-surface stability, inherited from Edmund Suess and William Morris Davis, had become so ingrained in models of geomorphic change as to appear almost unchallengeable.

At the same time, one thing that Eric Brown, who became my Ph.D. advisor, taught me very clearly was that in rejecting the whole set of ideas that was denudation chronology, they were throwing out a very promising baby with the bath water, so to speak. The bath water could certainly be got rid of, but the basic idea of denudation chronology—that surfaces existed in the landscape and could well be regarded as legitimate stratigraphic markers in much the same way as gravel beds—was a basis for understanding landscape evolution, which I believe remains valid. Although I didn't really realize it at the time, that really set me on the path to becoming something of an anomaly. Eric Brown was the last of Wooldridge's students, one of the last academics to have been trained in the era of denudation chronology, and I was Eric Brown's last Ph.D. student. He used to joke to me sometimes, "You are number 51, Paddy, and that's it." He was a man who had done his Ph.D. work on erosion surfaces in Wales. One of the first things I did when I became his Ph.D. student was to go to the University of London library and look at his Ph.D. thesis, which is awesome, and it led me to respect Eric Brown thereafter.

In the United Kingdom at this time, a Ph.D. was not gained by coursework, only by a thesis, written over a period of at least three years. The Ph.D. topic he had selected for me was an unusual one. He wanted me to look at river terraces in central London. He read me very well. I was the kind of person, and I remain the kind of person, who doesn't really like someone breathing down my neck and saying, "What are you doing? How are you doing this?" He was a first-class supervisor in that he left me largely to my own devices. But when I needed help, I went to him. He was there, and he gave the help and the direction that I needed.

The topic that he had selected for me was a very challenging one in many ways. People knew about the history of the Thames Valley represented in its river terraces upstream, and they knew about it downstream, but they didn't know about it in central London because they had this sprawling city on top of the terraces. So I spent three and a half years going around to building sites in central London, lowering myself down holes, taking samples of sand and gravel, and getting a lot of borehole records.

If they would dig a well?

Or put up a building . . .

You would sample it first?

Yes, that's right. I also gathered archival data relating to borehole records. I took all these data and plotted them using a computer mapping program called SYMAP, which, at the time, was very innovative. I never felt intimidated by computers. I've always seen them as an important means to an end but certainly not an end in themselves.

In three and a half years I finished the thesis. I discovered 17 pairs of river terraces in central London, on either side of the river, and found, in at least a very general sense, a diagnostic set of gravels [drifts] overlying terraces at different levels. I also found, as many people do for three years on a narrow Ph.D. topic, it to be very difficult to stay focused. Eric Brown and Claudio Vita-Finzi both tried to keep me on the straight and narrow on my topic, but I know I got distracted. I got very interested in the history of London's rivers, for example, and the way in which these rivers had changed course, not through geological evidence but through historical evidence. So I spent a lot of time in the British Museum and the London Library looking up all the evidence of where these rivers had once flowed. But, when I came to the end of it, I realized that I didn't want to spend the rest of my life as an urban geomorphologist, an urban geologist. It was very interesting, but it certainly wasn't mainstream, and there were lots of other issues in landscape science that fascinated me. I really wanted to become part of those.

I had the purely fortuitous opportunity to go to the south Atlantic in 1981 to visit the islands of Ascension and St. Helena, where Napoleon was exiled. With a team of about six undergraduates and another postgraduate, we spent about four weeks on St. Helena. It took nearly two weeks by boat to get there. We made a geomorphological map of the island, and we sampled the geology. When we came back, we wrote a huge report, which was very well received. Really, after that, I became interested in islands. I did a lot of reading on other south Atlantic islands, in particular, to see if there were analogs or correlates to the kinds of things that we found on St. Helena and Ascension. I realized that really no one was looking at island geology or geomorphology as a specific subject, in its own right. I realized it really was an untapped field and there was potentially a niche, if I wanted, for me to fill.

Then it really just goes back to the day in south London when I was just sitting in the middle of winter, in my little apartment. The windows wouldn't close properly, and there were icicles hanging on the inside of the windows. Oh, it was so cold! I was reading the newspaper. I saw this job advertised at the University of the South Pacific based at its main teaching campus in Suva, Fiji. I looked in my atlas and I saw where this was. I saw all the islands and I thought, *this* is the place for a budding island geomorphologist. So, I applied, and I got the job.

In 1985, I moved to the University of the South Pacific, which serves 12 different island nations, each of which are too small to have its own university. I'm based at the main teaching campus in Suva, in Fiji. I went there, ascended the hierarchy, and I am now head of the geography department. My interests in islands have burgeoned. Certainly, I'm still interested in central London, but you can't keep up in so many fields. The Pacific islands have been very good to me, in terms of my career, in that very few geomorphologists and very few people with interests specifically in island geomorphology have worked in that region before. Most of the environmental geoscience research that has been done in the Pacific islands has been by people who are not primarily interested in the long-term evolution of these environments. Many marine geological surveys have been done in the islands with resource potential in mind: hydrocarbons, manganese nodules, those sorts of things. There has been quite a lot of work done on coral reefs, but I've been very critical of a lot of it because I think many scientists uncritically transfer the ideas they have developed on reefs elsewhere, particularly on the Great Barrier Reef, to island reefs, which I think are different in many ways. I think one of the big problems with a lot of the environmental geoscientific research in the Pacific islands is that it is being done by people from outside the region who are bringing their expertise, which they have developed in another geographic region, to bear on the Pacific islands. That doesn't really allow the Pacific islands to be recognized as unique environments, or different environments, from continental environments, in particular.

When I first got to the Pacific islands, to Fiji, I began to do fieldwork without really knowing what I was looking at. There had been very little written on what I was really interested in about the Pacific islands, on any islands, really. So I was doing what I am now criticizing other people for. I was going to the islands, and I was trying to analyze their landscape history by looking at river terraces.

But you have to start somewhere.

Sure. Oh yes, I'm not saying I really could have done it any other way. But I think that over the 13 years that I've been there, I've come full circle in many ways. A lot of the interpretations I made initially I've since revised, which is, of course, a euphemism for saying "changed completely," and I think that is quite understandable. I really have very little time for academics who don't occasionally change their minds and who like to pretend everything has been crystal clear to them from the word go. I admire people who can stand up and say, "I once thought this, but now I think that," because I realize that is what real life and research is all about.

I think the other point to make is that a lot of my own research on the Pacific islands has basically been done by myself. If you look at my CV, most of my publications are by me alone. That, I think, has not been very healthy, even though it has been unavoidable. I would very much have liked to have collaborated more with people than I have done, but I'm at a small and very remote university. I get plenty of opportunity to do field research, but I can't wait around for someone to come to the region to work with me. Notwithstanding that, I have worked with some very good people over the years in the Pacific islands. Particularly influential has been Cliff Ollier, who is emeritus professor of geography at the University of New England in Australia. He has been on several lengthy field trips with me in quite remote parts of Fiji. He has certainly challenged a lot of my interpretations, and he has gotten me to think about things that, in the normal course of events, I might not have been thinking about.

I also involve undergraduate students in my field research. In the last five years, I've had two major research trips to other parts of the region where people live. On the first one, to an island called Totoya, we took eight undergraduate students along. There was myself, an anthropologist, and two archaeologists. What we were really interested in doing was to try to find out something about early human-environment relations on this island. There has been an orthodoxy around for a long time which says that the Pacific islands, before people arrived, were covered in forest. When the first people arrived, they cut down all the forest, and they devastated the environment. I've never really bought that. I've always been suspicious of that kind of explanation. One of the reasons why the debate hasn't been resolved is because of a lack of data. So the purpose of the trip to Totoya was really to get the data which would allow this idea to be tested once and for all, at least for this specific island.

Under the direction of the anthropologist, the students went off and collected oral histories from the elderly people living on the islands. I went around and mapped landslides and coastlines. We got the environmental history that way. The archaeologists looked at pottery and bones. Basically, we put together a history of human settlement and a history of contemporaneous environmental development. Unfortunately, it didn't work as well as we had hoped. We only got back 2,000 years, and the record was very patchy. There were lots of reasons why people may have moved, and there were lots of conflicting signs about what people may have been doing to the environment and how the environment might have been changing. Not particularly discouraged by that, we had another trip about 18 months ago to a group of islands in northwest Fiji. The results of this are still being analyzed. We took cores through some swamps that we found. I'm hoping that the paleovegetation record will extend back to the last glacial maximum, about 18,000 radiocarbon years ago. We've got a fairly good handle on the human history about 3,000 years ago. We are hoping that, when all of the results come in, we will have data on those islands which will enable us to resolve this particular controversy.

I think it is important for everyone to realize the kinds of questions people are asking that cannot be answered by looking at a computer model.

I think I can give you a very good illustration of that. Ever since 1988, when the IPCC, the Intergovernmental Panel on Climate Change, was established by the United Nations Environmental Program, there has been an increasing interest in the potential impacts of climate change and sea-level rise on the Pacific islands. When you think about some of the atoll nations, which are just strips of land with massive coastlines, they are basically all coast, no more than two to three meters above sea level. You think about the potential impact of sea-level rise of half a meter on these islands. You can, I think, get some idea of the seriousness of the problem. The work of the IPCC as directed toward the Pacific island coastlines has encountered several difficulties. The main one is that they are using GCMs, global circulation models, to predict or to estimate the likely effects of climate change on various countries. But the resolution of these models is so crude, you can't even get New Zealand on them, let alone small Pacific islands!

A lot of my energies, and those of my students, has been directed toward working in the field and trying to find out exactly what is going on along modern coasts and exactly how those processes may change in the future. A project that I have just finished has been to look at artificial shoreline protection structures in the Pacific islands. Most industrialized countries have the financial resources to be able to protect the most important parts of their coastline from the effects of sea-level rise by building artificial structures. Most Pacific islands, because they have very low GDPs [gross domestic product], because they have massive coastlines compared to their land areas, are really not in a position to do that. A lot of my own work, recently, has been directed to determining what people did in the past to protect shorelines, what materials they used, what structures they built, and how effective these were against coastal erosion, sea-level rise.

We found that, basically, up until about 30 years ago, most Pacific island communities didn't feel the need, or didn't have the need, to protect their coasts. They could move. Population pressure was not so great. Land tenure or land ownership was not well established. People were fairly free to respond that way to sea-level rise. But now, of course, with much greater population densities, land divided up, it is much more difficult for people to respond like that. We are working with Japanese engineers trying to come up with a design for a sea wall that is effective but is also reasonably priced and uses locally available material, maybe bamboo. So there are these kind of issues.

In the Pacific islands, we often don't have long records of continually monitored data. One of the issues that has particularly concerned me is the long-term trend of sea-level change in the Pacific because that is intimately bound up with coastal erosion and flooding. We don't have long-term records. The longest time series data we have are from Honolulu. That is about 100 years of data. But Honolulu is far from the islands of the southwest Pacific, where we can only go back 25 years at most in terms of continuously monitored sea-level data. The sites where they place the tide gauges are unstable in a tectonic sense. So trying to get the eustatic signature out of the observed record is difficult.

On several occasions over the past 13 years, I have assembled groups of students from all over the Pacific islands. The 12 countries that the University of the South Pacific serves all send their students to the Suva campus in Fiji. At the end of the main academic year, in November, I get groups of students, maybe 30 or 40, from all over this vast region. I instruct them to go back to their homes, coastal villages and to sit down with the elderly people in the villages and, basically, to interview them, to get information about how the shoreline has changed. I've picked some places where I know the shoreline has been tectonically unstable. The vast majority are places where the shoreline has been tectonically stable, at least over short time periods, about a couple of hundred years. Those data have been coming in over the years. In a way, that is a fine example of a situation where fieldwork can be used to provide data that are not otherwise available, as they are in many other parts of the world. The students ask, "When you lived in the village and you were ten years old, where was the shoreline? Where did the high tide come up to? Were there houses where there is now sea?" We have a unique record of shoreline changes, which I believe is a crude proxy for sea-level change for this vast region, which we couldn't have gotten any other way. I find that the students generally do a lot more than I expect of them. I expect them to talk for two hours, and they'll go and talk for 12 hours. They bring me back pages and pages of notes and diagrams, and they are keen to follow it through.

Let me talk a little bit about the formal undergraduate field trips we run at the University of the South Pacific. Certainly, when I arrived, there were field trips operating, but there wasn't the tradition of field trips such as had been so important to me when I was an undergraduate in London. I can look at the 13 years I've spent at the University of the South Pacific as a running battle with the administration to get funds and basically to get them to recognize the critical importance of formal fieldwork. In a university that doesn't have much money, and in a university where there isn't a tradition of that, where there aren't vehicles for that purpose, it has been an uphill battle. It has been for year after year. Sometimes I despair.

Most administrators don't perceive field trips as being something intellectually helpful. They perceive them as being a holiday, really. Certainly, there are students who perceive them that way as well. But, in my experience, they pay tremendous dividends in terms of most students' appreciation of what they are learning. When I first went to the University of the South Pacific in 1985, I was assigned the first-year undergraduate course in physical geography to teach. At the time, it attracted 60–70 students, but now we have about 180. I've taught that course ever since, except the years I've been on leave. The first time I taught this course, I started a one-night field trip, effectively two days in the field. It has been continued ever since, and I get marvelous reactions from the students. Every single one of them, almost without exception, says that the field trip was the high point of the course, not just in terms of enjoyment but also in terms of understanding.

It is an intensive field trip. I bombard them from the word go. We get on buses and go right up into the mountainous interior of the island. We stay overnight in a Fijian village, which many of them haven't done before, so there is a cultural experience on top of the environmental one. We talk about floods, we talk about landslides, we talk about late Cenozoic subduction and the whole origin of the landscape. We stop and look at rocks. I make them walk through rivers, measure pebbles and cobbles, and start to think about the landscape in ways they've never done before they come to university. In fact, there is always a small percentage of students who are very anxious about going out.

Into the bush?

That's right. I make them walk through a rainforest and climb up a muddy slope. I say, "That is why it is called a rainforest, because it rains a lot." When they come back, you can sense the change in the way they approach the class, the way they do their assignments. They are inspired, in the same way I was inspired when I was an undergraduate.

The whole issue of whether models are intrinsically valuable or whether they require verification to make them so is a very important one. I've always been very firmly on one side of the fence, since the time when I was computer mapping for my Ph.D. in London in the late 1970s. Computer models can certainly aid our understanding of processes, but it is very important to teach, particularly to undergraduate students, that the models are not a satisfactory end in themselves. I certainly don't mean to sound derogatory about people who spend their lives developing computer models, because of course these things are very important.

I would never like to see field science, as a technique, fall by the wayside. I don't think that remote ways of sensing the landscape can ever replace that. A number of the aerial photographs that I've worked with in the Pacific islands and drawn contours from, when I get on the ground, I find them to be nonsense. The contours are of the treetops, not the ground surface itself. I've become a bit cynical of the value and potential of those kinds of aids. There is really no substitute for getting out on the ground, getting your hands dirty, walking around the landscape, and talking to the people.

You and I both have field experience in parts of the world that don't have a long history of observation. Contrast this to when I worked in central London. I had to reconcile my own observations with ones that went back hundreds of years. I had to incorporate those observations. But here, I can go to islands only a half day away that don't even have a map, let alone a geological map. There isn't the long history of observations. It is perpetually stimulating.

At an undergraduate institution, my goal has been to keep my research broad enough so that students with diverse interests can become involved.

CONSTANCE SOJA (b. 1955)

Associate Professor of Geology
Colgate University
Interviewed in her office 12 November 1997

Being commended in her tenure review at Colgate University for broadening the scope of her research program (as opposed to overspecializing) pleasantly surprised Connie Soja, but it reflects the academic philosophy of some good, small, liberal arts colleges. A product of such a program, Soja explains it still took several years to develop the confidence to overcome her uncertainties about pursuing science as a career. Soja relished the challenges of field camp. She passes the field tradition to her own students through their integral participation in her field research on remote islands off Alaska. She applauds the breadth of the liberal arts tradition and urges geoscience departments to develop course requirements so that their majors graduate with sufficient breadth to enjoy successful careers in the earth sciences.

Helping other students in the seventh and eighth grades was my first indication that I liked science and could explain it to others. An interest in science petered out or waned for reasons I don't recall in high school. By the time I went to college at Denison University in Ohio, I planned to major in Spanish, but my mother kept encouraging me to take geology or archaeology. I completed a course in geology by the end of my first year, and I was hooked—not just by the subject but also by the instructor. Ken Bork became a mentor during those four years at Denison and continues in that role today.

During high school, I hadn't envisioned going into science. No one in my immediate family was really science oriented. So in college I deliberated, Is this what I want to do? Is this what I can do well? Is this what I should do or is this just a side interest that I really shouldn't pursue? By the beginning of my junior year, after taking more geology and Spanish classes (I wasn't sure what I would do with either subject in a career), I realized that I was captivated more by geology than any other subject. I remember some family members saying, "I don't see you as a geologist or scientist; this seems like a very strange choice for you," whereas others said, "You are an outdoors person; you're learning about the things you've always loved. It fits." I did plan to have a career, but it still took me another two years after college before I decided to pursue geology as a profession. Once I made the decision to earn an advanced degree in geology, I never looked back. I knew at that point that I had enough geology under my belt, had tried out other possibilities, and geology was what I liked the most.

Now that I am a geology professor, having had trouble myself making that paradigm shift during college has helped me be a better advisor to my students. Perhaps I was so slow to see myself as a scientist because I grew up with the classic image that science is only for geniuses and/or nerds, who can readily encapsulate a whole range of factual information and make something of it. What one learns rather quickly in science courses at the college level is that science is a way of asking and answering questions about the natural world.

Field experiences were an important part of my geologic training from the beginning. At Denison, we usually had a spring break trip to splendid geologic sites in the eastern or southern United States, as well as weekend field trips in various classes. We were quickly introduced to the idea that some aspects of geology make a lot more sense (or can be more easily explained) in the field than in the classroom. This became crystal clear at Indiana University's field camp in Montana, which I attended during the summer between my junior and senior years at Denison. That field course was the hardest class I took as an undergraduate, but it was the most beneficial because it allowed me to apply all kinds of textbook knowledge to real field situations. I probably learned more geology that summer than I could have in two or three years of courses.

The fieldwork wasn't easy; in fact it was very difficult. But I encountered every imaginable rock type of every possible age, and we gained experience mapping both deformed and unaltered rocks. Traveling throughout the western United States, interacting with students from other colleges, synthesizing lots of geologic data, and getting to know the faculty really expanded my horizons in geology. It helped me re-

Connie Soja (*center*) and her Colgate undergraduate assistants Allison Gleason and Brian Flynn, staying dry on a boat near tidewater glaciers in the upper part of Glacier Bay, Alaska, in the late 1990s. Photo by Brian White.

alize that I could do field geology even though I needed to improve. By the time I was in graduate school two years later, there were opportunities to do primarily lab-based theses, but by then I knew how much I enjoyed working outdoors.

At the University of Oregon I was fortunate to work with Dr. Norman Savage, who had a large-scale biostratigraphy project in southeastern Alaska. To tell the truth, as much as I was interested in the paleontological project that he proposed, its location was also very compelling. I had never been to Alaska but was challenged by the idea of working in a wilderness environment where the geology had not received much scrutiny.

I've now been conducting fieldwork in southeastern Alaska for two decades [since 1980]. Doing field research there continues to be challenging and exciting—the terrain is mountainous, the coastal scenery is beautiful, the wildlife is plentiful, and the projects have been very rewarding. Because so few geologists have worked in southeast Alaska, it still feels like an unexplored area. The fieldwork is concentrated primarily along the coasts of mountainous islands, which are covered by temperate rainforest down to the shoreline. Except for mountain peaks above the tree line or where landslides have occurred, rocks are accessible for study only in the rocky intertidal zone, and even there they are partly covered by kelp or barnacles.

I quickly learned that fieldwork in coastal Alaska is quite different from doing field research at inland sites. The use of a small, rugged boat [a skiff] is mandatory, as is a rudimentary knowledge of outboard engines and their repair. Reference to accurate nautical charts and daily tide tables is essential so that safe access can be gained to rocks exposed along the shores of offshore islands. Even in the summer, it rains frequently in this maritime belt of northwestern North America. Both because of the rain and our work in the intertidal zone, we wear knee-length rubber boots and are completely enclosed in rubberized fabric (rain pants, jackets, and hats). Generally even in the summer it's cool enough to wear a wool hat and gloves most of the day. The rain and slippery algae create treacherous conditions when working on rocks that are exposed along the shore. Fortunately, because the rocks are usually dipping at slight to steep angles, we can sample a considerable thickness of rock on the beach without resorting to rock-climbing techniques.

As a paleontologist, the focus of each summer's field expeditions has been to map unexplored areas while measuring, describing, and photographing in detail stratigraphic sections. We extract as much fossil material from the rocks as possible, including hand samples with fossils enclosed in sedimentary matrix, and send our samples back to the laboratory for further analysis. Once back in the lab, sample material is analyzed to answer questions about both the fossils and enclosing matrix as well as about the tectonic implications of our research.

I've been working on rocks that formed 350–420 million years ago during the Silurian and Devonian periods. Understanding the paleoecology and paleogeography of organisms fossilized in Alaska has enabled us to develop models that explain where one of Alaska's largest crustal fragments originated. Scientists in the 1980s first recognized that the North American continent experienced growth through time with the accretion of "exotic" crustal material, or terranes, onto its margins. In Alaska and other parts of the western United States, colliding offshore islands, microcontinents, and uplifted pieces of ocean floor became welded to the continent, which added new geologic real estate and created a confusion of geologic relationships for geologists to sort out millions of years later. The magnitude and duration of terrane assembly is still under investigation by geoscientists hoping to unravel this giant jigsaw puzzle from many different angles.

In my research, we compare Alaskan fossils with those of the same age from other parts of the world to determine where a particular terrane originated. Collaborative work with a Russian colleague at sites in Alaska and Russia's Ural Mountains yield an exciting, but not unexpected, conclusion—that the Alexander terrane we work on was always located in the Northern Hemisphere. A competing hypothesis, unsupported by our fossil evidence, proposes that the Alexander terrane is a fragment of Australia, which was transported by plate tectonic motion to its present position in Alaska. Distinctive sponge and microbial fossils in Silurian rocks refute that idea and establish that an earlier Arctic Ocean, the Uralian Seaway, allowed organisms to transmigrate along its margins and to form reefs that can be traced today from Alaska to Russia.

In my dissertation research, I focused on Devonian brachiopods from Kasaan Island, Alaska. These fossils had first been studied around the turn of the century by

E. M. Kindle of the U.S. Geological Survey. He interpreted the brachiopods as having Uralian [Russian] affinities, but at the time he did not have a plate tectonic framework within which to explore the implications of that interpretation. In the 1980s, [Peter] Coney's ideas about "suspect terranes" began to congeal; geoscientists began to realize that the "oddball" fossils, distinctive metamorphic patterns, and abrupt changes in stratigraphy at contact zones exposed in many areas of the western United States made sense in light of accretionary tectonics. Because no one had reconsidered Kindle's fossils from the standpoint of modern taxonomy, Dr. Savage was keen to have a student revisit the Kasaan Island section. It was a wonderful dissertation project—to restudy Kindle's collections and field notes, to augment his and Dr. Savage's material with more fossils that I collected on Kasaan Island, and to make taxonomic revisions in order to address the questions Kindle had raised years before.

It turns out that modern taxonomy guided me to a different set of conclusions from Kindle's and to ask different kinds of questions. I determined that the brachiopods are neither strongly Uralian nor North American in character. Instead, the entire brachiopod biota can be characterized as having mixed affinities and, as such, the brachiopods do not give a clear indication of where the Alexander terrane may have been located with respect to other continents hundreds of millions of years ago. However, the mixed paleobiogeographic signature of the brachiopod communities supported the interpretations made by petrologists and geochemists: that this part of Alaska was once a series of offshore, volcanic islands located some distance away from continental margins. On Kasaan Island, because a significant proportion of endemic taxa coexisted with species known from many other continents, this indicated to me that colonization of these offshore islands was largely the result of chance and the oceanographic transport of larvae from distant sites.

Recognizing that this unusual suite of fossils represented an ancient island allowed an interesting set of evolutionary and paleobiogeographic questions to be addressed. For example, can ancient biotas be used to determine where an island was once positioned before becoming tectonically "glued" to a continent as an exotic terrane? Did island biotas in the past serve as sanctuaries during times of ecologic crisis for organisms that later pioneered recovery phases after mass extinction events? Or do islands act as repositories for organisms that persist as relict taxa before going extinct as isolated evolutionary dead-ends? Fossil biotas that once inhabited ancient islands are still a rarity in the geologic record. With the completion of my dissertation, it was clear that answering these questions would require more data and would also have implications for conservation research as our modern biodiversity crisis unfolds.

Involving students in research and fieldwork is an important aspect of the educational process at many undergraduate colleges. My NSF grants have been awarded largely through the program that funds Research for Undergraduate Institutions (RUI). Those grants stipulate that undergraduates must undertake substantive independent research with faculty supervision. These funding opportunities have been important because they have allowed me to encourage students to design, implement, and evaluate the success of projects they have chosen to pursue in the field with my guidance. Once students take ownership of a project and see the rocks and fossils as "theirs," my experience has shown that their hearts and minds become more fully engaged in the science itself. My role is to help students identify questions that can be addressed in either a semester or a year and to help them see the big picture. Physically, the students are a tremendous boost to my field research because managing the boat, coastal navigation, and sampling would be impossible without their help. From a research perspective, the more people involved in a project, the greater the likelihood that no rock will go unturned and that a diversity of ideas will be considered as the research progresses.

I usually select my summer research students from those who are just completing my paleontology course each spring. Naturally, I prefer to work with students who have already shown an interest in paleontology, but I do not expect necessarily that they plan to attend graduate school or pursue a career in that particular field. Working with students in Alaska on invertebrate fossils becomes the perfect educational venue. Not only are the students learning a tremendous amount by doing their own research on fossilized organisms, but they can also see an incredible diversity of marine life and make comparisons between modern and ancient members of the same phylum. For example, in my paleontology course, we study fossil echinoderms, and in Alaska we can observe living starfish, sea cucumbers, and sea urchins moving over rocks that are jam-packed with fossil crinoids [types of echinoderms that are common in the fossil record]. This diversity of past and present life in Alaska provides many wonderful opportunities to reinforce what we have discussed in the classroom and to integrate that with information from the field and lab.

As we attempt to answer a particular set of questions, invariably our research raises exciting new ones to explore. I encourage my students to consider questions they'd like to answer and to make sure before we leave the field that they have the appropriate set of samples to address those questions. When we're back in the lab, students use rock saws to do the preparatory work necessary to analyze specimens under the microscope. Concurrently, they begin a thorough review of the literature, and by semester's end a synthesis emerges that, with hard work and determination, is satisfying and fulfilling to everyone involved.

I also have learned from experience that the students who are the most successful at field research are outdoor enthusiasts. Camping in southeastern Alaska in inclement weather often involves living in very close conditions in a tent or RV [recreational vehicle], experiencing rain and cold for days at a time, a lack of privacy, and limited access to mail, telephones, and hot showers. Those of us besieged by email messages or the constant ring of a cell phone consider fieldwork a blessing where these technological advances have not penetrated (or are not affordable by research teams). Fortunately, simple but invariably tasty food is a welcome salve at the end of a long day in the field. Team spirit is a real plus in the field, but so is having an aptitude for science and being able to tolerate the moments of both tedium and serendipity that are an integral part of research.

Many of our students find that the research they begin in the summer and complete in a fall semester course is a real challenge. Successfully completing a research project requires a tremendous amount of self-motivation and dedication. Putting their ideas on paper and using appropriate scientific terminology are difficult for most students. As research advisors we provide support along the way, but the students who derive the most satisfaction from their projects are those who apply themselves and remain focused from the beginning to the end.

Students gain additional experience by preparing an abstract of their research and presenting results (usually in the form of a poster) at a regional geology conference each spring. That becomes the capstone experience—preparing an abstract for publication and at the meeting explaining to professional geologists the significance of their research. For many students, this is the moment when the excitement of discovery and of explanation crystallizes, and it's wonderful to see.

At an undergraduate institution, my goal has been to keep my research broad enough so that students with diverse interests can become involved. My field areas in Alaska have made that possible. And having a multitude of projects to supervise has made the science for me far more creative, new, and exciting than it might have been if I had focused only on brachiopods for my entire career. Certainly, if I'd done that, I would know more about brachiopods than I do, but I really like being challenged by the range of projects that we have tackled together.

In fact, when I came up for tenure two years ago, I was expecting some people to be critical, to ask why I hadn't published much recently on brachiopods, what happened to the brachiopod specialist? In contrast, what I heard was that some of the reviewers thought that my diversity of interests was a real strength. That is the wonderful thing about having a flexible curriculum here; we can teach the kinds of courses that we want, and we can pursue our research interests with tremendous support, including generous funding.

Our curriculum does a very good job of preparing our students for graduate school, for a career in geology and related fields, or for a profession in an area outside of the sciences. At a liberal arts college like Colgate, we have an obligation to ensure that our students are truly well educated in many subjects in the humanities, social sciences, and sciences. Across the curriculum, we emphasize the importance of critical thinking and the ability to frame cogent arguments well both orally and on paper. Our open curriculum in the geology department serves us well because it allows us to attract a broad range of students into the department through a number of introductory courses, like oceanography, physical or environmental geology, or evolution. We recommend that students then complete five core courses in mineralogy, petrology, sedimentology, structural geology, and paleontology. As students move to the advanced level, each of us offers a course in our particular specialty, such as tectonics, acid rain, or reefs. We also strongly recommend involvement in research during the senior year as well as a year of chemistry, physics or biology, and math. Many of our students, particularly those who intend to go to graduate school, complete some if not all of these courses.

New students and faculty have introduced other ideas to consider so that our curriculum continues to evolve in response to our own interests and changes in the profession. Our commitment to innovative research is stronger than ever. We are particularly gratified that a study by Franklin and Marshall College indicates that we are number two in the nation for sending geology students who matriculated at liberal arts institutions into graduate programs in geology. A more recent study in the *Journal of Geoscience Education* reveals that this success is related in part to our proactive research program, which has earned us the rank of either number one or number two for the total number of journal articles, pages, or abstracts published by an undergraduate geology department in the last 25 years. It's an exciting time to be in geology, and it's an honor to be a member of a geology department that is helping to train the next generation of geoscientists.

As you get further on in your career, you start to put things more in a larger contextual base; you start to integrate things on a different scale.

DENNIS HUBBARD (b. 1949)

Marine Consultant, Virgin Island Marine Associates
St. Croix, Virgin Islands
presently: Visiting Associate Professor of Geology
Oberlin College
Interviewed in my office 5 January 1998

Dennis Hubbard, entering the University of Massachusetts as a pre-med, was overwhelmed to find a biology department with more majors than there were residents in his home town. Finally discovering the small and friendly geology department, he was rapidly incorporated into all activities, including field research. Hubbard explains the dual advantages he experienced as a resident scientist at small West Indies Lab, owned by a nonresearch university: (1) teaching was ideal, with the philosophy that students learn how to ask the questions suggested to them by their personal experience in the field; and (2) with no pressure to publish, he found that he had time to work on research projects until he was satisfied with the product.

My involvement in geology, and where I am now, is the result of a number of unpredictable circumstances. When I started at the University of Massachusetts, I was a biology major, pre-med. The biology department had an extraordinary number of majors and seemed like a big cattle farm. All through high school I assumed I was going to be a doctor. At the end of my sophomore year in college, I decided that wasn't what I was going to do. Number one was academic performance. Coming out of high school, I had high grades and high SATs. Coming from a town of 3,000 people and into a university community that was literally 30 times the population of my home town, I was overwhelmed. I was in a very large, impersonal department, complicated further by the competition of pre-meds.

Here I was, halfway through college and still looking for what I wanted to do. My ultimate decision to go into geology goes back to a number of things. One is having been outdoors all my life. The Appalachian Trail is the main street of the town I grew up in. Another is somewhat cliché: when I was in my junior or senior year of high school, there was a regional premiere of Jacques Cousteau's black-and-white movie *Silent World*, and I got into sport diving. This was a time when you didn't need a certification card, and the YMCA course seemed like training for the British commandos. I was looking for a discipline that worked in the field. I pretty much ruled out marine biology because I was fed up with the biology department. There was a very active field geology program at UMass. Despite my misgivings based on a high school geology course, I checked around and found that it was a field-oriented program in its general training and also that it provided opportunities for undergraduates to work in the field helping graduate students and professors. And it was a small department. It had only 60 undergraduates. That was the other end of the spectrum from biology.

I went to summer school in order to be able to graduate on time. My advisor figured out a fast track, where I took 18 credits of lab courses a semester. In hindsight, I think it was a big advantage. Because I was taking mineralogy at the same time I was taking sedimentology, paleontology, and so forth, I started to see the interlinking among them. Otherwise, it might have seemed like an unrelated sequence of courses. It was a very friendly department. It had Christmas parties for the faculty and graduate students, and the undergraduates were always included in these things. It was a close, family-type environment, which was the sort of thing you'd get in a small school. I think the fact that it was such a small department had a lot to do with my success. Suddenly, I was in a program where everyone cared how I was doing. It kept me on track. So I went from being a nominal student to straight As. I just enjoyed it.

The third day that I was a geology major, I was dumped into a field somewhere, given a map, and told to work out the geologic structure. We were mapping a large area, and we were each given a piece of it. We had to talk with other people and find out what they were seeing. The interaction was encouraged rather than frowned upon, just like "real" science. I think that was the first sense of working with colleagues that I had ever seen, which was 180° from my experiences in biology, competition like in the pre-med program. In retrospect, I think the difference wasn't so much from the way biologists and geologists really work. It is sheer numbers. Going from many thousands of undergraduate majors to just 60 makes it easier to get more interaction. The UMass biology program was well funded, and they were doing lots of cell and microbiology before it was the hot thing. It was on the cutting edge, but also at the beginning of what I would consider a real decline in biology—forgetting that there are things that are larger than cells.

That summer I was a field assistant for one of the Ph.D. students. I spent 15–16 weeks with him on the coast. I was hired partly because I was one of the few people in the university, let alone the department, who had diving experience. He was doing underwater work, looking at sand transport. He had a reputation that no one could get along with him and that he pounded field assistants into the ground. I ended up getting along just fine.

After that, I mainstreamed into the coastal geology program. Although I was doing structural geology and all of the regular courses, it was pretty obvious to me that the coastal group was doing something with an interesting application. I ended up working the next two summers as a field assistant. My advisor had a lot of students, but most of them stayed around quite a while. He had it set up so that at least half of what you learned was from graduate students who had been in the program. It was like an apprenticeship. On the other hand, if I had a problem, I could walk into his office and he'd put down what he was doing to talk with me. He never went around to ask graduate students what they were doing or how it was going. The move was yours to make. When he took on graduate students, he would never look at their grades. He looked at recommendations, and he never took a student without an interview. He tended to attract people who were very like-minded. I guess we were kind of an elitist, snobbish group, but it seemed to work. Many things that I've tried to do as a scientist go back to my experiences in graduate school.

I tend to be the kind of person who likes to work things through and finish them. My advisor was just the opposite. He went all over the world, and he encouraged us as graduate students to take advantage of it. So I went to Iceland, but it wasn't part of my thesis work. I went to the Bahamas as an aside. That's how I became involved with carbonates, so I guess it points out the value of these graduate student junkets.

I still remember when my advisor came to visit my field area. He stayed for about six hours. I showed him my data, and he said, "Very nice. What problems are you having?" When I explained a few things I couldn't figure out, he said, "Let's see if there is an airplane available." So we went off to the airport and flew over the area. Within ten minutes he had the strategy all worked out to answer my questions. Sure enough, within two weeks, all of the questions were answered. He was the person who could orchestrate things and get people headed in the right direction. On the other hand, he was a brilliant field scientist who had an incredible observational eye, like a modern [Sir Charles] Lyell or Darwin, and that had a lot to do with how I do field science today.

You feel that having gone on these field trips with the coastal geology group was valuable experience?

Oh, yes. I think that is what gives me some of the perspective that I have today. You've been through it. It's easy to tell where someone is in the educational process just by talking to them. The master's student knows a lot about a very specific thing but has difficulty relating it to anything outside his field area. Recent Ph.D. graduates will take a specific subject and talk about it over a broader spatial and conceptual scale, but they are pretty much confined to that subject. As they get experience, they realize that there is something outside of their specific interests. They start to interact and complement or contradict one another. As you get further on in your career, you start to put things more in a larger contextual base; you start to integrate things on a different scale.

I've been fortunate that the philosophy of breadth has stood me so well. My graduate training was in tidal clastics. The first thing I did when I finished my degree was spend 20 years in carbonates. Being able to draw on both of those fields and recognizing the interaction among biology, chemistry, physics, physical oceanography, and geology are valuable lessons that some people never learn. So many researchers are just doing lip service to those other fields. Starting with graduate school, I have always worked in programs where, as geologists, we worked with biologists, with engineers, with physical oceanographers—and it just seemed to work naturally. It wasn't anything that I did consciously, although I'm sure my advisor did. He thought long and hard before he put together the coastal research group. It included a wide variety of people and disciplines. Other than these few people we worked with, we would have a very difficult time getting people outside of our discipline interested in what we were doing.

With students, I think I can honestly say that it's generally been a real two-way experience. In most cases, I've gotten as much from them as they get from me—and that may be the key to the success. I think, in general, students have gotten better and they've gotten more focused, which is both good and bad. Good in that writing skills are better, and I'm basing that on what may have been in some ways the cream of the crop coming through West Indies Lab near the end. Communication skills were pretty abysmal when I first got there and worked with student projects. I think the main thing I got from students, other than the cheap labor, was critical questioning. I think you would probably not get that from the broad cross section of undergraduate students. I think it was the caliber of students that we had. They would ask, "Why do you say that?" You start to explain it and realize that you had never really thought it through. You had always taken it for gospel. This forced you to go back to the basics, to the building blocks.

I remember back in graduate school we had a reading seminar that my advisor organized. It taught me more about critical thinking but also developed a real suspicion for the literature. It is not always as clear-cut as it seems. That is one thing I always remember from the lab students. You'd give them an article, and they'd read it as though it were a textbook, the gospel. Then they'd go through their field projects and have problems anywhere from boats that wouldn't start to an uncooperative octopus that wouldn't respond to stimulus. Their first thought was, this project is real crap, I haven't accomplished anything, everything has gone wrong, it's not working. At this point, you do two things. First, you encourage them to see what they do have rather than what they don't, which is the way, as you know, any research project really ends up. You rarely end up with the questions you started with. Then the second thing you do is explain to them that this is the way most research goes. When they read that paper six weeks ago in class, they thought it was gospel and clear-cut—this guy just went through all the steps and got the numbers! But now they are thinking, maybe his project looked a lot like mine. Then the light goes on. No one wants to read the paper that says, "The boat wouldn't start; the chemicals were bad; and the experiment didn't work like I expected." So, you focus on what did work. The end result is that the paper always looks clean. You can lose sight of this, unless you've had experience with research and remember that this is the way it often is. Now, you're more suspicious of those papers that looked so neat and clean.

I always made the analogy that we are giving them the opportunity as undergraduates to break glassware along the way before they break glassware that really counts—as part of their master's thesis. They got exposed to all of that, and I think that has a lot to do with why West Indies Lab had such a disproportionate number of students going on to graduate school and why the lab has a disproportionate number of people who are now out there in the business as consultants, faculty, environmental engineers, and successful geoscientists, who survived the ax when Gulf Oil cut half its people. I think it was because they had the chance to get a leg up doing independent research before they did their master's thesis, that first rude lesson that most of us get exposed to in graduate school. They got it a little earlier. I got it but in a different way. I worked two seasons as a field assistant, so I got to see the guy I was working with make mistakes. But that early exposure is critical.

Looking back, it is an odd juxtaposition of coincidences that put me where I am. Think of what happened to all of us since the lab closed [1990, a year after Hurricane Hugo]. It was devastating. My main regret now is the loss to the scientific community, but it was obviously a large personal blow to all of us. On the other hand, the last three or four years I've spent consulting have given me a different perspective. It looks like I'll go back into the academic mainstream again now, but those years have given me a different attitude. If I had applied for a job from West Indies Lab, from academia into the USGS or into another university setting, my perspective would have been very different. I listen to colleagues complaining about how hard they are working and how they have this or that project to do. But the bottom line is that you do it because you want to do it, and I guess you complain because you can afford to.

When I was in graduate school, I realized I had a level of academic freedom that I would probably never have again. At a party after I defended, my advisor said, "Enjoy doing your dissertation. It's probably the last decent piece of science you'll ever do in your life," meaning, of course, this is the last chance you'll ever have to say, "OK, it's done because it's ready and it's the way I want it." You come out of graduate school, and a year later you are part of a nameless group looking for jobs. Whether you're a consultant, a government worker, or a faculty member, you are obliged to turn out the one or two or three pieces of crap a year that are needed to keep your vita growing at an acceptable rate. You've got to finish a project up because you have to move on. You've got to write that next grant, you've got a technician to feed. You just don't have the luxury of time any more.

I was very fortunate at the lab. I think the one real positive side of being at a very well known facility that was administered by a second-rate university was that "publish or perish" was not the rule. I would never have been able to write the carbonate budget paper had it not been for the luxury of waiting until I thought that paper was right. Obviously, having access to the field was essential, but I think that with the same facilities working out of someplace else, I just wouldn't have been given the freedom or the time.

Take someone like Conrad Neumann at Duke. Here's a scientist who published two articles in his first four years after graduate school, and they both won paper of the year in the journals where they were published. But, when it came to tenure, he only had two papers. Today, you have a person whose first paper says, "I think it is this way," his second paper says, "No, it's this way," and his third paper says, " I guess it could be either way." Now, he has three publications. I think that's what makes it so difficult to do good science. It's a death spiral, because there is now so much crap coming out that, as a researcher, you've got to keep reading this stuff to stay current. It's sensory overload. Papers are not always done in a scholarly fashion any more.

Trends in science? We are becoming more introspective. Even though we are exploring the planets, I think that both biology and geology have tended to look smaller and smaller. I realize that it is important to study the building blocks, but I think that we often tend to forget that those are part of a larger system. The other problem is that so much of what we do today seems to be driven by getting the next level of funding. My success is going to be driven less by coming up with cutting-edge things than by whether the cutting edge is salable. It becomes a marketing thing. About the time the lab closed, the gravy train came around to what I'd been doing for 20 years. People were getting excited about things we were already doing, and I was very fortunate in not having to compromise and say, "Oh, it's a hot new thing, so I'd better declare myself an expert!" Too many scientists make their living by just jumping from one gravy train to another and setting themselves up as self-proclaimed experts. The problem is that the system never seems to weed these folks out. I was proud of the fact that most of us at West Indies Lab kept on with what we were doing all those years, and the gravy train came around to us. This was a kind of self-vindication that what we were doing was worthwhile.

Postscript (May 1999): When I worked at WIL, I would sometimes just stop and think, I can't believe they pay me to do this. I had forgotten that feeling until I got back in front of a class this year at Oberlin. We all complain about the workload we are under, but we should never lose sight of how incredibly fortunate we really are. As a former classmate put it, "Sure beats working in a felt mill!"

It is one thing to offer students questions, but it is another to have the questions originate with them.

Myra Shulman (b. 1954)

Lecturer, Department of Ecology and Evolution
Cornell University
Interviewed in her office 13 November 1997

An NSF-sponsored high school ecology camp introduced Myra Shulman to a variety of terrestrial ecosystems but, just as important, to bright youngsters from around the country whose knowledge of natural history was impressive. She emphasizes the importance of an interactive atmosphere within an academic group large enough to encompass a diversity of disciplines but small enough to foster dynamic exchange. For her, these included undergraduate years at UCLA with the ecology group, her grad school life at the University of Washington, and teaching and research field experience at marine labs (including WIL, where we first met). An early love of ecological concepts prompted Shulman to take every possible ecology course she could at UCLA in lieu of some physiology, morphology, and natural history courses.

When I talked with TOM CALLAHAN *at NSF, he deplored the fact that most schools are giving up the tradional "-ology" courses: mammalogy, ornithology, invertebrate zoology, and so forth.*

I've always felt that job candidates should be able to sell themselves as having expertise in a taxon as well as in a conceptual area, whether you're a physiological ecologist, morphologist, behavioral ecologist, or whatever. The problem is that when you get down to the molecular biologists, they don't have the -ologist training. But then, I never received that training either. I could call myself a coral reef biologist in the sense that I have fairly general knowledge about the groups of organisms which live on coral reefs, but I was never an ichthyologist, although I worked on fish. I think people should receive that training, and I've frequently regretted that I did not have it.

I did a summer program for high school students, a terrestrial program out in the deserts of Nevada. Descriptive ecology. It was great. It was the whole camping thing over again, which I really enjoyed. We did the desert; we studied Pyramid Lake; we studied alpine lakes. There were probably 20 kids from all over the country and that was what was really amazing! I remember there was this one girl with really long red hair from Texas. She had this Texas southern accent, which I had always associated with dumb. Dumb! But she was so smart.

So, meeting people from other parts of the country and having these prejudices . . .

Which I hadn't even thought I had. But when I listened to what she had to say, I said, not so dumb. In fact, smart, very bright! There were people there who knew every bird we'd see. They were really bright and they were really into natural history biology.

I got a lot of encouragement for my interests at UCLA. I knew I wanted to be an ecologist already, so as soon as I could I took ecology, and you could take other -ology courses, and I did. I took herpetology, I took invertebrates. When I was in graduate school I took entomology and ornithology. When I was a sophomore, I was taking ecology in the winter. For the very first time, they [Martin Cody and HARRY THOMPSON] were going to offer the field biology quarter (FBQ). I went into Martin Cody's office to talk about the course. We were going to do a lot of camping and so he asked what experience we had had, etc., and then he said, "We've had a lot of requests from seniors, and we have to consider seniority." I said, "Does it make more sense to you to accept seniors who want to go to medical school or a sophomore who wants to be an ecologist?" I got in.

UCLA really focused me, probably too much, in ecology. I took every course that was offered, every field experience, and graduate seminars. In retrospect, I would have told an undergraduate, "OK, you are going to do this. You are going to go to graduate school in ecology, but meanwhile . . ."

Take more physiology, more morphology.

Yes, a broader education would have been nice. When I got to graduate school, I would say that I knew a lot more about ecology than the graduate students who entered with me, but my breadth was definitely less. The field experiences were great, though. I did the FBQ twice. The second time was the first time they offered a course in marine ecology at Catalina.

Your first marine fieldwork?

Yes. For me, the choice of marine versus terrestrial wasn't an issue. For a lot of people, like [my husband], JIM MORIN, the ocean just calls to them and that's it. But for me, it was the ideas in ecology and now in evolutionary biology that always call to me. What system or what organism doesn't make that much difference to me.

When I applied to graduate school, I decided to apply for the best advisor and graduate school combination, not whether it was terrestrial or marine. I was accepted at a number of places. I decided to go with Bob Paine at the University of Washington. It was obviously a good choice. I had always wanted to go to the tropics. Somehow, I got that bug in my brain. It must have been something I learned at UCLA. When I got to UW, Bob said, "Every student I've seen here who has gone to OTS [Organization for Tropical Studies] has ended up to no good end. I really don't want you doing that." I thought, hmm, am I being given a choice here or not? But, it wasn't like a parent-child relationship. He could have afforded just to say goodbye. You know your parents love you, no matter what. Bob didn't want me to take the OTS course. In retrospect, it would have been a good thing. But, he may have been concerned that I would have ended up doing terrestrial stuff. I don't know what the issue was. So I said, "All right."

I was looking around to consider my options. I knew John Ebersole, one of Martin Cody's graduate students when I was at UCLA, and he was working in St. John. Bob Paine suggested that I write to John Ogden and I did. I found out that John and Bill Gladfelter were giving a fish course at West Indies Lab, so I went. It was 1977. You were just going off to graduate school at UCLA, and I had just finished there, so we met briefly and talked. I took this fish class. Tropical fish ecology. All of my knowledge of tropical fish identification, I owe to Bill. We sat there with those slides and went over them. He told us what to look for. This last fall, I taught a tropical fish ecology course at UCLA for the marine biology quarter held in Hawaii. The fish there were pretty easy to work with. I dragged out my notes from that class. Bill would have us out there counting fish on all of the patch reefs at Tague Bay.

There I was in St. Croix, taking this fish class, and I decided I wanted to work there. Basically the kinds of questions I looked at were the effects of predators, the relationship between predation and competition, competition for shelter, available shelter for various fishes, especially recruits. So my questions were generated both from the literature and being out there on the reef.

A number of the questions I've worked on have arisen through curiosity. Why do damsel fish change color from the juvenile to the adult stage? Do I have the answer to this question? No. It is just one of these things that jumps out at you. Any undergraduate who is out there will ask, "Why are the juveniles so brightly colored and the adults so dull?" Other things have been driven by certain ecological events, like the *Diadema* die-off in Panama. I was down there on my postdoc. In fact, I was there to study why damsel fish change colors, but while I was down there, these sea urchins started keeling over. So I found myself counting sea urchins. Ross Robertson and I started doing this stuff on sea urchin–damsel fish interactions. We could have done that research before this, but it was stimulated by asking, "Jeez, what's going to happen now that all these really important herbivores are disappearing?" There it is driven by natural events.

The nice thing about established marine labs, like Friday Harbor, Shoals, or West Indies Lab, is that you get the history of the site and the expertise in different aspects of biology and geology. For instance, when I came to the West Indies Lab initially, I hadn't had much marine experience. One of the most useful things for me was talking to Mike Robblee, a graduate student working there, who had lots of field experience. Any simple thing—how to mark quadrats, where to get surveyors' tape, basic field methodology—somebody has to guide you. I mean, you could make it all up yourself, but it would sure take a lot of time. Mike was incredibly helpful to me. Someone else who was really a positive influence on me was John Ebersole. We were down at WIL together one time after he had finished his degree. He said, "Remember, Myra, how Jim Morin had us keep field notebooks? We may be working on fish but we need to learn about the rest of the system. We need to learn about corals and algae." As a result, we worked each day on learning new things, learning the corals, for instance. I kind of accept that now as the obvious, natural thing to do.

When I finished graduate school, I had a Smithsonian postdoc. San Blas, the STRI

[Smithsonian Tropical Research Institute] research station, was living on a sand patch with very intense interactions, more intense than St. Croix. Back in St. Croix, there was lots of room, things to do, and lots of people with whom to interact, for instance, the whole geology group. DENNY [HUBBARD] was there and Ivan Gill, Arnie Miller, and Dave Jacobs. San Blas was a much more primitive field station than WIL, in terms of scientific facilities. In San Blas, everyone had his own boat; it was right outside your house, and you could just get in and set off. Other than watching the Indians or talking, there was nothing else to do but to do research or to read. The scenery doesn't change. It was interesting. A high proportion of the people on the island were assistants. There weren't actually very many people with whom to talk science. It was restricted.

That brings up a concern I have had. There are a lot of programs today, particularly government ones, with a lot of money for science. Many are directed by a Ph.D., but that person doesn't go out into the field any more. Basically, the fieldwork is given over to someone who has a master's or a bachelor's degree. These technicians are often conscientious, they can collect data in a set way, but they just don't have the perspective to make the crucial observation.

The worst part about that is that a course is set: we are doing this, and nobody is thinking about it. Nobody is looking at it and saying, "Wow, there are some interesting things happening. I should be looking at this; I need to adapt as I find things out." A lot of the fisheries research is like you described. They go out and sample things time after time after time. Maybe somebody thought about it at the beginning. But, you know, that isn't how you do research.

When you do research, you think about something, you go out there and try something, and you say, "This isn't what I thought, and maybe I better do things differently." It is an interaction between what you observe and how you interpret it. It is interesting to watch that process with students in proposals. They've made an observation. In fact, I do this myself. You remember something as happening a certain way, and you decide that you are going to do this, this, and this. Then you go out into the field, and it isn't the way you remember it. You started with one observation. The next ten times, the fish does something different. You realize that you have to change course and adjust. This is the way real life is—not just in research.

I was talking to a postdoc here in scientific education who is doing a study on large lecture courses: how and what students learn. I asked, "How do you get students to think, to come up with questions in evolutionary biology?" It is one thing to offer students questions, but it is another to have the questions originate with them. [Gordon] Orians at Washington would have a 20-question thing, probably not unique to him. He would take students somewhere, an aboretum, a natural area, and ask them to identify questions about what they saw. This postdoc said, "Nelson Hairston does it in the ecology course here, but she said she doesn't think it works too well, because it doesn't go anywhere." The students never get an opportunity to try to seek the answers to their questions. Whereas, in the field marine ecology course, I would have the students choose one, two, or three questions that they are interested in personally.

Then we would discuss these in a group: what hypotheses might account for this, how would you go about testing that, and so forth. Then the students would have two weeks to do a research project on it. So it was very clear to the students why they were doing this, why they were generating questions.

These weapons were built to be used, not dismantled, and it was a huge technical issue.

PHILLIP LOBEL (b. 1953)

Associate Professor of Biology
Boston University Marine Program
Marine Biological Laboratory
Interviewed in his office 23 October 1997

Phil Lobel was an undergraduate at the University of Hawaii in the heyday of Pacific field expeditions and a grad student at Harvard, where field research was (and continues to be) generously supported through private funds. He is clearly driven by a love of fishes cultivated since early childhood. He is an ichthyologist, a field that seems an anachronism in modern academia, although he extends his study of questions of fish biology (such as the presence and function of vocalizations in fishes) with the use of specialized tools. Lobel's work with the military, helping to dispose of stockpiles of chemical and biological weapons in an ecologically sound manner, foretells of ever-increasing numbers of environmental problems, which mankind will be forced to confront in the near future.

Some of the people from the Cleveland Aquarium invented Instant Ocean for use in saltwater aquaria far from a natural source of seawater. When I was 14, I started volunteering when they were developing the synthetic sea salts. My job involved lots of grunt work in the lab.

It gave you good formal chemistry training and practice.

Sure, at 14, 15 years old they were teaching me Winkler titrations. When I was about 15, Jerry Klay [a premier field collector of tropical marine fishes whom I had met at the aquarium], Glen Northcut [now a neurobiologist at Scripps], and Joe and Sally Bauer [both M.D.s] went down to Florida to some of the big shark tournaments to collect the brains of the landed sharks. This was some of my first ocean diving and the first time I was diving with scientists. Ray Cray, another contact through the aquarium, was one of the first aquavets. He had set up an animal-importing facility in Barranquilla, Colombia, and he also had contacts on the Amazon River in Leticia. He helped me arrange my trip to the Amazon and hooked me up with a local collector. From Leticia, the native collector and I would go off into the jungle for a few weeks at a time, collecting fishes. I missed a few months of high school, but my grades were good, so nobody was upset. Of course, my parents supported it. I knew how to camp and had spent a lot of time hunting, but my father had also taught me survival and combat skills. He had been a soldier during World War II in the Pacific.

When I applied to the University of Hawaii, very naively I wrote in my application letter that I wanted to bring all my fishes, my pet python, and my sea lion to Hawaii. I went to Hawaii to work with Bill Gosline. He had worked with catfishes and other freshwater fishes in the Amazon, as well as being an ichthyologist in Hawaii. I was really interested in freshwater fishes. At home I had about 30 or 40 aquaria with African fishes. Being an undergraduate at Hawaii was pivotal. I was there during the heyday of Enewetak and the Pacific expeditions. I spent a few summers at Enewetak as the assistant manager of the lab. There were two or three of us running the boats, and we got to meet all of the scientists coming through. The University of Hawaii was at its peak field research period. There was a lot of monitoring of the atomic bomb tests, and there was broad economic expansion, so there were expeditions going to various islands. Marty Vitousek and Jack Randall used to sail to Tahiti and pump air to one another with a hookah rig. Marty had a 70-foot boat and a twin-engine airplane. He was a UH physical oceanographer running all of the remote Pacific Island stations. He had an airstrip at Fanning and a tugboat there, so he could get out to the more remote islands. He set up the first temperature and tide stations on several of the remote islands. His base was Fanning, a Gilbertese island, where the majority of the 400 people spoke only Gilbertese. During my undergraduate career, I probably spent two months there assisting a postdoc student, Deetsie Chave, and I went back to do my graduate work there.

I started working on my honors thesis with Gosline during my freshman year. He started me by saying, "Pick a common fish, a Hawaiian endemic, and learn everything you can about it. These are the things you do; you look at reproduction." You have a focus. So, during the four years I was in Hawaii, I studied the pygmy angelfish, *Centropyge potteri,* and everything about its biology. One of the things I did, being an undergraduate with lots of energy, lots of time to go diving, and none of the worries of the world, was to collect lots of fish. They were very abundant. This was before commercial fish collectors, environmental degradation, and everything else that affects fish abundance on reefs. I ended up spearing hundreds of fish as part of my study.

At the time, no one knew where and when most reef fish spawned. We did know about damsel fish, parrot fish, and wrasses that spawn in the daytime. I started collecting *C. potteri* and looked at their gonads. Under Gosline's advice, I collected fish every single week, getting very high sample resolution. Prior to this, most biologists who worked in the tropics were trained in the temperate zone. The standard protocol for studying fish reproduction was to sample once a month, look at the gonads, and

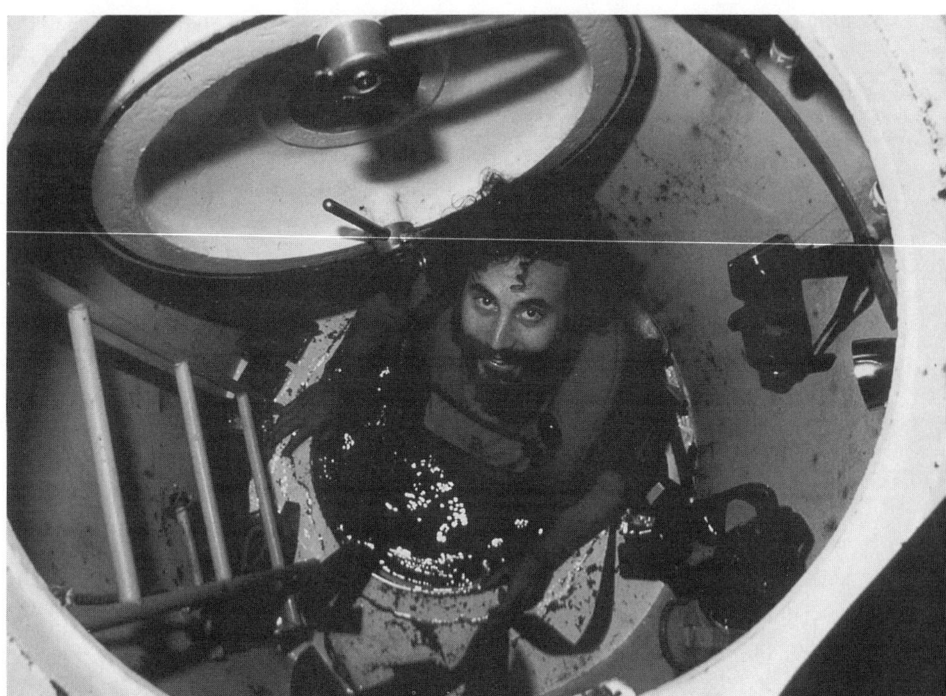

Popping up the hatch, Phil Lobel returns from a field excursion to the underwater saturation diving facility, the *Hydrolab*, in St. Croix, Virgin Islands, sometime in the late 1970s. Photo by Clem Bowman.

you detect a nice pattern. In the tropics, monthly sampling results in data scattered all over the place. The conventional wisdom was that, since temperature didn't change much throughout the year, and the gonad data were all over the place, spawning occurs randomly and pretty much all the time in the tropics. I was interested in this problem. I started analyzing the gonad data that I had collected every week. There was a clear lunar cycle and a clear annual cycle. I thought, I've got to get in the water and look at this. On one trip back home, I went back to Florida to visit Jerry Klay. He told me, "I have *Centropyge* spawning in my aquarium." He was a good observer. The *Centropyge* in Jerry's aquarium always spawned just before the lights went out around 10 or 11 P.M. I watched their behavior and thought, hmm, maybe they are spawning in the evening. Of course, I didn't know if the fish normally spawned at 10 or 11 or if it was due to Jerry keeping the lights on in the aquarium until that time, and they spawned before the lights went out.

Back in Hawaii, a friend helped me launch the University of Hawaii whaler at 3 in the afternoon with about ten scuba tanks loaded on board. I went out to a small reef where I had been studying these fish, out in front of Waikiki, in about 30 feet of water. Back then, that reef was gorgeous. I planned to spend as much time as it took in the water so that I could determine when the *Centropyge* spawned. Seeing the fish in Jerry's aquarium clued me in to what to look for in the field. I planned to get it on film. Fortunately for me, the fish spawned at dusk. In captivity, they were cued to the normal time the light went off. I got photos. I told Ernie Reese I had pictures of the fish spawning, and he said, "Sure, of course you do." No one believed me until I actually showed them the pictures.

By the time I met you in 1975, I had graduated from Hawaii, and I was at Harvard working with Karel Liem. He was really into African cichlids. He refined our understanding of their pharyngeal jaws, their morphology, and the capabilities these fishes might have to adapting to their environment. Karel was a lab scientist and, at that point, no one was studying these fishes in Africa, in the field. I had maintained my interest in African cichlids. I wanted to use the experience that I now had, diving and studying fish underwater, and take that into the African lakes to study the fish there. People had been doing this extensively on reefs, but at that point, nobody had done this in Africa. However, in 1975, Africa became a hostile place to work. Students at an African field station were kidnapped. Also, the funding didn't come through. Karel wrote a proposal to NSF and had me in to do fieldwork, but the reviewers said, "You are a great lab scientist, but you know nothing about fieldwork, and you shouldn't have a field component to your work." We were going to resubmit the following year, but I didn't want to wait two or three years to see if we could get the funding. I was already pretty advanced, and I didn't want to end up spending six or seven years on a Ph.D. I had already published three or four scientific papers based on work I had done as an undergraduate.

It was common to read in the literature that acidic stomachs couldn't possibly occur in marine fishes because of the buffering capacity of the environment. I thought, *this* is an interesting project. I asked the question: Is stomach acidity a possible mechanism for lysing algae in marine fishes? My first academic year at Harvard, I took Jane Lubchenco's ecology class. Of course, she had been working on grazers and the effects they had on communities in the rocky intertidal. I also remember [Larry] Slobodkin from Stony Brook giving a seminar: "Oh, the whole world is green, does that mean food is unlimited for herbivores?" I decided to do my paper for Lubchenco's class on herbivorous fishes, to review the literature and to think about this question.

Harvard is an incredibly great school. They had various in-house grants, for which you could write a proposal, which I did. I requested money to buy a pH meter, one that I could take into the field and test my idea about marine fishes using acidic stomachs to lyse algae. I hooked up with Marty Vitousek and went back to Fanning Island, where I spent the summer. Fanning was great, a beautiful tropical atoll with abundant fishes that I could collect right from shore. I learned that to get an accurate reading on stomach acidity you essentially needed to measure pH on a live fish. You had to be in the water, spear the fish, dissect it, and have the probe in its stomach within a minute or two of when you collect it. I had 19 species, showing that fishes with thin-walled stomachs that eat mainly blue-green algae and diatoms lyse the algae using stomach acidity and that mechanical tritering stomachs were basic. I came back and worked that up. John Ogden visited Harvard and gave a seminar. I talked to him about my work, and he told me about the seagrass project, and that's what brought me to St. Croix. He had told me about some preliminary work he had started along with Ileana

Clavijo with the bucktooth parrot fish. I came down and thought, what an incredible system! While at the West Indies Lab, I had the good fortune to be student and dive-slave for Ogden, John Lythgoe, and BILL MCFARLAND during their studies of grunt migrations.

I made several contacts in the military as a result of all the work I had done on the Big Island in Hawaii, doing the remote-sensing studies with Allan Robinson. A project came up: the U.S. government wanted to build a chemical weapons incinerator on Johnston atoll. They had to do an environmental impact statement. Jim Maragos at the Army Corps of Engineers was in charge. This was pre–London Ocean Dumping Convention. The government wanted to dispose of the scrubber brine from the incinerator filters in the ocean. Of course, today, they are almost done with the project. But at the time, this was the prototype, state-of-the-art, fully robotic system for getting rid of the mustard and nerve gases in bombs and rockets that were never meant to be taken apart. These weapons were built to be used, not dismantled, and it was a huge technical issue. It was also clear that environmental issues were coming online. The army never got enough credit, and still doesn't, for being forward-thinking enough to know that it needed to build pollution abatement systems where nothing leaks out. In this case, it is pretty obvious, if a burp of noxious material occurs, you get death for miles around. Another issue is that incinerators produce acid emissions and this is associated with acid rain. So they put in sodium scrubbers to neutralize the acid. What do you end up with? Salt. They were ending up with 5,000 tons of salt with some heavy metals and other things in it. They figured the cheapest way to get rid of it was to take it out into the ocean and dump it. In 1985, the lifetime cost estimate for the project if dumping were used, including the purchase of a dedicated ship and barges, was $3.5 million. To landfill the brine in the United States would cost $350 million. So, Maragos is in charge of doing this work and wants to bring the Hawaii submarine down for this part of the study. First they just wanted to bring the salt right outside the atoll and dump it. The submarine came to look at the slope, to determine if it would be affected.

I had known Jim for about 10 years and Deetsie Chave for 15. We couldn't bring a big oceanographic ship to help with the work, but we had tugboats on the island. They knew the work I had done on the Big Island with currents. I had portable oceanographic systems. They wanted me to come on board to do the EIS [environmental impact statement] since they had other projects they were busy with. They wanted me to write the documents and do the oceanography. I was the only one interested. I think physical oceanographers in Hawaii could easily have done it, but they weren't interested. They might have seen it as too applied. The Johnston project, through DOD [Department of Defense], has been my main funding source through these years.

We have our own lab on one of the smaller islands and have a few nice boats. Working with the military and the engineers through the years, I believe that I have been educating them about the environment and the role of scientists and science in environmental impact assessment. I help them to change their programs in ways so that they accomplish what they need to accomplish, but with a little extra effort, it doesn't cause environmental impacts. We've put together some of this information, laying out the doctrine in sort of a "lessons learned" format. Recently, the House and Senate passed a resolution that it is in the national interest to avoid unnecessary environmental degradation. In terms of reef environments, this also includes doing things like amphibious operations. That is some of the applied work we are doing.

I've been working on several problems through the years, but the single greatest obstacle in doing these coupled physical oceanography and biological studies is having comparable measurements. Today, with computers, airplanes, satellites, rapid data assimilation from expendable bathythermographs, rapid flow field, we can get at the oceanography. But how do we do the biology? How do we study fish reproduction? Well, we still kill fish and look at their gonads, drag plankton nets through the water, or go scuba diving and observe fish. As you well know, all of those methods are incredibly time consuming, manually intensive, incredibly limited, and you still can't get time series measurements. For example, with gonad measurements, you have to kill the animals as you go. And as the years have gone by, with pollution and overfishing, you just can't collect fish as I used to as an undergraduate. In Hawaii today, you just don't see the abundance of fishes I used to see. The difficulty with plankton nets is, well, you have to know the flow field where you sample. It makes a big difference if you are outside or on the edge of an eddy. Even so, there are many sampling problems, including net avoidance, aftersampling, and all the manual labor of sorting and identifying what you have. You don't know if you are sampling the right spot with diving. You can only be in one spot, and all the fish might be spawning 20 feet away right behind you. All of these problems don't allow for time series measurements of the actual spawning events that we want. We really need a way to detect where and when the fish are spawning. It would be great to sit back in the lab and listen to my monitor and hear, yes, the fish are starting to spawn, and then I could go out and look.

Pat Colin wrote a paper while in Jamaica describing Dancing Lady reef, where schools of parrot fishes spawn. During the spawning rushes, they swim at speeds up to 40 miles per hour and turn rapidly. Pat showed me the site himself when we co-taught a course at the lab. I thought it might be possible for us to hear the turbulent movement of these fishes through the water.

Just like a submarine.

Yes. I had no reason to think any differently. In January of 1988, I went down to teach ichthyology in Jamaica for the East-West Program, when Ken Sebens ran it, and took my cameras down. By that time, parrot fish were probably the most studied fish spawning on the coral reef—Ross Robertson, the Ogdens. Lots of us had studied hamlet fish and spawning too. As part of my class in Jamaica, I'd say, "OK, we'll watch hamlets spawn today, parrot fish tomorrow." They spawned so reliably as far as place and time of day that you could schedule it into your field plan.

I wanted to test my system. I was going to record parrot fish spawning and hamlets spawning. This involved taking this camera system and the sound system, going out into the field, recording the fish, coming back, and playing the data. Remember, in the field, I couldn't hear anything. I came back, played the recordings, and I thought, oh, my God, this isn't real. The hamlet fish make a courtship sound. The male and female each make a courtship sound, and then they make a distinct sound at the moment they are releasing gametes. Fascinating. Parrot fish make a hydrodynamic sound, as I predicted. What I didn't expect was the vocalization. The sound that a hamlet fish makes when it is releasing gametes cannot be confused with another fish sound. That was one of the great revelations of my life. Another was with Allan Robinson, when we studied the eddies in Hawaii and actually saw the drogues go 'round and 'round.

SUMMARY: CHANGING TIDES

How the field tradition is passed in person to the next generation is frequently discussed in these interviews. Nevertheless, at research universities today, scientists are penalized if they spend too much time teaching; rewards, including promotions, come to those who garner the research dollars. Interestingly, people who were taught, have taught, or do teach at smaller institutions (e.g., liberal arts colleges or field labs) clearly derive great satisfaction from working with young aspiring scientists. Furthermore, people who work at institutions where education continues to have a high priority are allowed to complete their research objectives without undue pressure to publish early and are not badgered to work on what someone else thinks is important. The result is that sometimes the funding gravy train comes around to the backwater field that a researcher has had time to adequately define and characterize, such as Hubbard describes.

Intense intellectual interaction was discussed as an element integral to broadening and strengthening one's scientific viewpoints, as opposed to working in isolation, which limits the scope of one's own effort because ideas aren't refined by critical questioning. The benefits of a stimulating atmosphere is described in several of the interviews, an atmosphere found at the institutions where these scientists were students or staff in the 1970s and early 1980s, including UCLA, the University of Washington, the University of Hawaii, West Indies Lab, and Friday Harbor Labs. Dacey contrasts the opportunities for the open, optimistic exchange of ideas, which he first encountered at WHOI in the late 1970s to the present day: "People stay at home. They are exhausted by the day's events." Not only Patrick Nunn, isolated by geographical distance from other scientists in the South Pacific, but also all of the other scientists interviewed, who are similarly isolated, but in their cases, by the lack of time to engage in meaningful communication and reflection with students and colleagues of all ages, bemoan this situation.

Time or, I should say, the lack of it, is an issue brought up repeatedly. Advice given to some of the scientists echoes that given to the previous generation: enjoy being in graduate school; you'll never again have the time and freedom to do research that way again, the way you think it ought to be done. Job and personal commitments consume increasingly more time and energy as the years go on. Time is also a problem for today's student. Field camps are being made shorter because it is felt that students need the time to get summer jobs or because they are too expensive for the sponsoring institution. Opportunities to work in the field are grudgingly taken by graduate students today, because they feel that they cannot afford the time.

Perhaps borrowing from the ecological theory dominating biology curricula and popular science at the time, many refer to finding their "niche" in science. Analogous to many successful organisms, they forged their careers by making the philosophical decision to avoid competition. They set themselves apart by combining techniques and approaches which had not yet been done or by comparative studies to detect the ways evolution has solved problems. To continue successfully in academia, when job opportunities were limited and funding was restricted, it was almost a necessity to distinguish yourself from the crowd. The downside of this approach is that the funding agencies developed programs along disciplinary lines. If a scientist's approach were to cross disciplines or if the funding agency were just blind to a comparative approach (as some NIH and NSF program managers appear to be), the proposed research often fell through the cracks in a highly competitive funding scene.

When this generation began college, introductory courses in biology had shifted from the classification, structure, and function of organisms to the molecule to ecosystem approach, which began in the late 1960s (see Kozloff). Not surprisingly, this is the first generation we encounter that does not emphasize "know your organism." In fact, only Soja and Donoghue mention the loss of taxonomic expertise as a real concern in the natural sciences. These scientists are now at midcareer. Hubbard presents a model of the ontogeny of a scientist: he learns specifics (as an undergrad and master's student), becomes an expert in a regional field (as a doctoral student), and with continued career experience, he can then approach broader, more integrative questions that require more synthesis.

The women of this generation were not hampered by expectations to follow the traditional female role of homemaking. However, this resulted in a dilemma. Either one attempted to do three full-time jobs at once (scientist, wife, and mother), with the enormous commitments of time and energy required to gain advancement in academia and to have a successful family. Or one compromised: had no children or accepted a job at the academic fringe, a lectureship or a soft-money research position (neither of which came with much security or financial compensation). Another disconcerting theme in the interviews of the women in this chapter were the subtle (or perhaps, not so subtle) hints that science as a career is somehow not suitable for women. Three of the four women interviewed began college choosing a major (philosophy, English, Spanish, sociology) other than one in the sciences, and they worked before deciding to go back to graduate school.

Many in this generation came into the field more through love of ideas (order in nature, ecological principles, evolution as explained by Darwin) than through an early love of nature, yet they each are passionate about their science. While they were undergraduates, the natural sciences were a respected discipline because of the public's awareness of ecology. Interest in the environment was high. New field institutions (including West Indies Lab, Catalina Island, Bodega, Shoals, Smithsonian's Carrie Bow Cay) built in the late 1960s and early 1970s, as well as older field stations, field camps, and financial support from NSF, provided opportunities even for undergraduates to be at field stations, to interact with older scientists. This generation was inculcated with the views that field experience is necessary to good science and that breadth in a discipline is important.

However, in an age when time and energy are limiting factors, how much effort is worth putting into creating the kind of institutional atmosphere and mission that one may recognize is important? These scientists seem weary of day-to-day concerns within their careers. They still love asking questions about nature, but they find other aspects of their jobs draining. Is promoting the value of field science worth swimming against the tide? Perhaps not, for them as individuals. What does that mean for us in the future, as a society?

6

The Lean Years: The 1980s

So many people think they've got it all right, and something new comes up and turns it all on its head.

Peter Clift

As is related in their lively accounts, the scientists of the 1980s generation recognize the importance, to their science and to their careers, of the time they have spent in the field observing nature. Their undergraduate coursework in the sciences is often described as remote from the reality of personal experience; they conclude that they lacked passion for science until they gained field experience. The message they were given, however, is that in modern academia, modeling and model systems are the mode of excellence. They have come of age in a period when society demands the quick solution. Nevertheless, they have discovered that fieldwork takes time, and the answers often take years to properly develop. Unfortunately, they have also discovered that modern academia does not reward these efforts, which require so great an expenditure of time and energy. Obtaining funding for doing fieldwork is difficult. In contrast, funds are easy to secure for modeling. On a positive note, I found as I interviewed them that each sees a challenge in passing on Agassiz's legacy, although the opportunities they have for personally passing on the field tradition are few and far between.

What was going on with these birds? Wanting to know more, to figure it out maybe a little more clearly, kept pulling me back.

GABRIELLE NEVITT (b. 1960)

Assistant Professor of Neurobiology, Physiology, and Behavior
University of California, Davis
Interviewed in my office 29 October 1999

Just prior to the beginning of her seminar at WHOI, I observed Gaby Nevitt going into the audience to thank octogenarian Betsy Bang for her detailed anatomical studies, which were done more than 40 years previously. Bang had compared the brains of more than 100 species of birds, concluding that (contrary to the dogma of the day) at least some birds must be able to smell because many species had large olfactory lobes. This attention to the historical development of her field and the breadth of her biological training make Nevitt a throwback to older generations. While discussing her salmon and Antarctic bird olfactory physiology research, she illustrates how time (both in the field and in the mental processing of the data) is needed to adequately develop an understanding of natural processes and how intellectual breadth (in her case, switching from an invertebrate to a vertebrate system) allows one to pose hypotheses unhampered by rigid dogma within a system, even though it may be a battle to get the work published.

It was ages before I started to talk about my bird work because, even though I had worked with the bird story for about five years, it took me a long time to actually get a feeling for what they were doing. The way I got involved in this project was by pure happenstance. To me, it has been sort of my [Hopkins Marine Station] spring course project for the last several years. It was an area of the world that I knew nothing about, and although I always really liked birds, I had never worked with them. There are a lot of personal challenges working in the Antarctic, working at sea, and also just being away. That whole lifestyle of going to sea changes one's life and perspectives quite a bit. I got seasick. We had to cope with a lot of problems like that, and lots of people have trouble getting along in that environment. For me, when I say this, it sounds hokey, but I was definitely driven by these questions about olfactory foraging in this new environment. They were really bugging me. What was going on with these birds? Wanting to know more, to figure it out maybe a little more clearly, kept pulling me back. I wanted to somehow portray that sense a little bit in my talk. The early results opened up all of these other ideas to explore. It was a personal journey.

In my first summer of grad school, I took a class from Karel Liem at Friday Harbor. He was awesome! In that class was a great guy named John Konecke, a student of Tom

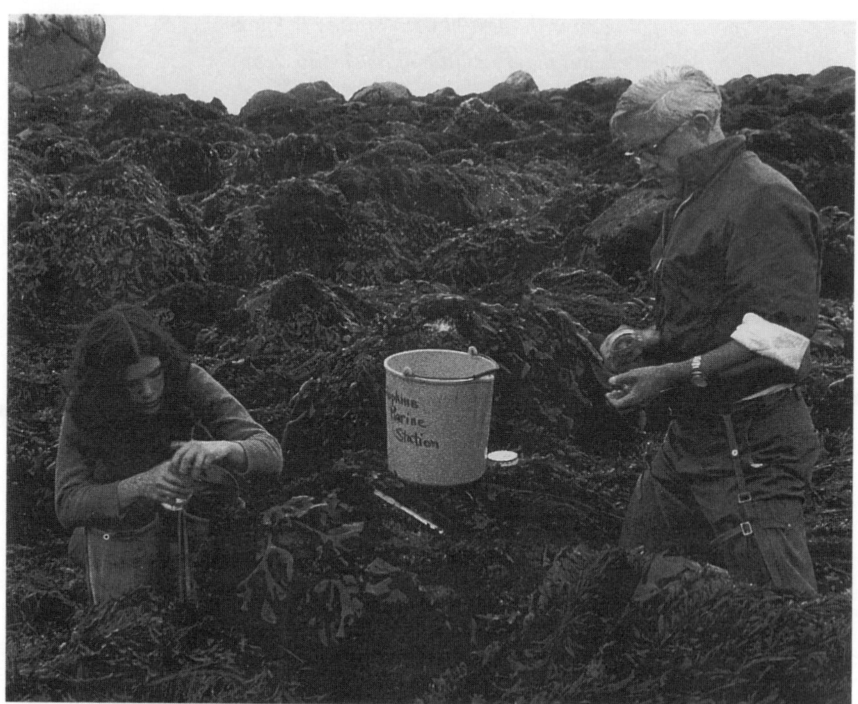

Gathering sponges in the intertidal off Hopkins Marine Station, Don Abbott gives advice about graduate school to his student Gaby Nevitt in 1982. Photo by Galen Howard Hilgard. Reprinted with permission from *Observing Marine Invertebrates: Drawing from the Laboratory*, by Donald P. Abbott (Stanford, Calif.: Stanford University Press, 1987), 376.

Quinn at UW, who was studying salmon. He had been the station manager at Palmer down in Antarctica. His wife played the fiddle, and I played the banjo, so naturally we became friends. So it was John who really gave me the bug to go work in Antarctica. During my second year, I started working on the salmon project. It was funny because Karel Liem had said to me, "You are going to go into cellular biology. You want to do molecular stuff." He always said the word *molecular* like it tasted bad. I did a flounder project with him at FHL. My paper was called "Do Fish Sniff?" and involved a lot of functional morphology with little pressure transducers implanted in fish noses.

Anyway, I did stay in Bill Moody's lab, and I did do a very cellular thesis on the cellular basis of olfaction. At that time, the patch clamp was used in our lab, and the technique was brand new. I read some papers by an old UW grad student named Dave Bodznick, who went on to work with Ted Bullock. Dave had found that salmon could discriminate between water types based on calcium concentration. I thought that maybe there was something in the peripheral olfactory system that detected the difference without a lot of secondary processing. This result of Dave's got me thinking because people had always thought of imprinting as a process that occurred way back in the brain. So, I started to think of the periphery. With an invertebrate background, I wasn't tied to the idea that things had to be integrated at the brain.

That is neat, to come in with a different background and not be prejudiced by the dogma of your field.

Yes. My advisor thought I was way too organismal to work in his lab, but I convinced him that I wanted to work on the salmon. I sold him on the idea of studying the cellular basis of olfaction using salmon as a model system. Physiologists love animals if you call them "model systems." One evening, a friend who had taken DPA's [Don Abbott] invert class with me was describing what her interests were, which were very applied, having to do with biomedical research. I was trying to explain to her about the project Bill wanted me to do. She said, "That is so obtuse, how sodium channels move around. That doesn't sound like you at all." I thought, I don't want to do this. It sounded horrible. I thought, that is not me. It was hard. I was in a lab where everybody was getting cover stories in *Science* because of their discoveries on sodium channels. I became the bastard child in the lab!

I started working on salmon and worked a lot with folks in Tom Quinn's lab in the School of Fisheries. I started working with Andy Dittman, who did a joint program in fisheries and pharmacology. We developed our own little operation, studying olfactory imprinting and homing in salmon. We needed to collaborate just to raise the salmon, and we raised tons of salmon, over 50,000 fish! It was insane what we did. We trucked them all over the place to imprint them with different odors at different times of their lives. We got to use a little hatchery in Steward Park in Seattle. It was tough though because we had no money to pay for fish food. At some point the lightbulb went on, and we figured that if we raised twice as many fish as we really needed, we could sell half of them to pay for all of them, to buy fish food. So we ended up doing that. We got a fish broker, and all of the money we made went back into the operation of the hatchery. We had quite an operation. It was neat to learn that aspect. We had to prepare the fish for market, and other times we sold them in the department, almost like a bake sale. It was really fun but a ton of work. I'm not sure you can do that any more.

Our big discovery was that there is some component of the olfactory memory retained in the nose, that the nose itself becomes somewhat tuned to the scent of the home stream. That was a different way of thinking, a new model for how salmon learn the scent of the home stream. It was way out there. It didn't make any sense based on what other people were saying about olfaction at all, because in olfaction, the thought was that olfactory receptor neurons constantly were turning over in a sort of haphazard way. But we had found that if we imprinted fish to odors when they were smolts that, later on, the olfactory epithelium would have an increased response to those odors, when the fish were maturing. That just went against the dogma, because those cells weren't supposed to be around. But, those were our results, and our model suggested a mechanism for how it worked. I did the electrophysiology, and Andy went on to show a biochemical correlate, an increased production of cyclic GMP [guanosine 5[1]-monophosphate] in response to imprinted odors, which correlated well with the behavioral and electrophysiological story.

We tried publishing the story in *Nature* and *Science*, but in the end, it wouldn't fly. When other people started to come out with similar stories in different systems, well, then we were accepted. It was so ironic. We got written up in *Science* this year for the work that it rejected ten years ago. When I was interviewed for the story, I had to laugh a long time over it. I told them, "I can't believe you are really interested in this. Do you know the battles we fought to get this published?" So, in the end, it all came through, but God what an ordeal!

In 1993 I went down to the Antarctic for the second time on the NOAA [National Oceanic and Atmospheric Administration] *Surveyor*. I did a series of slick experiments with the krill [to determine if they released the odor that attracted seabirds]. During the previous year, when I hadn't gone to sea at all, I started thinking that we had been looking at things the wrong way with the smelling experiments. It was as if there were a huge piece of the puzzle that was missing. I really felt there must be a way to measure odors over the ocean but convincing the funding agencies that we could do this was another story. We were getting great data, but I was a little puzzled. For example, birds that ate krill were responding to krill-scented slicks in the same way that they were responding to unscented slicks, and I thought that they should be coming in to feed if they were using olfaction to find food. Why weren't they doing this?

But then, another series of strange events occurred. We had a couple of weeks of work that were absolute hell. Then a storm picked up, and the ship rolled. I was entering data into the computer and the bungee cord securing my chair broke, sending me hurtling into a tool chest across the room. It was a bad accident. I got a tear in my kidney. I was basically fading in and out of consciousness for a week. That whole experience was life changing for me. I've never felt that kind of pain, but they weren't going to med-evac me off. I couldn't understand why they wouldn't do that but I probably wasn't good at communicating how much it hurt. As I was on the lower deck of the ship, there was no way that I could negotiate the stairs. Nobody suspected a kidney injury. They thought I had hurt my back. When we got into port, I could walk at that point, but it was just horrendous.

It was really bad, but it gave me time to think. And thinking about anything else was good to help cope with that pain. I knew something in the story was missing. There had to be a different way of looking at this problem, which I was missing entirely. When I finally did get off, we went into town, Punta Arenas, but I was still really messed up. My friend sat me down in a little cafe. Tim Bates, an atmospheric chemist, was at the next table, and I started talking to him. He was down for the next cruise. The next time I saw him I was back on ship. Since I was still in really bad shape, I asked if I could stay there while waiting for our plane to fly out. Tim came and started talking to me. I'm pretty sure he thought I was a complete weirdo, but I didn't care because by that time they were giving me pain killers. His group was doing all of this atmospheric chemistry. But, he asked me, "What is it that you are trying to do with these birds?" I told him that we were trying to determine what odors they used to hunt for prey. When I asked him what he did, he said, "What we do is measure this gas called dimethyl sulfide [DMS], which really stinks at high concentrations. Maybe your birds can smell it."

I thought, oh my God! I had read all about DMS, but I didn't know it had an odor that birds might be able to smell. A few months later I went to visit Tim's lab in Seattle. He showed me a map of DMS concentrations from a transect across the Drake Passage (the gas had low concentrations nearshore and a peak concentration somewhere in the middle). There are better plots available, but that was the original plot I saw, so I use it for my talks. I looked at that and I thought, oh my God, this is a completely different problem. These odor profiles were like mountain ranges; they could indicate areas of high primary productivity, but they could also be used as navigational cues to know where you are over the ocean. If a bird could identify a productive area by smell, then the search for prey would be much more straightforward. It just changed the problem in my mind.

It also changed my thinking about salmon migration. It wasn't an idea of looking for an ephemeral clue. We think a lot about visual maps and about how visual space changes as we pass through it, but we don't have the acute sense of smell that a lot of animals have, so we don't think of smell that way—a smell landscape you are traveling through. For me, that is when the lightbulb went on, and I think of things very differently now, including the way I approach the salmon question.

It is weird how things sometimes just work for you though. I only recently learned about JOHN DACEY's work detailing the release of DMS by phytoplankton during grazing by zooplankton. His data suggested to me how some species might use DMS to locate foraging hot spots. We have since extended these ideas to incorporate interactions between seabirds that use different odors to forage. For example, some species seem to respond to DMS while other species are more responsive to odors given off by macerated krill. I think that the DMS responders come in first, but leave before the bigger birds get there. In other words, they tend to avoid big mixed-species feeding aggregations that produce the smell of macerated krill. This is because some of these smaller species are subject to predation by some of the larger species that are sensitive to krill odors. I've actually witnessed predation events at sea at some of our experimental slicks, and it isn't pretty, let me tell you.

In the salmon work, we are continuing to test this model of olfactory imprinting and homing. We've had a hard time because many species of salmon have been listed as threatened or endangered in California, where I work. It changes your important questions. I really feel intrigued by the imprinting models, but it almost causes you to want to put your energies into saving these animals. Salmon are in sad shape in California. So my students have ended up doing much more conservation work than I have ever done. In a lot of this work, we can't work with hatchery fish, we can't work with wild fish, and we can't release fish into the wild. So a lot of my students' research has had to deal with looking at the behavior of naturally spawning fish. One of my students is interested in how fish choose mates. He has this idea that females will choose sneaker males to mate with rather than the larger guys. It has a lot of implications for management, because the smaller sneakers [jacks] will be culled and not allowed to spawn, but they are obviously critical in maintaining population fitness. He has gotten a lot of attention for his ideas, and it is exciting to see.

My students are great. I have a mandate that they need to have an organismal background to be in my lab. I take my students through the Animal Behavior Graduate Group. I ask myself, "What do they have in common?" They all, interestingly enough, have backgrounds in creative writing, and they all have strong backgrounds in organismal biology. They seem to want to be trained in integrative thinking. They are generally not neurophysiologists, at least they don't start out that way. Actually, my students don't come from a particular set of schools, but they tend to be older and have had a lot of field experience. I keep hoping students don't lose sight of organismal biology.

I worry that Bodega [Marine Lab] will become a molecular biology institute by the sea, like Hopkins is becoming, but with Susan Williams as the new director, I don't see that happening. When I went back to give a seminar at HMS a few years ago, I brought my whole lab group with me. I timed it so we could visit the intertidal. There was a great low tide right in the middle of the afternoon, but it was so strange because there was no one else out there working; they were all in their labs. I felt that we were ghost people there from a different planet, a different era. I chose to give a real organismal talk, emphasizing the bird story. What could be better for Hopkins Marine Station? But the audience just sat and looked at me with no response. It was the dullest crowd. They didn't know how to deal with an organismal talk, and I spent most of the hour wishing I had opted to talk about the salmon work. It is much more cellular. The question-and-answer period was sort of neat though. I told the story about Don Abbott, and I showed the picture I told you about. There was an audible gasp when I put up the picture of DPA and me collecting in the intertidal. I looked up and there were people in the audience literally with tears on their faces. All of these people came up to me afterward and said that they were so glad that I showed that picture and that I had gone on to become an organismal biologist. I was so moved by that. I still am.

In my humble opinion, interaction among scientists is one of the most important elements of science.

DINA FONSECA (b. 1964)

Postdoctoral Fellow, Molecular Genetics Laboratory
Smithsonian Institution
presently: National Research Council Fellow, BioSystematics Unit
Walter Reed Army Institute of Research
Interviewed at her office 2 November 1997

After experiencing "an American reality of teaching and interaction" during a summer at a field station, Dina Fonseca seized an opportunity to continue graduate studies in the United States, forsaking a guaranteed faculty position in Portugal. Watching live zooplankton at the field lab in Michigan convinced her of the need to understand the biology of an organism in relation to its environment, which led to Ph.D. research combining field, lab, and modeling work on stream insects. Her career path has tended steadily toward answering questions with applied applications, first in conservation biology and more recently in public health.

I wanted to be a scientist from when I was little. I cannot deal with a structured profession. I wouldn't be able to function if someone wanted me to be somewhere at 9 A.M. and to leave at 5 P.M. I am able to make my own choices when I do something. That is very important to me. I come from a middle-class family in Portugal. Initially I wanted to become a veterinarian because I loved animals. When I talked with my father about this, he took me to the meat market and said, "Do you want to spend your entire life looking at the quality of meat?" Because, if you are a veterinarian in Portugal, that's what you do. The difference between our views of veterinary science resulted from very different life experiences. I had read books—British, French, American books.

Did you ever read any of James Herriot's?

Oh, yes. Saw the TV series too. So I had all this literary knowledge about what a veterinarian was, and my father had this down-to-earth understanding of what it really was, in a city in the 1970s in Portugal. Talking to my father made me change my mind. I loved biology but fainted at the sight of blood. After 12th grade, I asked to be accepted in the biology department at the University of Coimbra as an undergraduate. In Portugal, after you finish four years of college, you can either spend one year taking teaching courses to become a high school teacher of biology and geology, or you can develop a senior thesis and go into research. In a cohort of maybe 100 people, I believe only five or six of us decided to go into research.

My undergraduate senior project was in community structure. We collected invertebrates from different areas in two streams. We also collected physical and chemical variables. In order to analyze that data, I learned basic computer programming and wrote a series of programs. At the time, we did not have access to more sophisticated software packages. So that was a fundamental part of my thesis, the development of the programs. I used them to analyze and compare the invertebrate community structures in the two streams. I wrote cluster analyses and used several different similarity indices. I also wrote ordination programs. It was a good experience because it made me understand what is behind some computer programs I now may use. I know I have a very different understanding from somebody who simply types in the numbers and runs the software package. I know in some cases what is behind the program, what it took to write it, what algorithm is in it.

When I finished my senior thesis at the University of Coimbra, I was hired by the biology department as a teaching assistant. I started doing my Ph.D. pursuing some of the questions I had examined in my senior thesis. But at that point, I became very unhappy with what I was doing. I felt isolated. My senior thesis advisor left to start his Ph.D. in England, and I felt that I didn't know enough to develop a worthwhile research project on my own. That feeling of isolation leads me to our discussion of biological stations. In my humble opinion, interaction among scientists is one of the most important elements of science, both among people in the same discipline and among people from different disciplines. As a fledgling scientist in Portugal, I felt isolated, not getting anywhere. People I spoke to thought I was doing well in my Ph.D. project, but I thought neither they nor I knew enough. Correct. At that point, Richard Bowker, a professor at Alma College, Michigan, and a visiting professor at the University of Coimbra, offered me the opportunity to come to the United States and take summer courses at a Central Michigan University biological station in Beaver Island, Michigan.

At this point I had really experienced a different reality—an American reality—of teaching and interaction. Also, I had become interested not so much in community-level biology but more in individual species and what the animals were actually doing. When I had done my senior thesis, we'd go into the field. We had all these fancy pH meters and other measuring instruments. We used a net to collect the animals, which we would put in formaldehyde and take to the lab, where we would count and identify them. The real puzzle was to connect the numbers of individual species with chemical and physical parameters. But, we'd lost completely all of the information about what these animals do. Most animals were dead by the time I first saw them.

Then I took zooplankton ecology at Beaver Island. During that course, we did a lot of watching of live animals—watching them behave—and a lot of other puzzles started to emerge. Puzzles connected to why organisms do what they do. I guess I started thinking that looking at things from a numbers perspective would eventually be interesting once I knew the background of the individual species. I became more attracted to the idea of understanding one species or two species than just going on counting.

I had a great time at the field lab. I was interacting with a group of knowledgeable, enthusiastic people—some were experts in reptiles, others in fishes, insects, zooplankton, or birds—in a very diverse and interesting environment. I didn't want it to end. I also learned that questions are not really that different in many cases across fields and also that people's views can be constrained by the animal they are looking at. I learned that sharing knowledge across fields can be very rewarding. People in other fields can come up with experimental designs or raise useful questions that you never would dream of. Many times a small detail in your system could be the emphasis of all of the questions in some other system.

And if you just stick with the literature for your taxon . . .

Right, it is impossible to read everything. Keeping up with the literature in your own field is hard. But my motto is "It's not the animal, it's the question."

One of the great things about biological stations is not only having all of these people working on many different aspects of the different organisms, but the fact that they are more available. There are common meals and field trips, which often lead to great discussions. Those two months at Beaver Island Biological Station changed my entire life. I did one year of coursework for a master's at Central Michigan University on leave from the University of Coimbra. That first December, I also took the GRE [Graduate Record Examinations]. I then applied for Ph.D. positions and accepted an offer from the ecology and evolution program at the University of Pennsylvania.

To be able to do a Ph.D. in the United States meant that I had to give up my position at the University of Coimbra. That meant exchanging the known for the unknown. I didn't have much information for a good cost-benefit analysis. My parents were not thrilled, but I was hooked. I really felt I had found my scientific habitat in the United States. During my Ph.D., I examined the behavior of black fly larvae in streams. Black fly larvae are passive suspension feeders and live in fast-flowing waters. They are fascinating creatures that fight each other for space, capture food, and escape predators in a very noisy environment. My Ph.D. advisor, David D. Hart, had been developing ways to measure the speed and other characteristics of the flows that occur in swift shallow streams. I became particularly interested in what factors make an individual larva decide to leave the area in which it is feeding and go find a better one. I should note that these larvae have no legs, they can loop like an inchworm on a rock, but water currents transport them between rocks. They trail a silk thread, which both helps them leave an area of slow flow—like a ballooning spider—and attaches them to new rocks.

I spent a lot of time exposing larvae to different flow conditions in the field and in the lab and observing their behavior. I also did lab analyses of their capacity to settle from the water column. From that basis I built models to try to examine the consequences of different individual behaviors to populations. In summary, I built my knowledge from the bottom up instead of letting correlations between variables be my guide to causality. Correlation analysis can mislead you. For example, during my Ph.D., I showed that the behavior of stream organisms reflects physical and biological constraints that ultimately result in unexpected correlations between densities of organisms and environmental variables. Ultimately, if during my senior thesis, I had measured black fly larval abundance and current speed and performed my top-down analysis, I would have concluded that current speed is not an important parameter in black fly larvae survival or black fly larval distribution in streams. My Ph.D. research and that of David Hart, however, showed that black fly larvae try to maximize feeding rates by actively searching for and feeding in fast flows. But high current speeds also decrease their probability of getting to those areas. The result of these two conflicting forces is an apparent lack of correlation between larval density and current speed.

My need to understand one species at a time doesn't mean I cannot make inferences across species. In some cases—in particular, in the case of endangered species—there is no time to collect all of the background information. That is why model species are so important. But it is fundamental for managers to do their homework about a particular species or set of species before using borrowed paradigms. Speaking of management and conservation, just recently I had a postseminar discussion with Gunila Ronsenqvist, a professor at the Norwegian Institute of Science and Technology. She is a behavioral biologist who works with female-female competition in pipe fish. We started talking about behavior and the effects of toxic substances. We discussed the possibility of using behavior as an index of toxicity, of habitat quality. She feels that animal behavior as a discipline is being underrated. People find it the softest of the soft sciences and are reluctant to fund projects in this area. If you start looking at it from a toxicological point of view, using it as an index, it could be one way to show how important it is to understand the behavior of the animal. A sublethal pollutant may affect the behavior of organisms. If the impaired behavior is linked to reproduction, a sublethal level of a toxic compound could mean extinction of a population or even a species. Therefore it is important to understand what is normal, healthy behavior. In principle you should be able to go from the behavioral mechanism all the way to the population and community levels.

Currently, as a Smithsonian postdoctoral fellow, my project is to understand the population structure of the vectors of avian malaria in Hawaii. Avian malaria is one of the most important contributors to the decline of endemic forest birds in Hawaii. Many people working at the Molecular Genetics Laboratory investigate similar conservation problems. Avian malaria, or any malaria for that matter, are parasites with a complex life cycle: they reproduce in the mosquito, the primary host, and use a vertebrate as a secondary host to grow, multiply, and infect other mosquitoes. Avian malaria is particularly lethal to Hawaiian forest birds because they only came in contact with the disease when the vector, the southern house mosquito, was introduced accidentally by ship from Mexico in 1826. To be able to control the disease in Hawaii, we will need to understand its vector. We know that trying to kill off mosquitoes with insecticides or the parasite with drugs is not efficient.

We know where the birds are, and we know their populations are few and far between. We need to decide where to build refuges and what populations are most endangered. Therefore, we need to know where the primary hosts are and what they are doing. The mosquito's behavior has the potential to influence the interaction between the parasite and the vertebrate. In Hawaii, if the mosquito cannot fly in high winds, then forest birds that live in windy areas may be sheltered from the disease. Also, in areas where mosquitoes do not disperse far from their breeding sites, due to climate or geography, it is conceivable that the malaria parasite will have a lower genetic diversity. An important consequence of low genetic diversity in the parasite is that it may be easier for the vertebrate's immune system to kill it. Understanding the population structure of the mosquito may allow us to identify areas where parasite strains are less virulent and possibly use those to generate vaccines. Therefore, I am currently examining the population structure of the southern house mosquito in Hawaii using genetic markers, and I went into the field in Hawaii to collect mosquitoes for my DNA analyses. The adult stages of insects, like black fly and mosquitoes, and many other invertebrates are like the larval stages of many marine invertebrates in that they spend most of their time in a fluid environment—air or water—and can disperse in a three-dimensional manner. That makes them hard to track.

The idea of open populations is quite recent, 15 years old or so. It is actually fascinating to analyze such a paradigm shift. We went from traditional ecology, where stability and equilibrium were omnipresent, to a concept of open populations, where immigrants can come from really far away and where local dynamics cannot be predicted neatly from what was present in the previous generation. In my mind, however, we are now coming full circle. Even when we know that populations are open, I feel it is important to ask, "How much are organisms really dispersing and what are the local and regional processes that mediate and control dispersal?"

Postscript: Since this interview, I have finished my tenure as a Smithsonian fellow. I am currently a National Research Council fellow working at the Walter Reed Army Institute of Research. I am still working on *Culex* mosquitoes, having now expanded the genetic analysis worldwide. Meanwhile, I have a new project, working on *Anopheles* mosquitoes and human malaria genetic diversity. This second project aims to test the hypothesis that the mosquito vectors are responsible for the high level of inbreeding commonly found in human malaria populations.

There are no equations in Origin of Species. *But does that make it poor science?*

Peter Clift (b. 1966)

Assistant Scientist, Department of Geology and Geophysics
Woods Hole Oceanographic Institution
Interviewed in his office 6 April 1998

Peter Clift, classically trained in geology at Oxford University, notes the disrespect shown toward field geology today by scientists who feel their disciplines, heavily defined by theory and modeling, are somehow superior to his. As we sat in his office doing this interview, he admitted that he himself took the skills and the art involved in doing field research for granted until he began to encounter both scientists and advanced students who were incapable of observing and interpreting data in field situations. Although some modern geoscientists would argue that the only way to understand a process is to study it "in the modern," that is, as it actually occurs, Clift disagrees. He uses two examples, the origin of ocean basins and the initiation of subduction zones, to illustrate the value and necessity of the historical approach, because these are processes that have resulted in features present on our modern-day Earth but that are no longer actively occurring.

I got into geology because I had a curiosity about fossils, and I was also interested in the oil industry. Offshore drilling seemed quite exciting. I thought geology was a good subject to do as a degree level because I wasn't sure that I was a sufficiently good mathematician or physicist to really be a good one. That said—I really didn't know much about geology. I took that on, thinking I would work for oil companies. My dad worked for BP but not as a geologist. He had a classics degree, actually, and didn't know anything about geology, but I got interested in the rocks by collecting fossils. I thought, good, there is a career path, something I can do with my degree. So I did that, but actually I didn't like it very much. It sort of seemed very fuzzy. I was regretting not having done physics and maths.

It was partly doing fieldwork that made me get excited about geology. At the end of my second year, I had to do an independent mapping project where we could choose the area. I chose the western Alps. I camped out there for a month and mapped on foot with my brother. Geologically, it's the north side of the Pelvoux Massif, between Grenoble and Briancon, in the French Alps. There is a plateau area on the north side and then a huge valley. There was this terrific vista of very high mountains, which I mapped the back of. I really enjoyed it, and it got me fired up, particularly about structural geology, faulting, and mountain belts. I thought, I would like to do some more of that. As it happened, the next year, when I graduated, the oil industry had gone to pot.

They said, "If you'd like to work for us, you'll need a Ph.D." Since my degree had been a very traditional course, I couldn't do anything too modeling related. I had forgotten much of my mathematics, and I wasn't interested in that.

I was interested in mapping structure and sediments. That was also of interest to the oil industry, so it was a happy coincidence that my geological interests also coincided with my intended career path. I found myself a graduate program in Edinburgh. I went there to work on the structure and sedimentology of southern Greece, which is a terrific place to map, despite the fact that it was a bit boring for long periods of time. I worked exclusively by myself. I spent six months in the field there, and I didn't speak Greek. I loved Greece and the village that I worked out of. I'm going to visit this year. I write to the guy who ran the taverna. We're still friends.

My advisor had been working in the eastern Mediterranean for a number of years, but he had never worked in southern Greece before. He had always worked in Turkey and Cyprus. He met a Greek guy at a meeting who said, "Please come and look at these rocks in my area." He went and thought the rocks were very interesting. They wrote a little paper together on it, but he thought there was a lot more to do in that region. He tried to get a student to do it, and he did. He got me.

My present research objectives are to study tectonic processes using the sedimentary record, especially in plate margins. So far, that has been looking at how continents break up in the north Atlantic and the central Atlantic. I've been looking at how convergent margins, subduction zones, evolve through time by looking at their sedimentary record. It provides you with a long-term picture of how the margin has evolved rather than the snapshot you get today. This is especially true in interoceanic settings, where you get these little tiny rocky volcanoes poking out through the top of the ocean. There is nothing there except igneous rock, so you've no idea what it looked like. So I think the sediment record has a lot of valuable information in it, which I'm trying to use to study the tectonics of these areas.

CLARK BURCHFIEL *said that there are some people who would argue that the only places to look at tectonic problems are active areas.*

I've heard that argument. It's a very popular one.

I understand basically what your field is, but I'm not familiar with the details and the major players. My bet is that there is an argument in this field right now, and people are taking sides.

There is, and they are.

Some people who work in active areas are being very dogmatic and saying, "This is where people should be asking questions because this is where the process is occurring."

That's absolutely right. That is exactly what is happening.

As opposed to the other side who say, "If you look at the history of places which have gone through this in the past and some of the old crust has reached the surface, then that will help us understand the process."

That's right. Well, I'm typically of that latter group. Although I like working in, and I like to see, the active systems. You can't see very deep though; that is the problem. And some things, I believe, are just not happening today. Two important things, actually. In the north Atlantic, for example, there are these enormous amounts of volcanics [rocks] which have erupted during the origin of this basin, up and down the margin here [points to a map on the wall], all related to probably Iceland. Now, there are new ocean basins opening today, the Red Sea and the Gulf of Aden, for example. But they don't seem to be nearly as volcanic. It seems as though these may be more catastrophic than the regular spreading process. If that is the case, you may not see that in the present day, in which case ignoring all of the older examples will mean cutting out a whole type of continental margin from your studies entirely.

The other thing that I'm particularly interested in, from my arc point of view, is the initiation of subduction zones. It is not clear to me that there are any initiating subduction zones in the present day. If we understand the rock record correctly, the volcanism that occurs shortly after the subduction initiates is very violent and has a very unusual chemistry. There are no active examples of that today, but it is very important.

There is no reason to assume that everything that has happened in the past is happening today.

I don't think it is. It seems to happen in bursts, like the whole of the subduction in the western Pacific from New Zealand at least until Japan all originated about 45 million years ago. Well, that's great, but it's not happening now. I'd like to know how subduction zones start. It's a pretty fundamental question. I don't think you can answer that very effectively in the modern oceans. That's a problem.

In fact, this document I'm brandishing here is a new initiative by NSF to work on continental margins on presently active margins. So, for both of those important problems—the origin of ocean basins and the initiation of subduction zones—this is worthless. Strangely enough, the people who wrote this want to study their own favorite system.

And they feel that, if necessary, it should be studied to the exclusion of other systems.

Yes, that's exactly what is happening. They are basically saying that there are limited resources. We need to concentrate on those systems that we can understand best, and we need to put our money into studying these particular ones.

How did you learn how to map?

I took a class. We were taught in the Lake District for a week and then we practiced on two or three other field trips. My undergraduate course at Oxford was quite field oriented. It has changed enormously, actually. They hardly teach that any more. We were taught how to map. There was an option if you wanted to do a lab project instead of a field project. But I don't think anyone ever did, at least in those days.

You were taught the techniques first time out, then you were given problems to work out on your own?

That's right, with graduate students and lecturing staff who came along. That is a rare skill these days. You know, I used to think anyone could do this. If you couldn't map, you weren't much of a geologist. But I now know that this is an increasingly rare art. If you were talking to Clark, he is a great practitioner of this. But there are not that many Clark Burchfiels being trained any more.

Several people have told me that MIT has the best field program in the country, if not the world. But Clark has to put a lot of effort into maintaining that program.

Yes, and he needs to. They are under pressure there to cut that back and to reduce the field time that they have. This is unfortunate, because I find people who are dealing with, for example, modeling or lab work, and they have no firsthand experience with the rock. It makes me worried about whether they really understand what they are doing. The classic example I enjoy is a story about a professor of geophysics at Cambridge, who is a very theoretical fellow. He doesn't know anything about field geology at all. He proposed that the mantle and the cross mantle boundary, the MoHo, was a sheer plane. He went to Iran to look at the ophiolite and the MoHo there with a structural geologist, a friend of mine. They went there, and the guy said, "Well, there you go. It's sheered." My friend replied, "No, it isn't! Just look at it. Look at it." He didn't know how to look at it, really, and he was modeling and publishing it. When they went out and saw the MoHo, there was no evidence for his model. He just wanted to believe it, therefore it must be correct. The physics said that it ought to be.

Just because your equations say it will work, doesn't mean it will work. Equations are not superior to observations, far from it. We have to take both things. It has been the trend over the last few years to denigrate fieldwork as a valuable mode of science. It's like observational science isn't real science, only computational science is real science. That's rubbish. I've never heard so much rubbish in my whole life. So, Darwinism is rubbish. There are no equations in *Origin of Species*. But does that make it poor science?

It worries me a little bit, not only the devolution of field programs at the undergraduate level, but the difficulty in getting field programs funded at any sort of level. A lot of lab and modeling work is very important, but unless you base it on a firm foundation, you can spend years on modeling what you think is a good observation. All it takes is for someone to make the correct set of observations, and all of your work is destroyed. It is all pointless if you are basing it on a false assumption. I don't like the superior attitude that modelers have regarding field observations, because it's stupid.

Why is it that undergraduates now at Oxford are not trained in field observation?

Because it is principally run by geophysicists and lab people, and they consider the field program to be largely outmoded and something that is not required any more for a successful scientific career. They felt they had been left somewhat behind by the plate tectonics people. Much of those advances had been at Cambridge.

I kick myself. I haven't been very bullish about the positive nature of fieldwork. When I first came here, I thought, I really ought to do more modeling and mathematical work. I tried to do that. I did fine, but it really wasn't all that great because I just don't have the training for it. But I just felt that the mode of excellence at Woods Hole was doing that sort of work, which was stupid. Now I've decided to be more bullish about the importance of fieldwork, not to deride other people's work but to say, "Look, this has value." I may not be able to differentiate, but this has value. This is important stuff.

I think there is a perception among the general public that everything that can be known is already known, there is just a glossing over that needs to be done.

That's usually a sign that something good is about to be discovered. I used to worry as an undergraduate. I'd think I had missed the bandwagon. Why couldn't I have been an active researcher during the plate tectonics era? They've found all the good stuff. There is only the dotting of i's and crossing of t's left. But, I don't believe that. We are not making the same sort of bounds at the present as they did in the 1960s, but so many times, people think they've got it all right, and something new comes up and turns it all on its head.

Look at this map of the world. People don't know anything about the southern oceans. There are only two or three bore holes on that whole area, the size of the continental United States. It's pathetic. We don't know anything about it. We don't know what's there, we don't even know our own planet. They spend billions of dollars on the space program, and they don't even know what the planet Earth is made of. They deserve funding, but it is a bit lopsided. I mean, when you argue that $25 million spent on the drilling program is wasted money, it is a drop in the bucket compared to particle physics or the space program. I can understand how Joe Public might be interested in the space program, in finding out what is way out there, but particle physics is extremely abstract. I wouldn't even pretend to understand it. Most people couldn't. How do these people manage to wrangle huge sums of money for really abstract science? I really don't know.

The start of the decay in the earth sciences was, I think, the difficulty in getting funding and support to do paleontology. It was seen as cataloging, boring, and not real science. But when it comes to it, if you need a fossil dated, you need a fossil dated. Without dates, you can't do very much. Perhaps it is boring doing the cataloging, but someone better know the catalog, otherwise you are all going to be in trouble.

You hear that you can use isotopes instead.

Not always, you can't. I wish it were the case, but it ain't. If we could date everything by isotopes, I'd be happy here at Woods Hole. There are so many mass specs. There are plenty of rocks I'd like to have the age of, but we don't. This one from my collection [on his office windowsill] would be a nice start. We have a pretty good idea but not a really good idea, mostly based on the fact that it looks fresh, actually.

So that means . . .

Well, it's probably not very old. One or 2 million as opposed to 45 million. The rocks in Tonga are of two ages. We think this is the younger one, but we don't know. We can't date it. It would be nice to have some bugs on top so a paleontologist can date it. Otherwise, we won't know.

It made me acutely aware of how little we know about fish biology, especially tropical fish.

JOSHUA SLADEK NOWLIS (b. 1966)

National Research Council Postdoctoral Fellow
National Marine Fisheries Service, Woods Hole, Massachusetts
Presently: Senior Scientist, Center for Marine Conservation
San Francisco, California
Interviewed in my office 17 February 1998

Josh Nowlis and I met on the street in Woods Hole, not having seen one another for more than a decade, since he had done his graduate field research in St. Croix. Ironically, he is best known for his scientific modeling, yet he forcefully explains how both a field science and a policy background have helped him develop realistic models about conservation-related issues. He initially chose field science because it was something to which he could relate; he didn't feel remote, as he did in other biological disciplines and in physics. He also discusses how the culture of a discipline, be it modeling, academic (as opposed to applied) research, or policy, often erects boundaries inhibiting potentially important collaborative work.

I started out at Brown University with every intention of being a physics major. I had really enjoyed the physics I took in high school. It was very conceptual and explained basic parts of how the world worked. But college-level physics really frustrated me because it was incredibly theoretical and felt very removed from the real world. That lasted a semester, and at the end, I just decided that I'd take a range of different classes to figure out what interested me. Among those classes was an embryology class, which was the first biology I'd had since 10th grade. It was definitely an interesting class, although it was so high-tech, it felt really removed from my personal experience. I've never been one who enjoys looking through a microscope, which is all we did. It was cellular, histological, developmental, and really focused on vertebrates. It was a good course, but it didn't catch me the way I was hoping something would.

That summer, I got a job working for NASA in its exobiology department. It is the group of people who look for extraterrestrial life and, as a corollary to that, they look at how life might have evolved in the solar system, starting from the very basic chemical compounds. What I found most frustrating with that job was that once again it just seemed very abstract to me. The work I was doing was on these invisible gases. The only indication I had that the reactions were actually taking place was by watching the level of mercury in mercury pumps and the graphs coming out of the gas chromatograph. And I was in a building with no windows in the summer in California!

I started getting a little nervous about what I was going to major in. I thought back to when I was a kid and what had really sparked my interest. I remember being just fascinated by PBS [Public Broadcasting Service] documentaries on the natural world. In the first semester of my junior year, I took an animal behavior class, which really focused on the evolution of behavior. I was hooked. Jon Waage, who has since moved on from Brown, was renowned as this challenging, engaging professor who really made you think. His research was on the evolution of mating systems in damsel flies. I loved his course and was really inspired by the material.

It was something you could see.

Exactly! I worked for Jon in the summer between my junior and senior years. I definitely found that kind of research so much more appealing than the work I had been doing earlier with NASA. I was outdoors, I was studying behaviors I could see with my own two eyes, and I was thinking on a basic level about the way the world works.

I had to decide whether to go to Cornell, to Duke, or to the University of Washington. I still have mixed feelings about that whole decision. It was a tough choice. What finally swayed me was DREW HARVELL, who said, "I have a need right now for a tech to work in St. Croix for me this summer." I had never done any diving before, and I had never been anywhere tropical. It was definitely an eye opener. The atmosphere at West Indies Lab was just spectacular in having all of these people with different interests there, but being a small enough place where you really had a lot of interaction among groups. In doing some experiments that Drew wanted done, I came across some interesting ideas for my thesis work as well. I ended up focusing primarily on the causes and consequences of host preferences in marine gastropods, working in the field in St. Croix and later in the Bahamas.

Actually, I shouldn't say at that point I was completely hooked. There have still been several times from grad school on where I really contemplated getting out of science altogether or reducing the field aspect of my research, or even just getting out of research altogether and getting into teaching. I guess the debate with me, especially in grad school, was that one of the things that really inspired me from the documentaries on ecology was the link to human impacts and environmental issues. In grad school at Cornell, at that point, there was virtually no support anywhere for studies that dealt with environmental impact. It was such a new thing, especially on the marine side. It wasn't really respected as an academic discipline.

One of the key experiences for me while I was in graduate school, trying to sort out what I wanted to do with my life when I finished, was a year that I took more or less off. I went to Washington, D.C., for about five months and found a volunteer position with Congressional Research Service [CRS], a group paid by Congress to advise it on technical issues. Two things amazed me about D.C. One was that, as somebody who almost had a Ph.D. in ecology with a real interest in environmental sciences, I had the hardest time finding a volunteer position working for somebody on the Hill. So many people want to do that and had more policy experience than I did. There is a split between applied and theoretical ecology; I think there is also really a split between the academic side of things and the policy side of things. One of the biases involved with that is that policy people tend to think of academic scientists as poor communicators. I had to prove that I could really write for a general audience before anybody would even have an interest in talking to me about being a volunteer.

On the Hill, I interacted with people across an incredible spectrum—from people who basically consider anything environmental to be a complete waste of Congress's time to the most extreme environmentalists. When I was at CRS, I focused mainly on the Endangered Species Act. Both sides were looking for ammunition for their points of view, with one side saying that the Endangered Species Act needed to be strengthened and the other wanting it gutted. What amazed me, though, in that environment, was how apolitical the Hill was compared to my academic department back at Cornell. I guess a small environment with a lot of people with very different interests is bound to be political.

And a lot of competition.

A huge amount of competition. Yes, it's the competition that is really the key.

On the Hill, people can argue over a point, but they can leave that behind and socialize whereas in a small department it doesn't work that well.

Again, these people's careers aren't so much on the line. As long as they've fought the fight that their constituents want, they can go home and feel OK about it, but if you've fought the fight and didn't get your grant proposal funded, you're in trouble. But I was naive about that.

I spent some time in the Philippines. That was the first time I got introduced to the whole concept of marine reserves, which has been the major focus of my research ever since. Silliman University was involved with setting up Apo Island, which is the textbook case study on marine reserves. They closed off 10% of the island's coastal waters to fishing. It's a very small island; only 300 people live there. By doing a lot of outreach, the lab finally was able to convince the fishermen there that it was worth their while to close what is a really small area.

So it was both a biological and sociological success?

The sociological part, actually, in my experiences, is much more important. One of my colleagues who also works on reserves has this saying, "Sociology trumps biology," and in many situations I think it is very true.

Doesn't it go back to "Give a man a fish, feed him for a day; teach him how to fish, feed him for a lifetime?"

Exactly. Very good wisdom. Incidentally, in Apo and everywhere I've since worked to create these reserves, a bigger selling point for the fishermen has been tourism rather than the potential for them to either rehabilitate a bad fishery or enhance one that is not in such bad shape.

That exposure to the reserves made me start thinking, how did they come up with a design? Everybody was talking about the fact that the reserve included 10% of the coastal waters. It seemed to be an important number for some reason. I started asking, "Where did this number come from?" Nobody knew.

It goes back to your math background.

Exactly. I hadn't done much math in college and had only toyed with it in grad school. But here was a great issue where I could just do some basic population biology to hopefully shed some light on the question: Is 10% really a good number? At that point, people were throwing 10% around a lot in the literature, what little literature existed on reserves.

It turned out to be just perfect timing. I was about a year or two ahead of the curve with that whole question. Since then, lots of people have looked it, but I had a head start and ran with it. In many circles, I'm the token theoretician who will come in and talk about the theory of reserves based almost entirely on mathematical models. It's actually a little bit ironic, that we are talking about field science when right now what I'm known for is theory.

But you have field science in your background.

I do. I think that it is actually so important that I have not only the field science, but also the policy experience in my background. There really have been probably a dozen other people who have done very similar models of reserves, but two problems arise. Either people have been bogged down in worrying about all of these little details—which, as a field scientist, I was able to say that I know these things are important, but I also know that I can ignore them for now and get to something that is much more general—or people don't ask targeted questions about the key design challenges. I've been able to keep focused through my policy experience.

I'm still a devout supporter of fieldwork, even more so now, having done this very theoretical project, because it made me acutely aware of how little we know about fish biology, especially tropical fish. I scoured the literature for anything I could find about life history information for species that are of commercial importance. You'd think that would be such an obvious thing to do, that people would have done it, but there is so little known, in tropical systems anyway. I think that kind of information is so important.

The other thing that I find a little frustrating in terms of field science is that it is relatively easy for me to get money to do more theory but not so for fieldwork. But, it's theory. It needs to be tested. I mean, I don't trust it. There are enough assumptions that are built into any kind of theoretical study one does. It's not worthless, but it's practically worthless until somebody gets out there and does some field validation of it. I think part of the problem is that people who do the theory are a little bit guilty of overstating the importance of their side of it. It is not often that you'll come across a theoretical paper that is transparent about the assumptions that are in the modeling, how those assumptions might influence the outcome of the model, and how field science could really help by clarifying whether the key assumptions are good or not. I try to do that a lot, and I've actually had papers sent back to me from editors saying, "We'll consider this, but you have to take out all this stuff that's making your work look less important." You just have to come out and say what your implications are and not question yourself. There is that culture among the theoreticians. With theory, you're supposed to prove something and then you state your proof, which to me just adds to the barrier that prevents field scientists from going out and testing the theory, because the theory isn't presented in a way that is accessible to people doing field science. That is something we have to work out among the people who are doing the theory.

I actually did the absurd in giving up a pretty secure faculty position at UVI [University of the Virgin Islands] to come here for a one year postdoc, with a potential for three years. I decided that I was well enough along with the reserve work I was doing. My secondary project was looking at land development and sedimentation impact on reefs. I was having absolutely no luck in getting funding for that. I was getting great reviews from everybody I submitted it to, but it was either a little bit out of their jurisdiction, or it was a little bit too applied, or it was just a little bit too weird with the land-water interface. I'd end up getting not quite the highest priority.

So I'd given up on that and decided, through contacts with Tim Smith here at NMFS [National Marine Fisheries Service], who at the time was branch chief for the protected species work, that it would be really interesting to come in and tackle the problem: What impact are the marine mammals having on the fisheries? The history behind this is that with the buildup of the marine mammal populations, primarily seals, as a result of the Marine Mammal Protection Act, I think there is no question that there are increased incidences of direct conflict between fishers and marine mammals. The marine mammals come around and harass the fishermen when they are trying to fish. The direct conflicts at least are increasing. A lot of people think there is also indirect competition going on because fishermen and marine mammals are going after fish, the same resource.

The same prey species?

Good question. We don't even know if it's the same prey. People just tend to say, "the fish"; they are all eating "fish" so there must be conflict.

The fishermen have called for programs to cull the seal populations. They want permission to shoot them when they see them, so they won't harass their gear, but I think a lot of them also firmly believe that the seals are decreasing their potential for catches because they are eating all of the fish. About two years ago, when I was first looking into this postdoc, Massachusetts Governor [William] Weld actually gave a speech in support of this. He said that there are all of these fisheries problems out on Georges Bank and the Gulf of Maine, and it is because of the seals. I've come into this a little skeptical that that is the case but definitely with an open mind. I've been doing some ecosystem modeling and looking at it from a couple of different perspectives to try to get an idea of how much conflict is really happening, not direct conflict with gear but indirect conflict through the food web.

Is much known about what fishes are eaten by the seals?

No, and that's the hardest part.

Has there been a switch in what the seals eat due to the changes in the fisheries?

Very interesting questions, and there is definitely very poor information. There is some information about the diet of seals and other marine mammals, which comes from gut contents of animals either that got stranded in some fishing gear and got drowned or animals that beached themselves. So there is growing information about that, but really what you need (especially to do some modeling with this, and to see how things might change with changes in fisheries regulations) is information about what a marine mammal species eats relative to what's available.

What's available must have changed because what the fishermen are catching has changed, the landings have changed?

Exactly. We didn't used to catch nearly so many skates. No question that that is the case. At this point, there is very poor information on what is presently available in areas where marine mammals feed, and historic information is even more limited.

One of the things that I saw as my role in this postdoc was to try to highlight what some key field needs would be, what information we need to answer some of these questions, to really come up with the answer. In fact, I don't suspect I'll be able to come up with an answer that I'm convinced about, and if I'm not convinced, I'm not going to convince anyone else, I hope. But my goal is to say that this is a key set of information that needs to be obtained and then also come up with some predictions that are based on a variety of assumptions. If those predictions turn out to be somewhat accurate, it might be that some of those assumptions are in the right ballpark. So that is the general idea. What's been nice about this is that there is incredible information about the fish, especially having come from working with tropical species. Every single species has really good information about its growth rates, its fecundity, its estimates of survivorship, and pretty good estimates of its fishing mortality. There is even, in a lot of cases, information about larval stages. There is nothing like that for the tropics, for anything of interest to fisheries. So, going from the bottom up, I'm finding it is pretty exciting how much information is available; it is the top-down part where the information is tricky.

Postscript (March 2000): I chose to forgo an interview for a job at Brown to take a job at the Center for Marine Conservation, a nonprofit group dedicated to marine conservation. At times I am frustrated by a lack of science colleagues within the center but also enjoy having a chance to weigh in on so many issues. My original science plans have suffered, but I am finding more time to catch up with those plans now. I value the role of fieldwork in general and good field stations in particular. I hope that through my work as a scientist and behind-the-scenes influencing of policy, I can help to build stronger links among these disciplines and encourage others to build similar bridges.

Those two years of just looking around and doing the wrong thing over and over again is a learning process.

KARLA PARSONS HUBBARD (b. 1961)

Postdoctoral Fellow, Rutgers University
Presently: Assistant Professor of Geology
Oberlin College
Interviewed in my office 5 January 1998

When Karla Parsons Hubbard was in college, she didn't think biologists went into the field any more. I'd like to think that her view has been altered since we've worked together on the effect of hurricanes on coral reefs. Hurricane Hugo caused unexpected interruptions in her career, prolonging her Ph.D. work and influencing future opportunities, but she concludes that it probably was beneficial as it gave her time to think and to make and learn from her own mistakes. She has finally realized her dream of becoming a professor in a department (at Oberlin College) that values teaching and the tradition of fieldwork, a dream inspired by the role model of her graduate advisor, Carl Brett.

What got me into field science was that I had the opportunity to take a high school summer school class at Northfield Mt. Herman School in Massachusetts. My sister had gone for a summer, and she took a writing course. I got this brochure and saw "marine science in St. Croix." I decided I didn't want to do anything that would be good for me. I wanted to go to St. Croix and learn how to dive. The summer after my senior year, I went back and took the advanced course, and for my first two years of college, being a TA for those courses was my summer job. It was great. I got into coral reefs and diving, and I went into college thinking I wanted to be a marine biologist.

I went to Beloit College in Wisconsin. There was no such thing as a marine science major there, but I thought I'd be a biology major and go on to graduate school. I quickly decided—it may not have been true but it seemed like—biologists don't work in the field. They look into microscopes. It's cellular, it's genetics, and I hated genetics. It was mostly pre-med types. In my second semester of my sophomore year, I took Geology 101. The department was all field oriented, field trips, and that's what I wanted.

I was so glad that I went to a small college. Those were my two choices, University of Wisconsin or Beloit. All of the professors were concerned with everybody. You'd have a class of ten people. For some of the lectures, you'd just go to the professor's house for dinner. I wouldn't trade that for anything. Watching friends of mine at the University of Wisconsin—a big university, big lecture halls, some of the lectures on tape. There wasn't anybody in the room; you'd go buy your lecture notes. What's that all about? How do you learn if you go buy your lecture notes? Isn't the material in the textbook?

Would you say you had a typical Ph.D. experience?

Ha, no! Well, actually, I probably did—it took forever.

With unexpected interruptions though?

That was a godsend! I'm the world's biggest procrastinator. It took me three and a half years to get my master's—under very good conditions. I mean, my field area was right there, I could drive out to it. It's just me. I'm slow. A lot of it is that I wasn't really sure I wanted to go on. So you postpone graduation, you postpone decisions. For the first two years at Rochester, I behaved as I did in college. I allowed myself to be taught and led around by the nose. When my master's project began to come together, I started to think for myself. That probably was the turning point in my motivation to continue. I finished my master's when I realized, yes, I did want to go on in research. I got it done and started something new.

I guess I was two years into my Ph.D. when Hurricane Hugo hit us down in St. Croix. I was really flailing around. I lost all my field notes in the hurricane. Two years of field notes fit into one notebook—that's not so good. But we found them. Really, I started to get something done after Hugo. I'm sure it's typical. I flailed around and looked too narrowly. I really wasn't getting any experiments going. DENNY [HUBBARD] helped a lot in getting it organized.

But isn't it true that those two years of looking around and then six months of doing something different, as you participated in research projects assessing the effect of the hurricane on the reefs, helped in the long run?

Yes. Without this experience looking around at different aspects of reef science, I never would have known how to structure my research. I'm suspicious of somebody who just goes in there and gets a Ph.D. in two years. Either their master's led directly into their Ph.D. or someone chose a topic for them, because those two years of just looking around and doing the wrong thing over and over again is a learning process.

Denny's perspective, which is more geological than I had, helped. I mean, I didn't

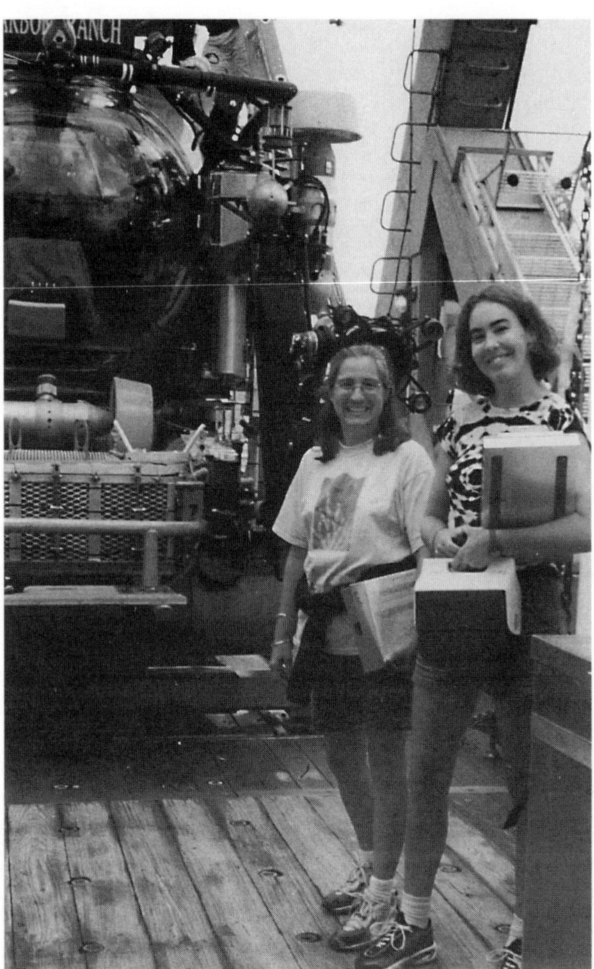

Preparing to board the submersible *Johnson Sea Link*, Karla Hubbard (*left*) enjoys the opportunity to pass the art of field science to her Oberlin undergraduate student and assistant, Bekah Shepard. Photo by Sally Walker, University of Georgia.

know anything about the geology of tropical coasts or coral reefs. The fact is that how shells are preserved is so dependent on where they are in relation to the sediment coming in. So all of the experience that Denny had was so helpful in determining where I should place my natural experiments. I really was better at the biological side of my research, even coming out of a very good geology graduate program. My sedimentology was weak, and my carbonates were totally weak. My undergraduate training was hard-rock oriented. I was really pretty clueless. Just helping Denny on his projects, which delayed what I was doing, but helping him with drilling reef cores and learning about the whole aspect of reef development, solidified what I needed to do.

The writing process was so painful. Carl is really into his students' projects, and his editing is complete, almost too complete. The best thing for me was to leave, because I would have let him do it. So being in St. Croix and writing, I wrote what I wanted to write, and Denny overhauled it completely. But, that brings up another point—your husband shouldn't overhaul your work. It's not healthy, but fortunately in our case, it worked out. We fought through it, tears and things, but he was right.

I've become a better writer. Denny's help also guided me toward the literature on carbonate systems, and he got me to interact with other people—Ivan Gill and Susan Williams, for instance—who have worked with carbonates. He provided critical editing, and he knew the sedimentology side, but most of what I was doing was paleontology. Denny didn't have background in that. That side was mine. Then I sent my dissertation up to Carl as a much more solid piece of work. His editing wasn't quite so harsh, and it wasn't so completely rewritten. That was pretty healthy for me.

The postdoc has given me additional writing experience so that now I have so much more confidence in writing a paper. Finally now, in 1998, I'm about to submit the first paper out of my thesis. That's because I've had to write papers for the project that I'm working on now, the Shelf and Slope Experimental Taphonomy Initiative. It includes seven different scientists from four different institutions, and I'm the only postdoc. The amount of data we have is phenomenal, and so I've given maybe five talks on the research, and I've got maybe four or five papers that I've mostly written.

In your field, do you feel people can generate realistic, testable models with never having had any field experience, just having read the literature?

No, absolutely not.

Why not?

It would be easier to answer that by thinking about people I know who have little field knowledge. What they come up with in a model, it makes sense, but it's not the way it is. You can see where they are coming from, if they are going by the books and papers, but if you haven't been out there, there are just so many factors that come into play. It is like standing in a bay and not knowing what the upland topography is, and you might not know that from reading a paper. You are not going to get the run-off and the sedimentation patterns from the literature. You have to be there and see it. Denny and I joke about how it seems that biologists, no offense, are just starting to come around and figure out that the physical environment has an influence on the fish or whatever. And they've been out there!

When I was in college, the geology department was really field oriented. At the time, the USGS was talking about ending its mapping program. According to the government, everything had been mapped, we don't need to go out and map geology any more. Everybody was horrified. Who are these idiots? Well, they are government administrators who have never been out in the field, and they have no use for it.

They pull a map out of a drawer and say . . .

It's been done. All the little squares have been filled in.

Postscript (September 1999): I always did want to teach at a small college and here I am at Oberlin College. It is great. I really think that my emphasis on fieldwork and research for undergrads helped me get this job. Unfortunately, last year, I was too overwhelmed to take the students into the field as much as I had wanted to, but this year I hope to do more.

It is not the technology that excites the kids; it is the person using the technology with them who excites the children.

Ellen Prager (b. 1962)

Marine Scientist and Freelance Writer
Arlington, Virginia
Presently: Assistant Dean, Rosenstiel School of Marine and Atmospheric Sciences
University of Miami
Interviewed in my office 9 April 1998, with Arthur Gaines

Ellen Prager is a WIL undergraduate alumna, a bright, capable, and enthusiastic student who was invited back to work as lab supervisor for a year. She illustrates how spending time in the field with other scientists has innumerable benefits, including acquiring new skills, learning how others approach and solve problems, and seeing new sites. She also points out that among the necessary ingredients for a successful field scientist are a good sense of humor and the expectation that things will probably go wrong. While Prager enjoyed teaching in a field program and directing a field facility, she has chosen to attempt to reach a broader audience through explaining the work of scientists to the public by various means, including books, videos, and the internet. She aims to create personal bonds among members of the public and natural scientists by highlighting the passion and conviction of researchers.

I went to college at Wesleyan University in Connecticut. It was a great experience there, the sense that you can do whatever you want. You make your own life. Independent thinking is very important at Wesleyan. I was interested in a wide range of sciences, especially biology and geology. I took a marine biogeology course, and I just loved it. That's when I decided to take a semester and go to the West Indies Lab. That was where I got hooked.

I was set for life after that; I couldn't change my path. West Indies Lab was really important for me for a couple of reasons. One, I had already been doing field science. At Wesleyan, I worked for the environmental science department. I would go out with them doing things, like sampling on the Connecticut River. At West Indies Lab, I got exposed to experts in the field who were really passionate about what they were doing. They absolutely loved it. They could pass it on to the students they were teaching, and not only that, they were really interesting and funny people whom I wanted to be like and have the kind of job they had. They were the kind of people with whom I'd like to spend time. That was fortunate, because many of those people are now my friends and colleagues. That was something I had sort of missed at other places. I didn't see role models.

It was the mentoring and the excitement of being in the field! The way the WIL program was set up, we'd have a lecture in the morning and go out and see things in the field in the afternoon. I am convinced there is no better way to get somebody excited about the environment than to actually bring them to it, especially the marine environment. You can do it justice to some extent through video, but you have to get them out into the field. For me, I was hooked. I knew there was a way to have a lifestyle that would be challenging, that I would enjoy, and I could do all the things I love. I could be athletic. I could be intellectually challenged. I could be outside. I could be in the water. I could work with the marine environment, which I found incredibly fascinating.

While I was a student at the West Indies Lab—this is one of my favorite stories to tell students—I got on my bike and I went over to *Hydrolab* [the underwater habitat research program] one Saturday. It was about ten miles away. I asked, "Do you have any summer jobs? Can I be a diver for you?" They asked, "Can you pick up these tanks?" "Sure." They said, "Great, we'll see you in the summer." I rode back to the lab and told the other students. They asked, "How did you get that job? How did you do that?" I said, "I asked." It was the beginning of taking a very proactive stance in my career. It is not going to hurt you to ask or to pursue something. The worst that is going to happen is that somebody is going to say "no" to you. So what? To get in, you've just got to be a little adventurous and be willing to put yourself at risk for rejection, but you've got to do it. That was one instance where it was great; it worked out. I went to *Hydrolab*, and I went back again another time. I made contacts with the people doing what I wanted to do. They were not only professional, but they were really wonderful people at the same time. They cared about training me, teaching me, giving me the chance to have the same passion for the ocean that they had.

In both my master's and my Ph.D., I was really lucky that I was invited by people to help them in places around the world. That was hugely beneficial to me in my career. You don't always know where you want to go, you don't always know what your interests are until you see a lot of things and can narrow it down—this is what I want to do research on. I've seen a lot of things, but this is where I want to focus my energy. That helps you.

For me to do work in the Pacific, to understand the differences between the Caribbean and the Pacific, is crucial. If I want to generalize in my writing, I need to have the perspective of having seen other places. I also think it makes me a better person. Some of the things I've done on some of these adventures, I've had to face challenges myself, personally and physically. I've had to deal with places in the world which are very different from what I am typically used to. I am much more understanding of the world and the way it works. In working in a number of Third World countries, I appreciate the United States 100% more than if I had never done those things. I appreciate things that I consider luxuries that most Americans think are just standard ways of living—electricity, fresh water.

While continuing to do research (here using the underwater habitat *Aquarius* in the 1990s), Ellen Prager devotes much of her time and energy to bringing scientists' passion for science and the environment to the general public. Photo from the files of E. Prager.

I finished my master's in carbonate sedimentology, which I loved, I absolutely love it, but I flew through the program, I was burned out of academics. I was offered an opportunity to bypass my master's and just stay there and get a Ph.D., but I was so stressed from trying to crank it through. I thought I would be very limited in my perspective if I did that. I went to Shell Oil and found that I definitely didn't like the corporate world. I was bored.

Then I went back to do my doctorate. We tried to do work in coral reefs, but at the time reefs weren't getting any funding. We put in a big proposal to work in Fiji on physical processes in reef environments, which is one of my big interests. We didn't get funded. There was another project, circulation in a coastal bay, using a model and some observational data. It wasn't my real interest, but it was interesting enough and, to me, it was a means to an end. The data were there, I only had to collect a little more. Somebody had a model I could use, and I just had to learn it, which was a tough thing to do—to learn how a model worked and what it all meant. It was a good experience. I learned what a model can tell you and what it cannot tell you. I am using a wave model now, and I am comfortable using it. I just have to see what it does. It wasn't exactly what I wanted to do for my Ph.D. research, but it was a way for me to finish, get out, and do what I really wanted to do.

After graduate school, it was very interesting for me. It gets back to the field science issue. People said, "What you have to do is go to a big university and develop a big research program." I said, "That may not be what I want to do." I had an opportunity to work for Sea Education Association [SEA]. I was told, "You'll never get back into research if you go there." At SEA, we took students into the field and taught them through being there and through hands-on field research. I couldn't go to West Indies Lab to do that, but here was something very similar. I thought, here is a place I can do for people what you guys at West Indies Lab had done for me. It was great. I loved the teaching at SEA. It was half a semester teaching on shore in a typical classroom style and half a semester out at sea in a tall sailing ship teaching how to do oceanographic research at sea. We also stopped at places on shore and went snorkeling, where I could teach students about reefs. As a program to teach students about the ocean, to excite them about the ocean, and to give them confidence to get rid of a lot of the insecurities that students at that age have, SEA is fabulous. I even published a paper while I was at SEA, so I completed some research. But the sea time was exhausting. To do it right, I just put so much into each trip, six weeks out at sea, it was exhausting. The deep-sea stuff, while it was interesting and I enjoyed teaching it, was not my area of research. It was hard for me to do any sort of research. I got sort of burnt out. That is when I got the offer to go out to the Lee Stocking Island Lab in the Bahamas and be the director, and when I left Lee Stocking, the USGS offered me a position.

The other thing in field science—and how that lends itself to what I am doing now and where I want to be—is the emotional reaction to being there. The best way to excite somebody, if you are talking about ocean sciences, is to bring them to the sea. The best way to get somebody to want to protect the ocean, to support conservation, to support ocean science, to appreciate the ocean—like I do and I think most marine scientists do—is to bring them there. But for the majority of people, that is not possible: it could be economic; it could be cultural; it could be where they live. For most people, they are never going to have the opportunity to go scuba diving in the Florida Keys. A good part of the population will, and I would promote getting them there to do that. But, for a variety of reasons, a lot of the population will never have that opportunity, so my perspective is that you've got to bring it to them, not only with pictures and videos but through personalities—through the eyes of people who are already passionate about it and who can explain why the ocean is so fascinating to them. Then maybe you have a chance to stimulate the interest and the curiosity that I got from the films of Jacques Cousteau and *National Geographic* specials. That is why I have been tending toward the idea of bringing science to the public. I don't see people getting the same kind of exposure to these popularizing programs that I had. There is so much junk out there that they do get exposed to.

I feel that, as scientists, we should not only be doing a good job at getting out the science information, but we should also bring our passion for the science and for the environment to the public. I want to take that challenge on. I never want to give up doing research and continuing to be curious to try to find the answers to a few more questions. I want to keep asking those questions and to try to better understand the physical world. But now, I also want to put a lot of my time into bringing that to people. Someone recently asked me, "Why don't you apply to a university and teach?" I said, "I love to teach, and I am tempted at times to do that. But, at this point, if I can do things that are much broader, I can be more effective." Maybe I can lead someone to go on in school like I did or to take a class from a professor in marine science. I can be a visiting lecturer at a lot of places, but I think if I work on a bigger scale I can have more of an impact.

Of course, that is all part of fieldwork—the adventures you get yourself into while you are out there. Part of it is that these adventures can be humorous. In the Florida Keys, everyone says, "So, you do your research in the Florida Keys. Aren't you lucky! You'll be on the beach; you'll get to swim with the fish—how lovely." Then there is me and Bob Halley walking to our research area in waist-deep mud, with crabs and worms, sitting there taking these very tedious measurements of the mudbanks and how they change through time. It is hot as can be out on the banks. There are turkey vultures circling overhead wondering, hmm, how much longer will they last? That is my Florida Keys. That is fieldwork. It is hysterical out there, if anyone ever saw us! In fact, one of the things I tried—someone in the National Park Service suggested the idea to me—was snowshoes. Snowshoes across the mud bank because you won't sink. Great, I thought. So I got these snowshoes and I said, "Oh, I'm ready." I put them on. The first part of the mud bank is usually lined with seagrass. So it was great walking over the seagrasses, bending them down a little bit, staying on the surface. I stepped off the seagrass into a mud hole. The mud squeezed through all the holes in the snowshoe, and I was stuck for life! The snowshoes just sucked in, and I went face forward into the mud! So much for the snowshoe idea. We reverted to our method of crawling on our hands and knees. Or you can do a knee walk, if you are carrying equipment, core tubes.

I think as you get more experience in the field you come to expect that at least half of what you try isn't going to work. That is normal. You'll either get weathered out or your equipment fails. People who don't do a lot of fieldwork don't understand that. They'll say, "OK, I have three days and this one piece of equipment, and I'm going to solve the world's problems."

Arthur Gaines: *When I was at SEA, I used to tell students, "The world does not want to know why you didn't do your project. The world does not want to know why you failed. The only thing the world wants to know is what kind of success you had." When you go out there, there are going to be thousands of reasons not to do your project—you forgot the batteries, you are seasick, it's too rough, somebody stole your chemicals, something broke—but you've got to keep laughing, you've got to keep laughing. They'd look at each other and then look at me and say, "We're going to sea with this guy?"*

The other thing I would tell students at SEA is that coming up with a different conclusion is as important as finding what you predicted. That was always a hard sell. For you to come up with something other than you predicted is really important.

Arthur Gaines: *They think they've failed.*

Exactly. Also, I see people now coming out of school who were not pushed, like Dr. [Robert] Ginsburg pushed his students; he pushed us to our limits. I see people now who don't get pushed, who aren't taught how to write, who aren't taught how to give a talk. No one has pushed them to their limits, and they are unprepared. I can't believe this intern I had. He was great on the computer but had no common sense. He was terrible in the field. He almost burned down the building.

How are we educating students? We are so tied into technology—CD-ROMs, interactive units. I'll tell you what, and I'm sure this will come out—it is not the technology that excites the kids; it is the person using the technology with them who excites the children. They have to have the content as a context to use the technology. Technology is so new. People say, "give them the interactive CD-ROM, and they'll learn it." You need a teacher who is enthusiastic, motivated, and has knowledge and can say, "Oh, look at this!" A kid using a computer can get from screen to screen, but that doesn't mean he is learning what is important on each screen. I think that is a big problem. With the project that I am doing over the internet, one of my big problems is how do I bring it to people who don't have access over the internet. I'm not promoting this as something so great because of the technology, I'm saying this is how you can use the technology to bring the stuff to your students. That is a big difference and is analogous to field science versus remote education

When you think of fieldwork, this was a contrast compared to what I had done before, which is constantly writing down numbers, counting things, and looking at things.

Grete K. Hovelsrud-Broda (b. 1959)

Marine Policy Fellow, Woods Hole Oceanographic Institution
Presently: General Secretary, North Atlantic Marine Mammal Commission
University of Tromsø, Norway
Interviewed in her office 2 February 1999

Grete Hovelsrud-Broda and I first met as colleagues at the Marine Policy Center at WHOI. She has used the skills she developed while exploring the countryside in all weathers during her Norwegian childhood and through ecological and geological fieldwork in the Arctic to do a scientific study of her first love, the peoples of the Arctic. In her account of her year spent in a small Greenland village investigating the social and economic ramifications of seal hunting, she compares and contrasts natural science and social science approaches.

When you go in the field and start living in a small, very isolated community, you make a big impact, obviously, just by coming there. You have to be very careful. That is one thing I found when I started my fieldwork—what a difference it is to be an anthropologist compared with an ecologist. My first field experience was with animals, reindeer, which go off, and you have to follow them, then dig in, and watch them. With the reindeer, we did a complete ecological study, their feeding habits and their feeding habitat, their reproductive cycles. My friends have records of populations up there for 20 years now. We also had permission to kill animals to do all sorts of biochemical studies and fat content. We had radio collars on the reindeer to see what their habitats were and their range. Svalbard reindeer are different from other reindeer that migrate. These are more or less stationary. In the end, we had done a quite comprehensive ecological study of these reindeer.

Then, I went back to the Arctic, to Spitsbergen, as a geological field assistant for a few summers. We were self-propelled, we kayaked and skied up the glaciers, camped and did Quaternary geology. I was working with Scott Lehman, who used to be here at WHOI. That was yet another way of doing fieldwork. We were looking at old beach strand lines, and we were looking at something that you don't even know that you are looking at because it is in the ground. We dug for shells and so on. That was a very interesting, different kind of field experience. I loved that, too.

Then, the people. I landed in a village in east Greenland. Here is a picture of the village. It is very small and very remote. I got out of the helicopter, and it left. I looked around and had no clue; I didn't know the language; I didn't know a person there. I had a place to live, actually, and Jim [Broda], my husband, was with me [for the first couple of weeks], so we were two. It was very interesting, but it was very different. How do you start? You can't start by walking into people's houses and talking to them. Also, anthropologists, in my opinion, should not do interviews in that sense, at first. They need to be part of the culture before they start doing interviews. So I decided to sit back, and I realized that this was now out of my hands. It was up to the community whether I was going to be able to do fieldwork or not.

I spent three weeks getting organized, walking around, and saying hello to people. People were friendly; there were just 160 people. I started baking bread, and I traded loaves of bread for seal meat. When you think of fieldwork, this was a contrast compared to what I had done before, which is constantly writing down numbers, counting things, and looking at things. All of a sudden, there was none of that. It was pure qualitative descriptive research, I would say, at first. I wrote field notes every night. I wrote down as much as I could, and I worked hard on the language.

I had done a pilot study in east Greenland prior to my arrival in the village. I was taught the basics of the language by a French anthropologist. Everyday in the village, I studied new words, and I went out into the community. The problem was that there is no published textbook of the language. There is no standard. So every word is written differently depending on to whom you speak. So time went by. My husband, Jim, who works here at the oceanographic, is so used to the quantitative—you collect data, and you talk about it. He was so frustrated. It seemed to him that I was just wandering around, which is exactly what I was doing! It was very important that I didn't—this was intuitive, but it was also through learning, I suppose—that I didn't come and push myself on the people.

It was also probably a very intense experience.

Incredible!

Did you have a preconceived set of things you were going to do?

Well, I did, but I didn't know if they were going to pan out. I had written this grant to do a pilot study. I said I was going to work with the seal hunters and their families. Well, I got there, and I realized that I couldn't work with the hunters. I was the wrong gender. I was a woman. I couldn't go out there with them. I had to go with the women. If I went out with the men, the connotations were that I was an available female. I figured that out in a day, so I immediately had to switch. That is one thing I think anthropologists always will have to be—incredibly flexible. If you are rigid, if you have said, "This is what I am going to get out of this," you are not going to make it.

I think in most science that is true.

I think you have to adjust all of your proposals once you go out into the field, in field research, but I think that when you deal with people, you are up against totally different variables that you don't even know about.

In my grant proposal to NSF, I said that I needed to go in the field for six to eight months, come back home for two or three months to write up what I have and to get a sense of the direction of my research, and then go back to the village. My intuitive sense, and also listening to other people, was that it is not a good idea to go into the field, stay for a year, and then you are gone and never come back. I didn't want to do that. I wanted, at least, to go back once, if not more. I am hoping to go back again, but that remains to be seen. NSF accepted that and even thought it was a great idea to go back home, work through my data, and see in what direction I wanted to continue.

So I went back. When I came back to that village, I was treated like the lost daughter. It was really nice. I came back in the middle of the winter. They were tucked into their houses, you couldn't really go anywhere. My questions were much clearer. The dynamics of the village, all of a sudden, sort of came together. I collected data from participant observation, which is what we call it in anthropology, which is the predominant mode of collecting data. You sit and drink a lot of coffee, you work on the seal skins when they come in, you eat the meat when that is cooked. You are with them all the time, you do stuff with them, and in that way you learn. You learn a lot about the dynamics and the culture.

The other thing was the interviews, which I did at the very end of the whole stay, the second stay. I wouldn't have known what to ask at the beginning. I'd have asked impossible questions. I still asked impossible questions, but I had knowledge of what to ask about and what not to ask about. I could also verify. It was interesting, because I had these standardized questions, and they would answer contradictory to what I had observed. I would say, "But, I saw you, so and so, with so and so, and you were sharing meat." And they'd say, "Oh, yes," and so I could get another explanation for why they were sharing meat. That was very useful, I thought, that kind of dynamics.

I want to talk a little about my field research. My feeling was, when I was there, that I was basically a two-year-old. Judged by my level of language ability, at the end, maybe a three-year-old, maybe four. I was in many ways treated like a child, and I knew basically nothing, except for my Arctic experience and my reindeer-slaughtering experience. That came in very handy, of course, as well as all the knowledge about the cold and the Arctic. But, as far as the social organization, how to behave, and what to do, so I didn't really mess up—I was a child, and they treated me as a child. I accepted being a child. I realized that, if I were trying to impress them that I was actually a grownup, they didn't have any use for any of the things I knew. They liked it when I baked bread and when I made certain things. I knew how to knit, I knitted all the time, and I knew how to sew, so I knew some things, but basically I was a kid, a little child.

I'm wondering if they were more open to someone who is childlike, appreciative, and learns, and they can see the person is learning.

I think that that is true. I mean, if you were constantly trying to resist that, I mean, we had to laugh together. That's one of the things you have to do. When we were struggling through our language barriers and telling jokes and so on, we were laughing like crazy. We would roll on the floor. I could share their laughter and their humor. Basically, I felt—sometimes, I would get anxious about this—I felt as if I were totally blank and that I just had to go with it, float with it, and then just write down what I saw.

Sometimes I felt that I was kind of spying, but they knew why I was there, and they wanted me to learn. When we were eating funny foods, like intestines or very old blood, they would say to me, "You shouldn't have this, this will not be good for you."

Was that true?

It was horrible, yes. If I had had that blood, I'm sure I would have been very sick. It was richer than a European stomach could ever imagine. So, basically, I ate everything they ate, and some of it was less interesting than other things. But I think it really helped me in that village. It was said that the villagers weren't interested in outsiders, and they basically didn't want anybody.

That was also a survival strategy for me, of course. I mean it is incredibly hard to do fieldwork like that, because you are always in the field. When I was studying reindeer or doing geology, I could go into my tent and I'm not doing fieldwork any more. But there, I was always doing fieldwork. The break I took on Sunday was to go skiing in the mountains with my notebook and write poems.

You are going off to a new position now, but you may come back to this?

Oh, absolutely. I'm not done with being a field anthropologist. I've taken a job with the North Atlantic Marine Mammal Commission as its general secretary. I'll basically be an executive. But part of it has to do with hunters, and I'll be working directly with the hunting communities. I could probably take control as to how much I want that contact or how little, because that is partly up to me. Since this is my field, I'll probably want a lot of interaction. Next week, I'm going to a workshop in Greenland on hunting methods, where many hunters will show up. I probably won't see any of my women friends at all, but I'll meet new hunters from new communities.

The problem, of course, is always funding. I've been lucky to go back to my village two times. I have been back in Greenland since, but not in the village. They keep asking me, when am I coming back? The woman who was my interpreter and friend died recently, last summer. That is really sad. I wanted to see her again. I have a mother, a father; I was adopted by a household; they called me daughter. Obviously, being in an area for only a year and a half, you don't get any depth.

At the anthropology department at Brandeis, I'd say I got a lot of support. But not anything, absolutely nothing about what to expect when I went out into the field. I asked several professors whom I knew—one worked in Palau, one in the Amazon—and they said, "Bring a lot of ziplock bags, for your notes, your underwear, whatever else you want to keep dry," because they went to the tropics. I said, "Well, that is a practical suggestion." I was told nothing. Methodology, very little. How do you actually go out and collect data? What I did was to write in my little books. Then, once a week, I sent a copy of my field notes out to my father. One of my colleagues had lost all of his research data in a fire, and he almost lost his life trying to go after them. So, I sent all my field notes out. When you look at the literature about how to do fieldwork—it is there—they don't really tell you about how tedious it is or how lonely you can feel or how helpless you are.

My dream is to go into the field with a colleague anthropologist, work together, see what we come up with, and see how much our perceptions determine the data. That would be really scary. Sometimes a couple goes out, and so you get the man's perspective and the woman's perspective, and it is very gender divided. It goes with the local gender division. But I think that would be fascinating, and you could really fill some gaps that way, I'm sure.

SUMMARY: CHANGING TIDES

Every person in this generation became curious to figure out how nature worked through direct interaction, rather than the remote approach they felt was being presented in college science classes. Fortunately, almost all found inspiring undergraduate and then graduate settings led by dynamic professors and equally enthusiastic colleagues and were able to develop their skills in field research, often at field stations. Unfortunately, many of these opportunities do not exist today: marine labs have closed or shifted emphasis away from fieldwork; geology departments have turned away from field science and reduced the length or even the requirement for a field camp; the "Don Abbotts," the generalists, have retired. A disturbing trend of recent years is for departments and also field institutions to emphasize the reductionist approach; they've switched from field-based to the more highly funded theoretical and lab-based research.

Critical career decisions were made at an earlier stage than in previous generations. Each was expected to specialize, to know the kind of research he or she wanted to do before going to graduate school. However, even if a subject were found (e.g., Prager's desire to study nearshore oceanography on reefs), the funds weren't available to do that kind of field research at that point in time. Following graduate school, everyone in this generation has had one or more postdocs (often involving administrative duties as young, inexperienced scientists), prolonging the time before they can settle in and formulate long-term career plans, and removing them from day-to-day research and field experiences.

The tendency for the experienced field researchers of this generation is to turn to alternative occupations away from traditional academia (which removes them from a continuation of personal field research); these vary from inspiring the public about natural science issues through public media (Prager), to working at NGOs [nongovernmental organizations] (Nowlis), to working as a government administrator (Hovelsrud-Broda). In the latter occupations, the scientists do not intend to give up field research permanently, but I suspect they will have to carefully craft the opportunity to return to it, to work on those questions that nature poses. Additionally, most of them are not in positions, nor would they have the time, to spend extensive periods doing fieldwork with younger colleagues, passing the art of field science in a personal manner to the next generation.

On a positive note, unlike previous generations, there is no sense that the women in these interviews were either implicitly or explicitly discouraged from careers in the natural sciences during their formative years. On the contrary, they seem to have been given encouragement from their parents, teachers, and peers that they were capable of doing whatever it was they set their minds on accomplishing.

Finally, the strong message running through these interviews is the continued relevance and immediate need for good field scientists. It is clear that environmental issues will be of major importance to our global economic systems in the near future. Conservation, as a justification for field support, has been an emphasized issue for this and the previous generation. Nevitt points out that her research animals, several species of salmon, have been put on the endangered species list; Fonseca worked on the malaria endangering endemic Hawaiian honey creepers; Nowlis has worked on several different fisheries issues. Fonseca discusses how knowledge of animal behavior, requiring detailed field studies, may be a tool in the future to monitor toxicity in air and water environments. To adequately address these emerging environmental problems, we will require scientists who are experienced in (preferably comparative) field situations, field techniques, and field sites.

7

Natural Sciences at the Beginning of the Twenty-first Century

J'admire la Science, bien sur. Mais j'admire aussi la Sagesse.

Antoine de Saint-Exupery

Are we currently producing the next generation of Agassiz's heirs? In these pages, we had glimpses of the lives of some of the people who have discovered something new about our natural world, from before World War II to the present. While they have been fortunate to have inherited Agassiz's legacy themselves, the world has changed in their lifetimes. Curtis Cate, in *Antoine de Saint-Exupery: His Life and Times* (New York: Putnam, 1970), pointed out, "For it is one of the great (and greatly overlooked) ironies of our age that while the last war [World War II] was fought by the western democracies in the name of liberty and the 'dignity of the human individual,' the values which are now everywhere in the ascendant are unindividual, quantitative, and collective" (571). Yet, we have seen that good science requires individual creativity, a qualitative appreciation of the world through all of one's senses, an emotional connection to one's work (as Mimi Koehl states, "being passionate about wanting to understand it"), and plain hard work. A good mentor cultivates these traits in the next generation. He leads, he listens, and he continues to learn as well.

How was Louis Agassiz remembered by his students? One of them, David Starr Jordan, the father of modern ichthyology and the first president of Stanford University, encapsulated the essence of a true education in his memoirs, *The Days of Man* (New York: World, 1922), as he describes the Anderson School of Natural History six months after Agassiz's death:

The following summer we gathered again at Penikese. . . . Eager new faces now appeared. . . . Wise teachers were present as before, the work was stimulating—but a sense of loss was felt above everything else. One evening, therefore, we met in the lecture hall, and each spoke as best he could of the absent Master. The words which longest remained with us were those of Samuel Garman:

> He was the best friend that ever [a] student had.

On the walls we put several mottoes taken from Agassiz's talks to us:

> STUDY NATURE, NOT BOOKS.
>
> BE NOT AFRAID TO SAY, "I DO NOT KNOW."
>
> STRIVE TO DETERMINE WHAT REALLY EXISTS

These striking phrases, written on cloth, were left for fifteen years in the empty building, whence they were then carried by my student . . . to the Marine Station at Woods Hole, in some degree the natural successor of Penikese. (1:117–18)

Students and masters learning together. Respect for and love of nature. The transfer of knowledge from generation to generation. The institutional setting to facilitate access to nature and to prior knowledge. The cultivation of an atmosphere of scholarly inquiry and mutual respect among students of all ages. The search for truth in the natural world. Agassiz's legacy.

The Pendulum Swings

What is the life of a scientist at the dawn of the twenty-first century? Don Abbott, methodically collecting sponges in the intertidal with his student, made the point, "Science can provide a great life for a certain kind of person." What drives the field scientist of today not only to work long hours (for which he receives reasonable monetary compensation, but certainly not a great deal in relation to that given other professionals in law, medicine, or finance), but also to deal with the frustrations of academic and government bureaucracies, which erode more and more of his time and energy? We find that the motivation for a field science career is still much the same today as it was in the 1940s. Inspired by nature, prepared and encouraged by mentors, one relishes the physical, emotional, and intellectual challenges of fieldwork. Using creativity, one integrates knowledge, experience, and observations of the natural world to discover something new. It is that simple. But is today's scientist able to share this passion with the next generation?

We have seen that, by the 1960s, the composition of faculty and graduate students in biology and geology programs began to alter with some new, not classically trained additions. In geology departments, physics and chemistry majors were brought in as graduate students to help unravel the mysteries underlying the evolution of the Earth. In biology departments, physics- or chemistry-trained molecular biologists were hired to unlock the secrets of life within the delicate strands of DNA. Both of these were and certainly still are worthwhile objectives. These specialists were welcomed into the academic departments. The generalists (geologists and biologists trained in the 1940s and 1950s) recognized the expertise of their new colleagues and saw these new approaches as an additional way to help understand the natural world.

Problems arose, however, because many specialists, unfortunately, are not broad enough in their thinking to appreciate the diverse roles a generalist brings to a department. Some quite arrogantly dismiss a natural history approach as old-fashioned, not appropriate in today's modern world. In his autobiography, E. O. Wilson recounts the battles fought at Harvard when the young, brash Nobel Prize–winner James Watson joined their faculty. Harry Thompson found that when he asked some new specialists in molecular biology on the UCLA biology faculty to contribute to the general teaching assignments, they couldn't. Thompson sadly points out, "They didn't know any biology." Young Peter Clift recounted a story of when a famous geophysicist from Cambridge, whose models suggested the MoHo was a sheer plane, was brought face to face with contradictory data in the field, "He didn't know how to look at it, really, and he was modeling it and publishing it." Clift, from the 1980s generation, presently endures an academic climate in which a visiting geophysicist is comfortable to dismiss the discipline of geology as "a blast from the past." Friends of mine have heard colleagues argue: It is simply no longer necessary to teach field biology since people don't do it any more. It doesn't need to be done, and we therefore ought to emphasize cell and molecular biology. Parallel arguments are heard in geology departments.

These arguments ignore the reality that many scientific discoveries of tomorrow will still be dependent on natural science fieldwork to provide basic data on which to build more and more quantitative understandings of how the Earth, the atmosphere, the oceans, and life evolve. Students need to have a framework of basic information and practical experience upon which to build their own particular expertise, to provide a context in which to place their own unique contribution. It is important to remember Clark Burchfiel's statement, "It's always been very clear to me that no matter how sophisticated your science gets, most of the sophistication is in the tools, not in the science. Therefore, if you don't understand the very basic relationships between things as they exist in the field, you can apply all of the most sophisticated tools in the world to those problems, and you won't come up with the right answers."

Even in the limited selection of scientific work presented in this volume, it is evident that basic research incorporating fieldwork has contributed substantially to medical research, for example, neurotoxins from cone snails are used as pain killers and in other medical applications; we are learning more about how to treat obesity through studying fat metabolism in hibernating bears; GFP (green fluorescent protein, first discovered and studied in hydroids and jellyfish, is used to tag other proteins in cellular research); we now have increased knowledge about transparency of the human eye lens, by comparing the proteins in the eyes of different but related species in all vertebrate classes living in temperature extremes; and we are learning more about problems with the visual pigment rhodopsin in humans, which has been studied by comparing the rhodopsin molecule in related fish species living at different depths.

Basic science has also contributed to environmental research on interpreting rockfall and other geological hazards correctly; on understanding natural abiotic and biotic cycles by correctly interpreting signals in nature (to help decipher climatic patterns, for instance); on using animal behavior as an indicator of environmental toxicity; and on gathering data during our planet's various upheavals, including hurricanes, volcanic eruptions, earthquakes, and disease outbreaks, to better predict the extent and duration of impact.

In natural resource management , field science contributes to knowing the structure and function of ecosystems and the natural history of organisms therein well enough so that changes can be detected (e.g., polar bears, coral reefs); finding and procuring in an environmentally friendly manner petroleum and mineral resources and developing alternative energy supplies (such as methane-producing algae); and conserving and restoring habitats and protecting endangered species. The responsibility of the field scientist is to provide the framework of natural history knowledge that underlies and grounds truths, even in the most sophisticated molecular and cellular biology results, geophysical and geochemical analyses, satellite remote sensing, ecosystem modeling, and medical and environmental studies in the complex, real world of nature.

The dichotomy of viewpoints—questioning the value of field science or a natural history approach versus a reductionist approach to science—was not sufficient in itself to cause biology departments to drop organismal biology or geology departments to drop stratigraphy as central focuses from which to explore and to relate other areas of the curriculum. The government, stepping into academia, particularly making scientific research into a big business, was the vital blow. John Dacey recalled what his father, a distinguished and respected academic, had explained: once you depend on the government, you are ruled by it, because you fall subject to its whims. For governments, the notion of academic excellence is way down on the list of priorities.

University administrations are driven by overhead from grants. They don't like fieldwork, but not only because it brings in less money to the institution. As budgets are reviewed, field courses and fieldwork are often the first to be eliminated. They are easy targets because there are liability risks; the travel and logistics are unusual; the student-to-faculty ratio is unacceptable. Ironically, field research and field education is actually far less expensive than most laboratory work, which requires an enormous input of financial resources, not only to purchase the expensive equipment, but also to support the technical personnel required to keep these items operating. I am not suggesting that we eliminate laboratory research at all. I am merely suggesting that we have an equitable balance among the field, laboratory, and theoretical approaches to both education and research in the natural sciences.

Unfortunately, the interviews of the 1980s suggest that not many who began to accrue field skills and experience in that era continue as university faculty. Many are now working in administrative jobs far from the side of undergraduates or graduate students pondering the mysteries of nature in the field. Even if they do join a faculty, many senior scientists of the 1950s and 1960s agree that, in modern academia, a junior faculty member simply cannot afford to waste time teaching a field course, taking students to the field; otherwise, he would have no hope of receiving tenure. This has two consequences. First, the young faculty member has no opportunity to hone his own skills as a transmitter of the art of fieldwork, and second, he (and all of the students, I might add) is given the subtle (or not so subtle) signal that somehow this activity is not important. A skewed balance, heavily favoring the laboratory and theoretical approaches, has unfortunate implications for the future of field training; as ten more years go by, the junior faculty member who has been discouraged from wasting his time teaching field courses assumes a more senior position within the department. Not only will he not have had the important experience of teaching in the field, he will have forgotten its importance in his own development. As a result, he will not defend the value of teaching and learning field skills. He will not insist upon their inclusion as a necessary part of the education of the next generation of scientists.

We have seen that the presently active faculty complain that they do not have sufficient time today to think and reflect about their work as they should. Nor do they have many opportunities to take exploratory cruises or field trips with colleagues and students. Research geared to providing quick publications is rewarded, while lengthy studies or risky questions are discouraged. The pendulum has swung away from the natural history approach in education and in research. Specialists dominate the faculties of the natural sciences and the well-rounded generalists are few and far between. And, of course, confronted with closed office doors, an undergraduate student of today cannot glimpse the enviable lifestyle of a field scientist, as many of the 1940s through 1970s generations of students did. It does not exist any more, although sometimes, at field labs and camps, particularly even grant-driven faculty will get into the spirit of shared interaction with students and colleagues of all ages and will relish their teaching and their research, just like in the old days.

Why do I emphasize that Agassiz's legacy, a tradition that has stagnated in the past half century, must not be allowed to disappear? Some scientists whom I interviewed predicted that the twenty-first century will witness the domination of the environmental sciences, a position held by physics in the twentieth century. Humanity will be forced to confront the consequences of its actions, particularly those of the twentieth century, not confined to, but including, polluting every space on the surface of the Earth, the oceans, the atmosphere, and even on our planetary neighbors. Humanity has often been reckless in its interactions with nature. It requires humility to recognize that people are not superior to nature; indeed, humans are an integral part of nature and are subject to nature's rules. Perhaps now is the time for humanity to view its world as one of a cooperative partnership, a symbiosis, and to practice how to listen to nature instead of futilely attempting to dictate humanity's own terms. We might win some battles, but in the end we have no hope of winning the war.

These oral histories illustrate that a basic scientific understanding of our natural world is necessary to begin to deal with these environmental problems. They range from the molecular and cellular scale, relating to medicine and human health; to the regional scale, relating to agriculture and fishing, other land- and water-use practices, and geological hazards; to the global scale, relating to global warming and atmospheric and oceanic pollution. They are most effectively investigated by correctly interpreting the experiments that evolution (in the broad sense of Earth, ocean, atmosphere, and life) has done, and that requires direct experience with nature, a natural history approach to science. These signals are found in many forms, as we have seen, including the distribution and abundance of organisms through time and space; the chemical and physical signals in annual bands of ocean rock or mud, polar ice caps, coral skeletons, fish otoliths, loess and stream terraces, and river deltas; the behavior of organisms—snails, crabs, corals, seals, fishes, insect larvae, birds—in geographically different sites, under different environmental conditions; and comparative studies of visual pigments of fish, vertebrate eye proteins, cactus productivity, glacial histories, deltas, flying frogs, carbonate muds, mountain formation, serpentine belts, sea anemone body-wall construction, oceanic trenches, coral reefs, or polar ecosystems. The taxonomy of the organisms needs to be known, which means we need to produce scientists who know how to classify organisms. Nature's language must be deciphered by natural scientists. The art and discipline that is field science must be developed in some students, so that future generations will know how to communicate in nature's language as well.

Study Nature, Not Books

Not everyone is going to be a scientist. The number of field scientists will continue to be a small fraction of our professional population, but we want to ensure that we produce the scientist who will do the very best science. As Yossi Loya noted, "There are, fortunately, still some students whose basic drive to do coral reef research is their love of nature. They may be less brilliant, but they turn out in the end to produce better scientific work."

In our undergraduate education, we must teach the general public adequately about the basics of the natural sciences, so they will be informed decision makers; we must produce good technical people to support scientists, including good science teachers at all levels; we must provide a breadth of basic background information to those scientists who may choose to specialize later in a narrowly focused geological or biological discipline, so that they can place their work in a broader context; and we must also provide a conduit for the potential field scientist of the twenty-first century to start on that career path.

It is not merely the presently active field scientists who have benefited from field experiences. Specialists in cell biology, geophysics, geochemistry, and other subdisciplines of the earth and life sciences often credit their field experiences and breadth of training for the enhancement of their ability to ask relevant questions in their present specialized field. For instance, even some of the scientists included in this book have not considered themselves field scientists for much of their careers, yet it was precisely their field experiences that enabled them to ask significant questions. Carl Bowin stated, "The way that I approach the solution to the problem of the core-mantle boundary is through the empirical geologic way of thinking." Consider that Carl Bowin and Peter Molnar are best known for their geophysical studies, which require extensive computer modeling. Similarly, Eugene Kozloff, Alan Kohn, Mimi Koehl, Maggie McFall-Ngai, Gaby Nevitt, and other biologists study their organisms as much in the laboratory as they do in the field. Yet, all can place their results in a context because they know intimately the environment from which their data have been collected (for the biologists, where their organisms live), and they possess broad experience in other, perhaps contrasting, environments. Many currently successful cell biologists, of the 1960s and 1970s generations, also are appreciative of their knowledge of the diversity of life forms, which was obtained through field courses such as invertebrate zoology given at Hopkins Marine Station, Friday Harbor Labs, or other marine stations or campuses across the country. As a friend, Rocco Carsia (from the University of Medicine and Dentistry of New Jersey), recently told me, "I presently use cellular and molecular techniques only to solve more broad and comparative physiological problems." Finally, there are many students whose field experiences did not result in a vocation, but rather in a life-enriching avocation, an appreciation of some aspect of nature revealed to those who have learned how to observe it.

It is clear from the interviews that few of these scientists entered college with the clear intent to become field scientists. What drew them into science were some inspirational professors teaching about subjects to which the students could relate directly—science at the scale of human experience; extended field trips, a field station or camp experience, or a field assistant job; a professor who took a special interest, who provided research and/or teaching opportunities; a small group of collegial faculty, grad students, and undergrads providing an interactive atmosphere, reinforcing the importance of the field tradition. If the student had that inner drive, a love of nature, or a desire to understand the intrinsic order and beauty in nature, he was on his way to developing into a field scientist.

The Field Experience

In the mid-1800s Louis Agassiz encouraged students to participate in the learning process: to make their own observations of live organisms in natural settings, to formulate questions about them, and to struggle to determine how to develop an answer. Let us add to this an element in education stressed by Steve Wainwright: the student describes to others his discovery or creation and explains why it is important. Then we have a technique that could and should be employed in any subject one wishes to investigate: nature, medicine, economics, history, law. The key is that every field should base its theory in the reality of that discipline. Observe the entire system; be frustrated at the complexity, but learn to look for patterns; ask how those patterns may have emerged; collect the information needed to ascertain if your hypothesis is correct. Natural science:nature; medicine:the patient; history:the battlefield and the documents; economics:the company; law:the courtroom and the prison. In each case, this is a process that requires time; it is an experience that might make a person feel uncomfortable and frustrated, but he will be a better scientist, doctor, economist, historian, lawyer, or judge for having gone through it.

Through field experience, facing the panic of feeling uncertain, making mistakes, and learning how to adjust, a student gains confidence in his ability to collect valid data. Even an experienced field scientist, confronted with a new situation, is confused and momentarily frustrated. Even if he has looked at ostensibly the same thing for 20 years, he will delight himself by noticing something new. The value of extensive field experience is that he can quickly and decisively appraise the situation. He can rapidly determine the relevant observations he must begin to make.

Unfortunately for today's student, we live in a society that demands immediate results, the quick solution; it focuses on the bottom line and the quarterly report; people do not have the time to accrue experience in real-life settings. Attention spans are limited. People are focused on the next item in their busy daily schedule; they are looking forward to the next ephemeral pleasure in life, one purchased with money, not with experience. They are "chasing the ants and ignoring the elephants among us." Life seems to be getting faster. But is it? Nature doesn't hear this message and continues to evolve on its own schedule.

It is particularly disturbing to find, in many undergraduate departments, that even when practical experience is available (e.g., a laboratory session), it is often a different, separate course from the lectures on the subject. Some of the students in a lecture may have experienced the reality of the topic on which the professor is lecturing by attending labs, but most have not. Even those who do sign up for the additional course, the lab course, may take it in a different semester than they take the lecture course, or even if they take the two courses concurrently, the lecture and the lab course may well be taught by different professors, and the topics in the course do not parallel one another. Therefore, the kind of learning that occurs when practical experience (seeing the reality of a phenomenon) reinforces what one hears in lecture simply does not occur. Introductory courses in biology and geology should have some practical experience tied to lectures, including time spent in both the field and the laboratory with knowledgeable and experienced faculty. Majors in those disciplines should have extensive practical experience as part of the graduation requirement.

Often today one hears that for a graduate student, there is no time to take a summer course at a field station. Nor is there time to learn how the rock samples are selected or to see the blood, skeleton, or eye protein samples collected or how the ice core samples are obtained. They arrive in the professor's lab by FedEx or UPS. Very specific aspects of these samples are analyzed and described by the student, and yet he has no idea of how to relate this information to the real world. Moreover, if an anomalous result is detected, a student (or soon-to-be Ph.D.-level scientist) will have no idea how to determine whether this is indeed a clue that something different is happening than what he had originally thought. Science has become more quantitative with more emphasis on concepts, theory, and modeling; the skills required to manipulate numerical data have assumed a position of greater importance and become a priority in ranking the acceptance of potential graduate students in the top programs. As this emphasis on quantitative skills has increased, the inverse is true of observational skills, that is, the ability to make good observations, gather appropriate data, interpret the data, and place them in relevant context. Obviously, both skills are needed to make significant advancements in science, and both skills should be considered in the evaluation of our future scientists when they are selected for graduate programs.

The busy senior scientist increasingly depends on students and technicians to collect the data to answer the questions he has formulated because he doesn't have time to go into the field. He must be successful at writing and administering grant proposals, securing funding (including overhead) and in many cases administering a large lab. This distances him from the reality of the natural world he is purporting to study. It also turns his graduate student into a technician; although this technical expertise may be valuable for the research, if it does not depend on the student's own creativity, it does not give the student an opportunity to forge his own unique scientific niche. After the enormous demands on his time of writing proposals, administering grants, and cranking out publications, the scientist tries to squeeze in a little new research. He has very little time to keep up with the literature, which is expanding exponentially.

No wonder time is given grudgingly to teaching, especially when time and effort spent in preparing for and then conducting a field or lab session is given no special consideration in formulating hours credited for various academic activities. There certainly isn't time left to talk with students and colleagues in a collegial, positive atmosphere, to share new observations and new approaches, or to form lifelong friendships. Another paper must be generated, another proposal written for the lab to continue for the upcoming year. If an interesting new twist appears, a new observation, a new way to integrate ideas, it must be put on the back burner, as one follows the science that is getting funded. And, often, the science that is funded is constrained by the limited intellectual scope of a preconceived program or a limited funding cycle (two years or less) or is made to fit within the mission of a granting agency. What results is often rather dull and unimaginative, although lots of papers are generated, because one's publication record helps one get promoted and get new grants. What a vicious cycle. What a waste of the creative energy of a highly experienced scientist.

Agassiz's Heirs

To increase (or, if need be, restore) the natural history approach in the teaching of and research in the natural sciences, the U.S. academic community and the U.S. public must make the same commitment today as Agassiz made in the mid-1800s. They must ensure that students and scientists have the skills and the experience to study nature in nature. They must strive to recreate the intellectual scientific ferment of the 1960s and 1970s, to produce an atmosphere that encourages scientists to ask significant questions, to not be afraid to question the dogma of the present day, to take risks, be creative, develop new fields. This is why Holger Jannasch, Daniel Stanley, Yossi Loya, John Dacey, Dina Fonseca, and others interviewed wanted to learn and to practice science in the United States. This is the atmosphere Loya and others of his generation were able to bring back from abroad and create in Israel. It replaced the atmosphere formed by the rigid German system (where the professor was God) with a system that encouraged creativity and the ability to recognize a new phenomenon. Combine that with discipline and hard work, then add the logistical support to be able to run with an idea and pursue the answer. What results is good, significant science.

Additionally, it is essential that we make access to nature logistically feasible for field scientists, their students, and their technical support personnel. This means maintaining good, well-equipped field facilities, which allow a scientist easy access to field sites and a means to keep field specimens alive in the laboratory. Even the finest institutions of this type, for instance, Friday Harbor Laboratories and the Organization for Tropical Studies, must constantly work to secure adequate funding for yet another year; of equal concern is the necessity to ensure that the mission of the institution—teaching and research in the field—is not compromised by focusing on narrow lab and theoretical research, which could as easily be done on the main campus. It also requires excellent logistical support for field expeditions and oceanographic cruises to the remote corners of our planet.

All of this requires time, good organizational skills, and some (but not a great deal of) monetary support. But the payoff in enriching the lives of the students and the professors who work along with them is well worth it. Society benefits. Good science will result. The expense necessary to support fieldwork should not be calculated in student FTEs [full time equivalents], faculty-to-student ratios, or whatever quantity the university administrators use for ease in tabulation. It must be balanced against what (both people and research results) is produced. It should not be dependent on an endowment to continue—the reason Clark Burchfiel tells us that the superb field geology program run by MIT is assured and Mimi Koehl tells us that vertebrate field biology will be continued at UC Berkeley. Don't misunderstand me, an endowment is nice, if someone is willing to make it. But field access should be an important component of the education of students and the research of some of the professors—and it should be adequately supported monetarily—at any institution of higher learning that purports to produce majors in the geological or biological sciences.

Passing the natural history tradition on to the next generation is the role of the teacher in the scientist. Successful teaching is as important to the advancement of science as research. It is clear, especially in the oral histories of the older generations, that this is a role they cherished and one that was shared by many of their mentors. The satisfaction with being a good teacher, inspiring students, preparing them with a solid framework in the basics, supporting them in various ways as their career progresses is, of course, not confined to field scientists. We all have special teachers whom we can recall from various stages in our education. However, the bond created when you are with a person in the field, day and night, eating, gathering data, and trying to stay dry or warm is a special one. Often these students become lifelong friends. One person might love the opera, a second marathon running, and a third architecture. Yet we all have had the shared experience of allowing nature to pose questions. We have connected emotionally to another person through our mutual respect for nature; we value the same thing. Their other passions merely make these scientists even more interesting as people.

The scientists interviewed for this book are obviously a selection of those men and women who have made important contributions to the advancement of science through their field research and teaching during the past 60 years. I think it is significant that another group of scientists emerges as having had great influence on the careers of many of the scientists whom I interviewed. Of course, these include Ralph Smith of UC Berkeley and G. Evelyn Hutchinson of Yale, whose direct academic descendants are among the people I interviewed. It also includes Cornelius van Niel and Donald Abbott at Stanford University; Tom Goreau at the University of the West Indies; Richard Strathmann at Friday Harbor Labs; Bob Paine at the University of Washington; Richard Foster Flint and John Rodgers at Yale; Karel Liem at Harvard; Ross Robertson at STRI; Harry Thompson at UCLA; Larry Slobodkin at SUNY Stony Brook; Steve Wainwright and Steve Vogel at Duke University; Bill McFarland at Cornell; Harry Hess at Princeton; Brackett Hersey at WHOI; Joe Connell at UC Santa Barbara; Arthur Hasler at the University of Wisconsin; Bob Ginsburg at the University of Miami; Fred Fry at the University of Toronto; and John Ogden at West Indies Lab. I am not saying that these are the only members of this special group, but they are repeatedly mentioned in the interviews. They have made a major impact on science as practiced in the United States today because they created opportunities and challenged others to continue to develop their skills. These scientists have created turning points in other people's lives. Furthermore, the waves they propagated often had far-reaching affects because the people they inspired through personal intervention in their lives, in turn, influenced many other people. They have fulfilled their responsibility to the passing on of Agassiz's legacy.

The Natural Scientist and Science at the Dawn of the Twenty-first Century

For a scientist, there is excitement in sharing new information and new ideas with others; however, this is balanced to a degree with the threat that ideas might be stolen and pursued by another scientist. As a grad student at UCLA, I remember wondering why any scientist would even consider falsifying data, or publishing results that did not exist. The question arose because there was a widely publicized case of a medical scientist doing this. Why? To gain a promotion? To get the next grant? I'm not sure, but as I thought of the colleagues with whom I had worked at West Indies Lab, Hopkins Marine Station, Friday Harbor Labs, and UCLA, I couldn't imagine any of them doing this. How could someone betray the friendships forged during long hours in the field, long evenings of discussions about observations and interpretations? I'd like to think that the human relationships developed in the field sciences would place a high value on truth, on honesty, but in today's competitive, noncommunicative world of academic science, this is one of the factors about which I worry.

Another major concern of mine is that I find some scientists, particularly those who work almost exclusively with numbers and computers, act dispassionately, as if they have no emotional connection at all to their research. I think they should care whether their answers and models are approximations of the truth or whether they have any relevance to reality. Unfortunately, not all of them do.

A third problem in modern-day science is that there are some scientists who work on issues with potentially great societal impact, for instance, global climate change, who become advocates for their cause. They become personally attached to a certain answer. Instead of critically evaluating the available data and being open to the possibility that their viewpoint may have to be revised, they cling to their cause; it is their belief, actually. Sometimes they do this to continue to get funding from support groups, and sometimes it is a mistake: "I know this is now wrong, but the general public needs something like this to believe in." This does no one any good. It is harmful to scientific progress (it is antiscientific) and harmful to society. Yet we live in an age when public relations and spin is in (or out of) control in every facet of our lives.

We want our scientist of the twenty-first century to be flexible enough—and humble enough and honest enough—to change his view should new data suggest that the first interpretation (even if it were his brilliant idea) was incorrect. This is charmingly illustrated in the story Fred Nagle relates about his Princeton graduate advisor, Harry Hess. Nagle asked Hess, "Harry, aren't you disappointed? All those ideas you had and published on in the old days don't seem to be quite right. Aren't you discouraged?" Hess answered, "On the contrary. All the questions now are new. You guys have opened my eyes. All the questions have changed. I have a reason to go on!"

New thinking must emerge to keep what's good of the old and to add what is needed to create a stimulating atmosphere where the mission of the institution (college, university, research institution) is the pursuit of truth. The truth of nature is found in nature, which brings me to my last point: the diversity of academic roles (taking advantage of individual strengths and weaknesses and individual time conflicts) in the workplace. I am convinced that it can be created and can make work a better place, but it will require a little imagination and a lot of commitment on the part of professors and administrators. A scientist must be committed; he must believe in what he does; ultimately, to be truly successful, he must derive pleasure from his work. I think you will agree that the scientists whom I have interviewed love their work. But is it necessary that every scientist at an institution do exactly the same thing? conduct the same amount of research? run a large, technician-filled empire? teach the same kind of courses? Must an academic position be full time? Is there not an administrator with the creativity and an institution willing to develop some flexibility in the present academic system?

My initial concern, in considering this point, is the mixed plight of women, as we have seen through the past 50 years. Women have really only since the late 1970s joined the scientific workforce as equals (my generation either delayed getting a graduate degree or took a long time on a convoluted pathway to get into a tenure-track position). A woman who chooses a career as a professional field research scientist, particularly one in her 30s or 40s, is often found juggling research and teaching, children and life partner, pets and home—none getting the amount of attention it deserves. Sometimes a woman will compromise and accept a non–tenure-track position, a lectureship or a soft-money research position, which allows her more flexibility in dealing with her busy life, but it also holds no hint of security or of upward career development. It is clear from the interviews that women have just as much to contribute to the advancement of science as men. In addition, I suspect there are others, men and women, who might want to spend part of their lives on some other activity, such as writing, climbing, philosophy, music, dance, working out for the Olympics, photography, parenting, and so on.

Wouldn't we have a more diverse and creative faculty if we could encourage part-time as well as full-time participants? Young assistant professors and old professor emeriti? More expertise would be available to students and colleagues. People would be more enthusiastic about their lives if they could devote a segment of time to something they hold dear. I know the reaction that administrators would have to this suggestion. What about fringe benefits? What about retirement? Or (even more crucial in these business-oriented academic times) what about getting continued funding? What about the grant overhead brought in by that department? How will we support our administration? "Ridiculous," they would say, "this can't be done." For a moment, let us just think about what we could do if we did choose to do this.

Related to the idea of diversity in a department is one that perhaps may not be viewed as quite so radical. That is, it is time to bring academic departments back to the balance of teaching and research. Research is important, but a little more or a little less isn't really going to matter in the long run. What will matter is if we graduate a generation from our undergraduate programs that has no basic knowledge of organisms or rock and strata, whose learning has been all conceptual and theoretical. The Cultural Revolution in China set that country's extensive intellectual capacity back for generations. Our current trend in academia, where the pendulum has swung so far to theory (and not just in the natural sciences, I am told by my colleagues in other disciplines), might produce such an intellectual gap in our country. In my experience, it is beginning to occur, at least as far as the field sciences go. Who will integrate this material for the upcoming generation? Who will even be aware of the information?

The time is now to make a commitment to teach the basics in the natural sciences to the general student body and to inspire those select few who will go on to become the research scientists of the future. Those who go on to university and college jobs should care about teaching as well as research. Faculty in departments who teach undergraduates should contribute in some substantial way to this education. However, there can be diversity as well in this contribution. Not everyone should teach a big lecture course; we all know some people are lousy lecturers. But that professor could perhaps teach a case study or a special seminar, focusing on a general but active topic of research. Every faculty member should pull his weight, but not necessarily in exactly the same way. What he should be encouraged and given time to do is to share his passion for scholarship with those of younger (and older) generations. There should be a role for generalists in a department. Someone should be able to integrate material for students and for more specialized colleagues. In the best case, a resident natural historian (who perhaps has a specialized research topic) should be accessible to students and faculty, especially at field institutions. Courses such as natural history of vertebrates or invertebrate zoology should be taught on a main campus and include good labs and field trips when possible. Are they as effective as those taught at field stations? No, of course not. But it is the only introduction some potentially fine future scientists (Mimi Koehl is an example) will have to these organisms. This might be the inspirational course (as invertebrate zoology was for Larry Madin, Maggie McFall-Ngai, and others) that starts someone on the road to field science. Lecture and practical work should be combined. It is the best way to learn. Research output among the faculty should be encouraged (and even required in departments with graduate students) but balanced with sufficient time allocated for teaching.

Finally, the funding agencies as well as the academic administrations should acknowledge that people are diverse in the ways they do research as well. Some are better working alone. Some can function in a larger group. Different kinds of questions require different approaches. Some will require a much greater period of time to generate meaningful results. Models and hypotheses sometimes result from and are tested through observations, and sometimes it is necessary—although time consuming—to gather basic descriptive data. Even in a larger group, a scientist must continue to act as an individual, contributing his own expertise and creativity to a broader problem. Both kinds of research produce significant results, good science, as it were, and both should be supported. Effort should be made to reduce the bureaucratic paperwork now necessary to support one's research. A researcher should be given adequate time and resources to develop an idea and time to reflect before he communicates his results to the scientific community. Writing more grant proposals than papers is a terrible waste of resources. Administrations should strive to reduce the time spent on these activities, rather than add to the burden, as appears to be occurring today.

Your Contribution to the Advancement of Science in the Twenty-first Century

I hope you have been inspired by the passion and effort expended by the scientists whose stories you have read. I hope that you have learned more about how science is done and about some of the exciting things we do know about our world at the present time. I hope that you recognize that today and in the future we need well-trained field scientists investigating how nature works. I hope you will continue to follow some of the exciting areas that will develop exponentially in the near future, including a theory of climate change (which will integrate approaches and data from the earth, ocean, atmosphere, and life sciences); geobiology (at many temporal and spatial scales); and symbiosis (especially, the role of microorganisms in all facets of the evolution of life, the oceans, atmosphere, and the Earth).

I also hope that you will contribute to the advancement of science in the future in a creative way, one unique to you. You need not be a scientist. You may be a relative or a friend who introduces a child to the wonders of nature; you may be an English teacher or a history teacher who inspires a student and prepares him to seek the answers to questions in a disciplined, scholarly way; you may be able to provide financial support to provide scholarships for students to attend field activities or be able to help in the endowment of a field program; you may be on a university faculty or in a government agency that is making decisions about field support. Many people are involved and must become involved in the advancement of science, in our pursuit of knowledge about our complex and wonderful world.

David Starr Jordan's memories of Agassiz's legacy at Penikese Island remind us that things really haven't changed much through time: "At the end of that second summer—that of 1874—the Anderson School closed forever. There was nothing to do but pay the debts and shut the doors. Agassiz being gone, even the small sum necessary to carry on the work could nowhere be obtained. In the eyes of the businessman for whom it was named, the venture was a failure. But while Penikese is deserted, the impulse which came from Agassiz's work there still lives, and is deeply felt in every field of American science" (*The Days of Man*, vol. 1, p. 118). It is our responsibility to ensure that we help produce the next generation of Agassiz's heirs.

Index